NANO BIOMATERIALS

NANO BIOMATERIALS

Editors

V. Rajendran
P. Prabu
K.E. Geckeler

BLOOMSBURY
London • New Delhi • New York • Sydney

First Published, 2013

BLOOMSBURY PUBLISHING INDIA PVT. LTD.
London New Delhi New York Sydney

ISBN: 978-93-82563-37-2

Published by Bloomsbury Publishing India Pvt. Ltd.
VISHRUT Building, DDA Complex, Building No. 3, Pocket C-6 & 7
Vasant Kunj, New Delhi 110 070

Laser Typeset by SHINU SIMON
A-5/56, Moti Nagar, New Delhi 110 015

Printed at Anvi Composers
1 DDA Market, Block A-1B, Pachim Vihar, New Delhi 110 063

डॉ. डब्ल्यू सेल्वामूर्ति
विशिष्ट वैज्ञानिक एवं
मुख्य नियंत्रक अनुसंधान तथा विकास

Dr. W. Selvamurthy
Distinguished Scientist &
Chief Controller Research & Development

रक्षा मंत्रालय, भारत सरकार
रक्षा अनुसंधान तथा विकास संगठन
338, डी आर डी ओ भवन,
नई दिल्ली-110011

Government of India
Ministry of Defence
Defence Research and Development Organisation
338, DRDO Bhawan,
New Delhi-110011

FOREWORD

The ever growing Nanotechnology is being the strong pillar of support for humanity in its pursuit for development in all walks of life, be it in the field of technology for medicine, biotechnology or devices and materials to improve the quality and fulfill the needs of our day today life.

I am happy to note that Centre for Nano Science and Technology is organizing an International Symposium on Macro and Supramolecular Architectures and Materials (MAM-12) with a special topic on Nano systems and Applications during November 21-25, 2012 at K.S. Rangasamy College of Technology (KSRCT), KSR Group of Institutions, Tamil Nadu, India.

Dr. Rajendran has made a commendable work on enlightening Centre for Nano Science and Technology, at KSRCT with the state of the facility under the private institution, in rural parts of India at Tiruchengode. The achievements made by the centre in a short span of three years with more than 11 patents, completion of several research projects, publications in reputed international and national journals, collaboration with foreign and Indian Universities reveal the significant progress of the centre. It is also to be noted that the centre receives supports from funding agencies such as DST-Nanomission, DRDO, BRNS, UGC-DAE and IGCAR.

I am sure that the conference proceedings will be an asset for the young scientists and the beginners to pursue the research carrier in Nanotechnology. I heard that the conference is enlightened by 4 plenary talks including two noble laureates, 36 invited talks and more than 350 delegates. The participants are from 24 countries like USA, UK, Russia, Germany, China etc. I am glad to state a foreword message that more than 200 papers are received and compiled under four broad titles of books. The contributed papers of 50 under the book entitled Nanobiomaterials encompass the development of biocompatible nanomaterials for implant and biomedical applications.

दूरभाष/Telephone : 23016976 टैली-फैक्स/Tele-fax : 23014259 ई मेल/E-mail : wselvamurthy@hqr.drdo.in

No doubt MAM-12 will be a better platform for the delegates from different parts of the world and from our home country to exchange their ideas and to explore the possible collaborations on their mutual interest. I congratulate Dr. KS Rangasamy, Principal Dr. K Thyagarajah and Director and Organising Chair Dr. V Rajendran and his team for their efforts to conduct the event more successful.

I wish the conference all success with fruitful deliberations.

Place: New Delhi

Date: 23 October 2012

(Dr. W Selvamurthy)
Distinguished Scientist and
Chief Controller (R&D)(LS & IC)

EDITORS

Dr. V. Rajendran FUSI, FASI, FInstP, is Director, Research & Development, K S R Group of Institutions and Centre for Nano Science and Technology, K.S Rangasamy College of Technology, Tamil Nadu, India. Under his able guidance, 11 scholars have completed and 14 scholars are pursuing their Ph.D. degrees. He has published more than 150 research papers in reputed international and national journals, 50 papers in conference proceedings, 28 refereed books, 4 R&D books and 11 patents. He has won many awards including the Prof. K. Arumugam National Award in 2011, Best faculty award in 2010, Raman-Chandra Sekhar silver medal (2010), Tamil Nadu Scientist Award, NDT Man of Year 2004, DAAD from Germany (2002), INSA, TNSCST Young Scientist, DAE/BRNS Visiting Scientist, Best paper award from MRSI, ASI, ASA and USI and Outstanding Organiser Award for the 7th National Symposium on Ultrasonics, 1996.

Dr. P. Prabu, Ph.D., Assistant Professor, Centre for Nano Science and Technology (CNST), K.S. Rangasamy College of Technology, Tamil Nadu, India. He has been working in the Centre for Nano Science and Technology, KSRCT for past 30 months. He obtained his Ph.D. in bionanosystem engineering from Chonbuk National University, South Korea. He has worked as a post doctorial fellow in Chonbuk National University during 2008-2009 and also at Nanyang Technological University, Singapore during 2009-2010. He works on drug delivery systems and electro spinning for tissue engineering and other biomedical applications using nanoparticles and nanofibers. Presently, he has to his credit more than 10 research papers in reputed international journals, 28 papers in international conference proceedings, one edited proceedings, one research & development book.

Dr. K.E. Geckeler is the Founding Chair of the Department of Nanobio Materials and Electronics, World Class University (WCU), Gwangju, South Korea. He is also a Professor in the Department of Materials Science and Engineering and the Department of Medical System Engineering at the Gwangju Institute of Science and Technology (GIST), South Korea and serves as Vice Director of the Gruenberg Center for Magnetic Nanomaterials (GCMN). He received his Ph.D. and M.D. degrees from the University of Tuebingen, Germany (both degrees: "magna cum laude") and spent sabbatical leaves at Harvard University, University of Montana, Clemson University (USA) and at the University of Montpellier (France). He received a series of awards such as the "Fonds der Chemischen Industrie", the "Fritz-Ter-Meer Award", the "Science Price of the President of Korea" and the "Outstanding Research Award of GIST". He has to his credit initiation and organizing the biannual MAM Symposia series. He has published more than 300 research articles and short communications, 12 book chapters, over 10 books, and over 80 patents.

PREFACE

The International Symposium on Macro- and Supramolecular architectures and materials (MAM-2012) with a special topic on Nano systems and Applications is organised by Centre for Nano Science and Technology (CNST) of K. S. Rangasamy College of Technology. This conference is jointly organised with World Class University, Gwangju Institute of Science and Technology (GIST), South Korea.

More than 525 abstracts have been received from India and abroad. Out of 400 full papers received, 260 have been selected and peer reviewed by the expert committee for the publications in conference books. All the received books are classified under four headings namely Nano Material Synthesis and Characterisation, Advanced Nanomaterial for industrial applications, Nano biomaterials and Applications of Nanomaterials: Electronics, Energy, & Environment.

A total of 260 accepted full papers, 65 of contributed papers for the inclusion in the book entitled Nano biomaterials after peer review. This book contains the contributed papers discussing about the appropriate applications of biomaterials for the benefit of humans in various eco friendly product developments for biomedical applications. The most effort has been taken in getting the review of all the contributed papers. All the contributed authors are extended by our sincere thanks for their timely submission and cooperation in carrying out suggestions by the referees.

The CNST has well established the start–of–art of experimental facilities and looking for industrial collaboration and partners for the technology transfer in thrust areas of nanotechnology. We are sure that the outcome of MAM-12 open up new avenue between the researchers and industrialists for the development of nano products.

We owe our special thanks to members of technical committee for peer review of contributed papers. The various government funding agencies, private organizations and industries are thankful for their munificent support and sponsor for the successful conduct the conference. The various committee chairs and members are highly acknowledged for making this event a grant success.

The support extended by Mr. Suresh Gopal and. Mrs. Jyoti Mehrotra of Bloomsbury Publishing India Pvt. Ltd in bringing out this book on time is highly appreciated.

V. Rajendran
P. Prabu
Kurt E. Geckeler

CONTENTS

Foreword v

Editors vii

Preface ix

1. Fabrication of Electrospun Paclitaxel Loaded Poly (ἐ-Caprolactone) Nanofibers for Drug Delivery 1
Elakkiya Thangaraju, Vaishnavi Thangaraju, Ramadhar Kumar, Natarajan Thirupathur Srinivasan and Sheeja Rajiv

2. Non-Covalent Modification is the Next Probable Step to Prepare a Collagen Based Biomaterial Using Di-Carboxylic Acids (Glutaric and Sebacic Acid) 5
Tapas Mitra, G. Sailakshmi and A. Gnanamani

3. Polymer Hot Embossing with Soft Stamps 9
D.K. Maurya, S. Kandasamy, F. Lapierre, K. Alameh and Y. Zhu

4. Influence of Molar Ratio on the Surface Characteristics of Ferrous Phosphate Nanoparticle Synthesis 13
P. Renuka Devi and E. Hemananthan

5. A Clinical Study of Advanced Aids-Ks with Liposomal Doxorubicin, Phleomycin and Vincristine in Nano Drug Delivery 17
Poornachari

6. Silver Hydroxyapatite Nanocomposite Prepared by a Novel Method for C-N Bond Coupling and Antibacterial Activity 23
Perumal Emayavaramban and Nallasamy Dharmaraj

7. Green Synthesis of Copper Oxide Nanoparticles and Their Antimicrobial and DNA Cleavage Studies 29
T. Abiraman, S. Malathi, M.D. Balakumaran, P.T. Kalaichelvan and S. Balasubramanian

8. Synthesis, Characterization and Electrochemistry of Ruthenium Terpyridine Complex Anchored to Gold Nanoparticle and its Anticancer Activity 35
T. Ezhilarasu, S. Malathi and S. Balasubramanian

9. Pharmacokinetics and Biodistribution of Negatively Charged Pectin Nanoparticles Encapsulating Paclitaxel 41
Anita K. Verma and A. Kumar

10. Development and Evaluation of Propellant Free Quick Break Foam Forming Clobetasol Propionate Nanoemulsion for Efficient Treatment of Psoriasis 47
N. Subramanian, P. Chandra Sekar, S. Abimanyu, S.P. Sharavanan, M. Nandagopal, D. Vijula Rani and N. Jamuna Rani

11. Development of Electrospun Poly (Propylene Carbonate) and Polycaprolactone Scaffolds for Tissue Engineering 53
Naveen Nagiah, Sivakumar Singaravelu, Lakshmi Madhavi, R. Anitha, Tiruchirapalli Sivagnanam Uma and Tirupattur Srinivasan Natarajan

12. Synthesis of Blended Fibers of Poly (3-hydroxybutyric Acid) and Poly (propylene carbonate) Scaffolds for Tissue Engineering 59
Naveen Nagiah, Giriprasath Ramanathan, Lakshmi Madhavi, R. Anitha, Tiruchirapalli Sivagnanam Uma and Tirupattur Srinivasan Natarajan

13. Design and Development of Artemisinin and Dexamethasone Topical Nanodispersion for the Effective Treatment of Age Related Macular Degeneration 65
Subramanian Natesan, Chandra Sekar Ponnusamy, Abimanyu Sugumaran and Venkateshwaran Krishnaswami

14. Formulation Development and Pharmaceutical Evaluation of Lornoxicam Magnetic Nanoemulsion 71
S. Latha, P. Selvamani, C. Prabu and M. Anis Kumar

15. Preparation and Evaluation of Prednisolone Magnetic Nanosuspension for Possible Use in Rheumatoid Arthritis Therapy 77
S. Latha, P. Selvamani, T. Silambarasi and M. Anis Kumar

16. Toxicity Assessment of Ormosil Nanoparticles 83
Dhiraj Kumar, C.K. Prashant, Amit Kumar Dinda and Susmita Mitra

17. Luminescence Studies on Er^{3+}/Yb^{3+} Doped Telluro-Borate Glasses 89
S. Arunkumar, K. Maheshvaran and K. Marimuthu

18. Studies on Biosynthesis of Iron Nanoparticles at Room Temperature Using Aqueous Extract of Aquatic Plant (Water Hyacinth) and its Application for Chromium (VI) Removal 95
R. Vasantha Gokilam, M. Anjali Nair, A. Kanakalakshmi, C. Sukumar and K. Shanthi

19. Studies on the Removal of Chromium (VI) by Iron Nanoparticle Synthesised from Fungi 101
A. Kanakalakshmi, M. Anjali Nair, R. Vasantha Gokilam, C. Sukumar and K. Shanthi

20. Comparison of Different Polymerization Protocols on the Molecular Recognition Performance of Molecularly Imprinted Polymers 107
Sushma Pardeshi, Rita Dhodapkar and Anupama Kumar

21. Hydrogen Bonded Supramolecular Architectures in Organic Dicarboxylate Salts 113
D. Sathya, R. Padmavathy, R. Jagan and K. Sivakumar

22. Synthesis of Gold Nanoparticles Using *Punica Granatum* Extract and Evaluation of Their Antimicrobial Activities 119
S. Lokina and V. Narayanan

23. Modification of Carbon Nanotubes for Bio-Applications and Toxicity Evaluation 125
P. Khalid, M.A. Hussain, P.D. Rekha, V.B. Suman and A.B. Arun

24. Growth and Structural Study of Cadmium Arachidate Multilayers on Solid Substrate Like Silicon and Glass Substrate by Langmuir Blodgett Technique 131
Dhrubojyoti Roy, Nayan Mani Das and P.S. Gupta

25. Intergrown and Microporous Zeolite-A Crystals to form Membrane Layer on Polysulfone Hollow Fiber Support 137
Sanjay G. Chaudhri, Ankit M. Kansara and Puyam S. Singh

26. Synthesis and Characterization of Pectin Based Silver Nanocomposite Hydrogels for Controlled Release 5-Fluorouracil 143
P. Ramasubba Reddy, N. Sivagangi Reddy, S. Siva Prasad, S. Eswaramma, K.S.V. Krishna Rao and Yong Ill Lee

27. Synthesis, Characterization and Investigations of Antimicrobial Properties of Silver Nanoparticles Via a Green Method Using Mulberry (Morus Alba L) Tree Leaves 149
R. Rubia, Yashika, Remy, Namrata, Rajeshwari, Indu Tucker Sidhwani, Kavita Vasdev, Mangala Joshi, Shubha M Gokhale and Vandna Luthra

28. Efficient Preparation and Size Evaluation of Gold Nanoparticles in Chitosan Suspension by One-Pot Chemical Reduction Method 155
R. Lakshmi Narayanan, M. Sivakumar and K. Venkateswara Rao

29. *Eucalyptus* Leaf Extract Mediated Synthesis of Iron Oxide Nanoparticles 161
M. Balamurugan and S. Saravanan

30. Novel Ring Like Metal-Organic Assemblies as Host for Small Molecule Recognition 167
Rohith P. John, Subashis Pradhan, Dohyun Moon and Myoung Soo Lah

31. Biogenic Silver Nanoparticles' *Bombyx Mori* Nuclear Polyhedrosis Virus (BmNPV) Inhibitory Mechanism 173
S. Tamilselvan, T. Ashokkumar, R. Geetha, K. Govindaraju, A. Mohamed Sadiq and G. Singaravelu

32. Biosynthesis of Silver Nanoparticles Using the Extract of *Alternanthera sessilis* 179
M. Jannathul Firdhouse and P. Lalitha

33. Self-assembled Cu (II) MOFs with N, O Donors Mixed Ligands 185
P. Mosae Selvakumar, E. Suresh and P.S. Subramanian

34. Gelatin Micro Particle Loaded with *Ephedra Gerardiana* Extract for Lung Tissue Regeneration 191
Deepti Singh, Dolly Singh, Soon Mo Choi and Sung Soo Han

35. Synthesis and Structural Properties of TiO_2 Nanocapsules Prepared by Hydrothermal Method: Growth Mechanism for Mass Production 197
M.A. Shah, S. Rubab and S.A. Akhoon

36. Cytotoxic Response to Carbon Nanotubes by Lymphocyte Cells 203
Seema Manchanda, Madhuri Sharon, Maheshwar Sharon and Amol Kakade

37. Temperature and pH Responsive *in situ* Gel Forming Nanodispersions for Efficient Ocular Delivery of Ibuprofen 211
Subramanian Natesan, Vinoth Jeevanesan, Vanitha Kuppusamy and Sivasankari Gomathinayagam

38. Enhancement of Modulus, Strength and Conducting Properties in Reduced Graphene Oxide-Graft-Poly (methyl methacrylate)/Poly Vinyl Chloride Nanocomposite 219
Rama K. Layek, Aniruddha Kundu and Arun K. Nandi

39. Biosynthesis and Characterization of Silver Nanoparticles using *Aspergillus niger sp.* 227
A. Sivasamy, K. Jagan, S. Nethaji and Sugania Josphine

40. Preparation and Antimicrobial Observations of Zinc Doped Nanohydroxyapatite 235
C. Deepa, A. Nishara Begum and S. Aravindan

41. Insitu Preparation and Characterization of Hydroxyapatite/PEG Nanocomposite and its Electro Chemical Studies 243
C.P. Dhanalakshmi, R. Suresh, K. Giribabu, L. Vijayalakshmi and V. Narayanan

42. Studies on Bio-medical Application of ZnO Coated on Plasma Treated Fabric 251
S. Inbakumar and A. Anukaliani

43. Evaluation of Surface Acetylated and Internally Quaternized Poly(Propylene Imine) Dendrimer as Biocompatible Drug Carrier Using Nimesulide As A Model Drug 259
E. Murugan and D.P. Geetha Rani

44. Synthesis of Quaternary Ammonium Functionalized Poly(Propylene Imine) Dendrimer and its Possible Application as Drug Carrier Using Piroxicam as a Model Drug 269
E. Murugan, D.P. Geetha Rani and V. Yogaraj

45. Facile Green Synthesis of Silver Nanoparticles and Their Antiviral Activity 279
T. Ashokkumar, S. Tamilselvan, R. Geetha, K. Govindaraju, A. Mohamed Sadiq and G. Singaravelu

46. Green Synthesis of Gold Nanoparticles and Their Anticancer Activity 289
R. Geetha, T. Ashokkumar, S. Tamilselvan, K. Govindaraju, A. Mohamed Sadiq and G. Singaravelu

47. Green Synthesis of Silver Nanoparticles from *Dodonaea Viscosa* Linn. Aqueous Extract and its Antibacterial Activity 299
Pannerselvam Balashanmugam and Puthupalayam Thangavelu Kalaichelvan

48. Biocompatible Titania–Chitosan Nanocomposites for Tissue Engineering and Orthopedic Applications 309
K. Kavitha, M. Prabhu, R. Suriyaprabha, P. Prabu, V. Rajendran and T. Jayakumar

49. Zirconium-Substituted Nanobioactive Glass Particles for Bone Implant Applications 315
M. Prabhu, K. Kavitha, S. Sutha, P. Manivasakan, V. Rajendran and P. Kulandaivelu

50. Antimicrobial Properties of Nano-Sized Silver-Doped Hydroxyapatite Composite for Biomedical Applications 321
S. Sutha, G. Karunakaran, M. Prabhu and V. Rajendran

Author Index 327

FABRICATION OF ELECTROSPUN PACLITAXEL LOADED POLY (Ɛ-CAPROLACTONE) NANOFIBERS FOR DRUG DELIVERY

ELAKKIYA THANGARAJU, VAISHNAVI THANGARAJU[1], RAMADHAR KUMAR[2], NATARAJAN THIRUPATHUR SRINIVASAN[3] AND SHEEJA RAJIV

Department of Chemistry, Anna University, Chennai, Tamil Nadu
[1]Department of Chemistry, P.S.G.R. Krishnammal College for Women, Coimbatore, Tamil Nadu
[2]Department of Bioproducts, Central Leather Research Institute, Adyar, Chennai, Tamil Nadu
[3]Department of Physics, Indian Institute of Technology Madras (IITM), Chennai, Tamil Nadu
E-mail: sheeja@annauniv.edu

ABSTRACT

The present work reports the preparation of Poly (Ɛ-Caprolactone) (PCL) and Paclitaxel loaded PCL (P-PCL) nanofibers by electrospinning. A series of PCL solution (25 wt %) and Paclitaxel loaded PCL solution containing Paclitaxel (0.1 and 0.2 wt %) were electrospun into nanofibers. The average diameter of PCL and P-PCL in the range of 50-500 nm was confirmed by SEM. The TEM images showed the dispersion of Paclitaxel on P-PCL. The decrease in crystallinity of PCL with increasing Paclitaxel concentration of 0.1 to 0.2 wt % was confirmed by XRD. The functional groups in the PCL and P-PCL nanofibers were confirmed by FTIR. The TGA results showed the degradation of PCL and P-PCL close to 400°C. The glass transition temperature, melting temperature, and decomposition temperature of PCL and P-PCL were studied by DSC. The water uptake percentage was found to be 15%. The percentage porosity and contact angle of PCL were found to be 92% and 112±3° with deionised water respectively. The storage modulus, loss modulus and tan δ of PCL were confirmed by DMA. The percentage cumulative release of paclitaxel at the end of 15th day for 0.1 and 0.2 wt% formulation was 67.80±1.6 and 70.31±1.5, respectively. The in-vitro biological cytotoxicity studies were performed using C6 glioma cells and NIH 3T3 fibroblast by MTT assay and SEM analysis.

V. Rajendran, P. Prabu and K.E. Geckeler (eds.)
Nano Biomaterials, pp. 1–4 (2012).

INTRODUCTION

The development of nanotechnology through the improvement of electrospinning methods has the benefit of biomedical and tissue engineering applications.[1–3] Electrospinning nanofibers has such great properties like high surface area to volume ratio and have received very good attention because of their potential applications for drug delivery carriers, wound dressings, and tissue repairs.[4] PCL is a synthetic biodegradable and biocompatible polymer which has been approved for use by the FDA since the 1970s. Paclitaxel is a natural product with antitumor activity and belongs to a class of chemotherapy drugs from plant alkaloids.[5] In the present study, P-PCL nanofibrous mats were assessed for their potential use as carriers for anti-tumour delivery of Paclitaxel.

EXPERIMENTAL

PCL (25 wt%) was dissolved in chloroform at 80°C and gently stirred continuously for about 5 hours. Paclitaxel (0.1–0.2 wt%) was then added to PCL solution and continuously stirred for 2 hours. The polymer and the drug solution was electrospun at a flow rate of 0.5 ml/h, applied voltage of 26 kV and tip to collector distance was kept at 13 cm. Surface morphological studies were performed using SEM, TEM, FTIR, and XRD. Thermal, Physico-chemical and Mechanical Properties were studied by using TGA, DSC and DMA. *In-vitro* drug release was performed using PBS at 100 rpm and 25°C. To observe the biocompatibility of the formulations, mouse embryonic fibroblasts (NIH 3T3) and rat C6 glioma cells were used. 3-(4, 5-Dimethylthiazol-2-yl)-2, 5-dipheyltetrazolium (MTT) assay was performed to compare the degree of cell proliferation and viability.[6]

RESULTS AND DISCUSSION

Surface Morphological Studies

The morphology of the PCL and P-PCL fibers was smooth and uniform as shown in Figure 1. The diameter of PCL nanofibers was found to be in the range of 50–400 nm while P-PCL nanofibers was increased after incorporation of Paclitaxel in the range of 50–500 nm. TEM images clearly confirm that Paclitaxel was well dispersed in the PCL nanofibrous mat. The two sharp diffraction peaks were observed for PCL at $2\theta = 21.41°$ and $2\theta = 23.64°$. The semi-crystalline PCL was converted into highly amorphous after loading of Paclitaxel, which confirms that the amorphous form of Paclitaxel dispersed in the PCL nanofibrous mat, has the benefit of increasing the solubility and bioavailability of Paclitaxel. The peak at 1646 cm^{-1} which is due to C=O amide stretching. By increasing the concentration of paclitaxel to PCL, the peak at 1646 cm^{-1} is present which confirms the Paclitaxel was dispersed in the PCL nanofibrous mat.

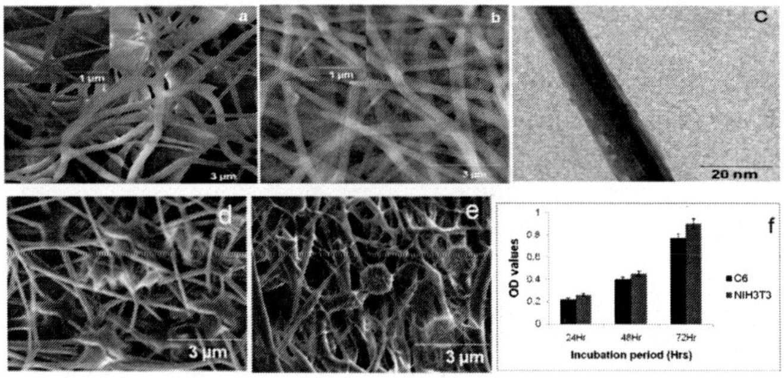

Fig. 1: SEM Images of Electrospun Mats (a) PCL (b) P-PCL (c) TEM Images of PCL and SEM Photograph for PCL and P-PCL (d) NIH 3T3 (e) C6 Glioma Cells and (f) MTT Assay for P-PCL Nanofibrous Mat

Thermal, Physico-Chemical and Mechanical Properties

The decomposition temperature of PCL starts at 393°C whereas P-PCL was found to be 385°C. The higher decomposition temperature of PCL shifts to lower temperature in the P-PCL which further confirm the conversion of crystalline nature to amorphous form as evidenced in the XRD. The low melting temperature of PCL however did not increase after loading Paclitaxel which was confirmed by DSC. The absence of crystalline peak concluded that the crystalline nature of Paclitaxel was fully converted to amorphous form which is advantageous for drug delivery. The TGA and DSC results confirm that PCL and P-PCL nanofibrous mats were stable upto 400°C. The percentage porosity of PCL was found to be 92%, the formula was reported earlier.[2] The water uptake percentage of PCL was found to be 15%. The contact angle of PCL was found to be 112 ± 3°, which confirms the hydrophobic nature of PCL. The inflection point of the storage modulus is at about (–) 60°C–70°C which corresponds to the glass transition temperature of the PCL.

In-vitro Drug Release Studies and Cytotoxicity Studies

The assessed viability using MTT evidenced more inhibitory effect of Paclitaxel on C6 glioma cells in comparison to fibroblasts at the end of 72h. The SEM pictures of the PCL nanofibrous scaffolds after 72 h of cell culture (NIH 3T3) show appropriate adhesion of the cells on the nanofibrous scaffolds and proves the biocompatibility of PCL as shown in Figure 1(d) while the C6 glioma cells grown on P-PCL nanofibrous scaffolds is as shown in Figure 1(e), were less dense the fibroblasts. Hence, the SEM pictures supported the observation that P-PCL nanofibrous scaffolds can be effectively used for cancer therapy. An effective drug delivery scaffold is capable of targeting the drugs at the right time period with controlled dose. At the initial stage, loosely bound Paclitaxel on the surface resulted in small burst release effect while later more controlled and sustained release was observed. The percentage cumulative Paclitaxel release at the end of 15th day for 0.1 and 0.2 wt% formulations was 67.80 ± 1 and 70.31 ± 1.5. 0.2 wt% formulation performed higher release rates than those of 0.1 wt%.

CONCLUSIONS

In the present paper, Paclitaxel, widely known for its anti-tumour properties, was added to neat PCL in various amounts (i.e. 0.1–0.2 wt % based on the weight of PCL). All the results confirmed the P-PCL nanofibrous mat was well suited for drug delivery applications.

ACKNOWLEDGEMENTS

The authors acknowledge the Department of Science and Technology (DST) for financial assistance (Project No. SR/S1/PC-56/2009 (G) dated 12.07.2011. The authors acknowledge the help offered by Late Dr. Praveen Kumar Sehgal, Chief Scientist, Department of Bioproducts, CLRI, and Chennai – 20 in assessing the biocompatibility tests of the present work.

REFERENCES

[1] Wenguo, C., Yue, Z. and Jiang, C., Electrospun nanofibrous materials for tissue engineering and drug delivery, *Sci. Technol. Adv. Mater.,* 11, doi:10.1088/1468-6996/11/1/014108, 2010.

[2] Elakkiya, T., Natarajan, T.S., Ramadhar, K., Sehgal, P.K. and Sheeja, R., Fabrication of electrospun Poly L-Lactide and Curcumin loaded Poly L-Lactide Nanofibers for drug delivery, *Fibers and Polymers*, 13, 823–830, 2012.

[3] Rahul, S., Velmurugan, T. and Seeram, R., Design Modifications in Electrospinning setup for Advanced Applications, *Journal of Nanomaterials,* doi: 10.1155/2011/317673, 2011.

[4] Hyuk, S.Y., Taek, G.K. and Tae, G.P., Surface-functionalized electrospun nanofibers for tissue engineering and drug delivery, *Advanced Drug Delivery Reviews*, 61, 1033–1042, 2009.

[5] Franscesco, I., Delia, M., Erman, P., Benjamin, N., Jean-Christophe, T., Sarra de, V., Giuseppe, F., Alessandro, M., Afksendivos, K., Robert, G., Michael, M. and Beat, H.W., Paclitaxel-Eluting Biodegradable Synthetic Vascular Prostheses A Step Towards Reduction of Neointima Formation?, *Journal of the American Heart Associaton*, doi:10.1161/CIRCULATION.AHA.109.848242, 2009.

[6] Van de Loosdrecht, A.A. *et al.*, A tetrazolium-based colorimetric MTT assay to quantitate human monocyte mediated cytotoxicity against leukemic cells from cell lines and patients with acute myeloid leukemia, *J. Immunol Methods*, 174, 311–320, 1994.

NON-COVALENT MODIFICATION IS THE NEXT PROBABLE STEP TO PREPARE A COLLAGEN BASED BIOMATERIAL USING DI-CARBOXYLIC ACIDS (GLUTARIC AND SEBACIC ACID)

TAPAS MITRA, G. SAILAKSHMI AND A. GNANAMANI

Microbiology Division, CSIR-CLRI, Adyar, Chennai, Tamil Nadu
E-mail: gnanamani3@gmail.com

ABSTRACT

In the present study, an attempt was made to engineer the collagen macromolecule using di-carboxylic acid (DCA) (glutaric and sebacic acid) individually and further evaluated the thermal, mechanical and biocompatible properties of the scaffold. A 3D Scaffold was prepared using collagen 0.5% (w/v) and dicarboxylic acid 0.2% (w/v), which, displayed an appreciable mechanical strength and in addition found biocompatible for NIH 3T3 fibroblast cells. In silico study revealed the non-covalent interaction between DCA and collagen.

INTRODUCTION

The existing problems associated with biopolymer materials on mechanical property have been challenged through suitable cross-linkers. The present study emphasizes non-covalent interaction, which offered the requisite property to the biopolymer material in a green manner. The study employs (i) collagen and glutaric acid; and (ii) collagen and sebacic acid for the preparation of biomaterial. These two di-carboxylic acids through its ionic as well as intermolecular hydrogen bonding interactions stabilizes the collagen and at the same time increases the mechanical property and in addition, makes the material biocompatible. Thus, non-covalent interactions satisfy the need of the hour with regard to biomaterial preparations is concerned.

V. Rajendran, P. Prabu and K.E. Geckeler (eds.)
Nano Biomaterials, pp. 5–8 (2012).

PREPARATION OF BIOPOLYMERS (3D- SCAFFOLD)

Collagen (0.5%, (w/v)) and DCA at different concentration was mixed with water and stirred at 4°C. Followed by stirring, the samples were subjected to centrifugation and was poured in Tarson (India) vial followed by frozen at –4°C for 2 h, –20°C for 12 h and –80°C according to Peng *et al.*[1] The frozen samples were lyophilized for 48 h at a vacuum of 7.5 militorr (1 Pa)

CHARACTERIZATION STUDIES

SEM analysis was performed to assess the morphology of the prepared scaffold. Degree of cross-linking was quantified using 2, 4, 6 -Trinitrobenzenesulfonic acid assay according to the procedure summarized by Bubnis *et al.*[2] For mechanical properties assessment, dried scaffold were used and tested using Universal Testing Machine (INSTRON model 1405) at a crosshead speed of 5 mm min^{-1} at 25°C and 65% relative humidity. Differential Scanning Calorimetry (DSC) analysis made using a differential scanning calorimeter, model -DSC Q 200 (V 23.10 Build 79).

For the docking study, chemical structures were generated using ACD/ChemSketch (ACD/ChemSketch Version 12, Advanced Chemistry Development, Inc., Toronto, ON, Canada, 2009). The three-dimensional structure of type-I collagen was generated using gencollagen program. To find out the interaction between type-I collagen with GA and SA, AUTODOCK has been used.

Finally, MTT assay has been performed using NIH 3T3 fibroblast cells to assess the biocompatibility of the prepared scaffold.

RESULTS AND DISCUSSION

It was obvious that most of the volume of the membranes was taken up by the interconnecting pore space as shown below (Figure 1).

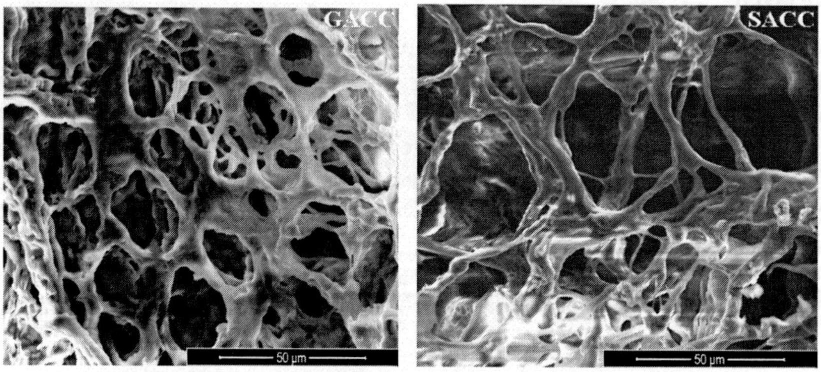

Fig. 1: SEM Figure of GACC (glutaric acid cross-linked collagen) and SACC (sebacic acid cross-linked collagen)

Results on the percentage of cross-linking degree suggested 60–65% cross linking of collagen with 0.2% GA and SA.

With regard to mechanical property, high tensile strength (MPa) values were observed for both the cross-linked biopolymers [Glutaric acid cross-linked collagen (GACC) - 9.34 and Sebacic acid cross-linked collagen (SACC)- 2.96] compared to the native collagen (0.13). Moreover, the Young's modulus of GACC and SACC was 75.02 and 18.61 respectively. The stiffness values (GACC- 1.68 and SACC- 0.69 N/mm) were also greater than that of the native collagen (0.3 N/mm).

The thermo grams of GA, SA, type-I collagen, GACC and SACC were shown in Figure 2. DSC studies recorded degradation temperature differences among GA (96°C), SA (135°C), type-I collagen (92°C), GACC (200°C) and SACC (125°C). The higher transition temperature suggests GACC and SACC had higher stability in high-temperature environment.

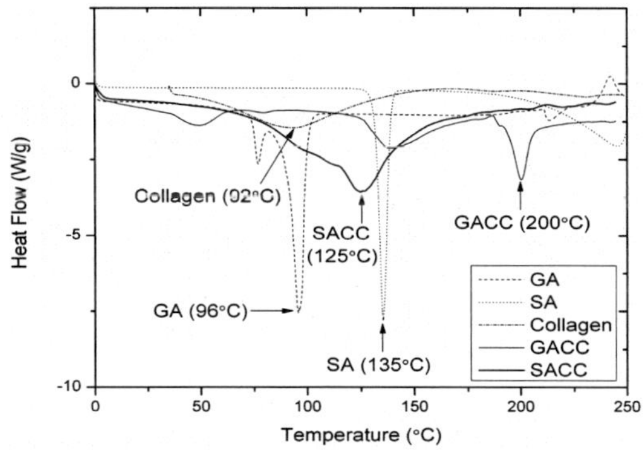

Fig. 2: DSC Analysis of GA, SA, Collagen, GACC and SACC

The intermolecular hydrogen bonding interaction between –COOH group of GA and SA and free ε-NH_2 group of lysine from type-I collagen were shown in following Figure 3.

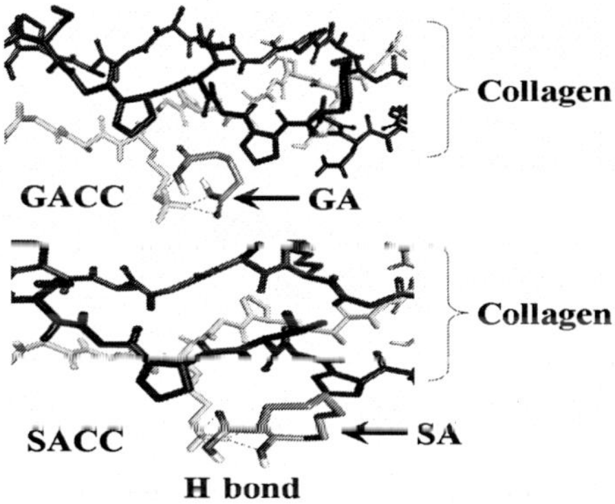

Fig. 3: Multiple Intermolecular Hydrogen Bonding Interactions between GA, SA and Collagen

MTT results clearly indicated that NIH 3T3 cells proliferated well on the surface of the GA and SA cross-linked scaffold.

CONCLUSIONS

The present study explicitly demonstrates GA and SA act as suitable cross-linkers for the preparation of biocompatible biopolymers with appreciable mechanical properties. The interaction between GA and SA with collagen was identified as non-covalent interactions. The resultant material found application as a wound dressing material or as an implant in clinical applications.

REFERENCES

[1] Peng, L., Cheng, X.R., Wang, J.W., Xu D.X. and Wang, G.E., Preparation and Evaluation of Porous Chitosan/Collagen Scaffolds for Periodontal Tissue Engineering, *J. Bioact. Compat. Polym.*, 21, 207–220, 2006.

[2] Mitra, T., Sailakshmi, G., Gnanamani, A., Raja, S.T.K., Thiruselvi, T., Gowri, V.M., Selvaraj, N.V., Ramesh, G. and Mandal, A.B., Preparation and characterization of a thermostable and biodegradable biopolymers using natural cross-linker, *Int. J. Biol. Macromol.,* 48, 276–285, 2011.

POLYMER HOT EMBOSSING WITH SOFT STAMPS

D.K. MAURYA, S. KANDASAMY[1,2], F. LAPIERRE[3], K. ALAMEH AND Y. ZHU[1,2]

Electron Science Research Institute, Edith Cowan University, Joondalup, WA 6027, Australia
[1]Melbourne Centre for Nanofabrication, 151 Wellington Road, Clayton, Victoria 3168, Australia
[2]School of Physics, University of Melbourne, Parkville, Victoria 3010, Australia
[3]CSIRO, Division of Materials Science and Engineering, Clayton, Victoria 3169, Australia
E-mail: d.maurya@ecu.edu.au

ABSTRACT

Microfluidic channels are fabricated in thick polycarbonate (PC) substrates by hot embossing using soft poly(dimethylsiloxane) (PDMS) stamps. The embossing force is in the range of 2.5–3 kN for a 4-inch diameter substrate, and the embossed microchannel is 70 µm deep. We investigate the influence of soft stamp precursor and curing agent mix ratio, and post thermal treatment on embossing conditions. Experimental results show that a soft stamp fabricated with 5:1 mix ratio and post-annealed at 150°C results in better embossing properties than conventional un-annealed stamps.

INTRODUCTION

In recent years, many polymer-based microfabrication techniques have been explored for application in bio-MEMS and microfluidics. Current major polymer microfabrication techniques include hot embossing,[1] injection molding [2] and soft lithography.[3] Among them, hot embossing is the most attractive as it offers high replication accuracy for micron-scale features. The basic principle of hot embossing is heating a polymer substrate above its glass transition temperature (T_g), *pressing it with a mold* or master stamp and gradually cooling it well below the glass transition temperature where patterns from the mold are completely transferred to the polymer substrate. The mold pattern is then permanently transferred to the substrate upon cooling. The hot embossing process exploits differences in thermomechanical properties between the stamp and the substrate. Typically, stamps can be either hard or soft. For hard stamps, which are made of silicon or nickel, fragility, thermal mismatch and cost are

V. Rajendran, P. Prabu and K.E. Geckeler (eds.)
Nano Biomaterials, pp. 9–12 (2012).

the main concern. On the other hand, soft stamps made of polymers such PDMS or epoxy resins can more practical.[4–7] The use of soft stamps made out of polymers as a replication tool would potentially reduce the cost and be more durable, compared to fragile hard stamps. Hot embossing using soft PDMS stamps has been successfully developed by several research groups in recent years,[4–7] and their practicality has been demonstrated.

PDMS is a common elastomeric material widely used in the fabrication of microfluidic devices because the devices made in this material can be readily fabricated by replica molding or soft lithography. PDMS is stable to a temperature of up to 300°C, it has moderately high physical toughness and relatively low modulus. In this paper, we report on the fabrication of microfluidic channels in thick PC substrates by hot embossing using PDMS soft stamps. The effects of PDMS precursor, curing agent mix ratio and post thermal treatment on the stamp performance are investigated for different embossing conditions.

EXPERIMENTAL

Soft lithography technique was used to fabricate soft stamp. Briefly, features on a chrome photomask were transferred onto a silicon wafer using conventional photolithography. The photoresist-defined silicon wafer was then anisotropically etched to a desired depth using Oxford Plasma Lab 100, which is an excellent machine for silicon deep reactive ion etching (Bosch process). After silicon DRIE, photoresist stripping was carried out by soaking the wafer in acetone followed by IPA and DIW rinse. Following photoresist stripping, anti stiction coating was applied, in a glove box, using silane (trichloro, 1H, 1H, 2H, 2H-perfluorooctyl).

The Sylgard 184 (Dow Corning) was used to fabricate soft stamps. First, the precursor and the curing agent were well mixed and poured onto a silicon mold fabricated as outlined above. The mold was generated having the negative replica of the desired features. Curing of the soft stamp was typically performed overnight at room temperature to allow effective elimination of bubbles in the precursor mixture. The flatness and proper leveling of the precursor mixture and mold while curing, was critical to ensure that the soft stamp has uniform thickness throughout. This was important to evenly replicate the structures during the embossing step.

The hot embossing process was carried out using EVG hot embosser (520HE). Table 1 shows an overview for the process parameters used for the hot embossing of microfluidic channels in PC substrates. The PC substrate diameter and thickness were 4-inch and 1 mm, respectively. Multiple embossing processes were carefully carried out without damaging the stamp.

Table 1: Hot Embossing Process Parameters

Stamp	*Stamp Ratio*	*Post Annealing*	*Embossing Conditions*			
			Force (kN)	*Temperature (°C)*	*Time (min)*	*Vacuum*
PDMS	10:1	No	3	165	10	No
PDMS	5:1	No	2.5	165	10	No
PDMS	5:1	Yes, overnight @150°C in oven.	2.5	165	10	No

RESULTS AND DISCUSSION

PDMS stamp hardness is typically sensitive to stamp mix ratio and curing temperature/post-annealing.[8] We exploited this PDMS property to fabricate soft stamp for successful embossing of micro-scale features in PC substrate. The soft stamp, which was fabricated using 5:1 stamp ratio and post-annealed overnight at 150°C, resulted in successful embossing of 70 μm deep microchannels in PC. Figure 1(a) shows microscopic image of embossed microchannel using 5:1 stamp mix ratio. The embossing force was 2.5 kN for a 100 mm diameter substrate. The microscopic inspection of embossed microchannels showed sign of partly connected microchannels. Figure 1(b) shows microscopic image of fully embossed microchannels using stamp prepared with 5:1 mix ratio and post-annealed overnight at 150°C. It is clear from successful embossing that post-annealing of PDMS stamp at 150°C improves stiffness/hardness of the stamp and results in excellent embossing properties. Figure 2 (a)–(b) shows a 3D optical image of the PDMS soft stamp and the embossed microchannels in PC, respectively. The replication accuracy for the embossed microchannels, which was measured by a surface profilometer, was better than 5% after 20 embossing cycles, confirming the practicality of the stamp for mass production.

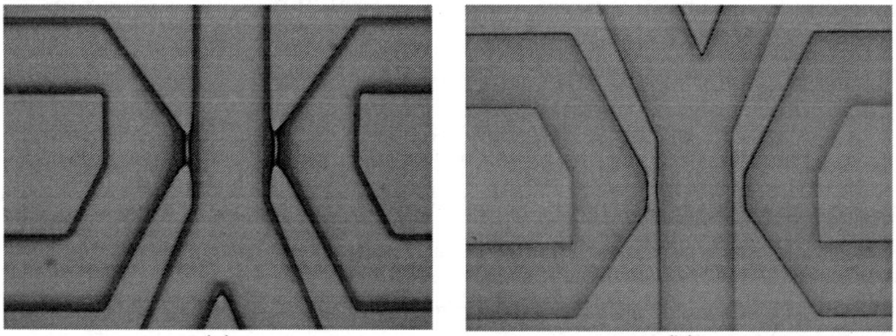

(a) (b)

Fig. 1: Microphotographs of Embossed Microchannel in PC Using (a) 5:1 Stamp Ratio, (b) 5:1 Stamp Ratio and Post Annealed Overnight at 150°C

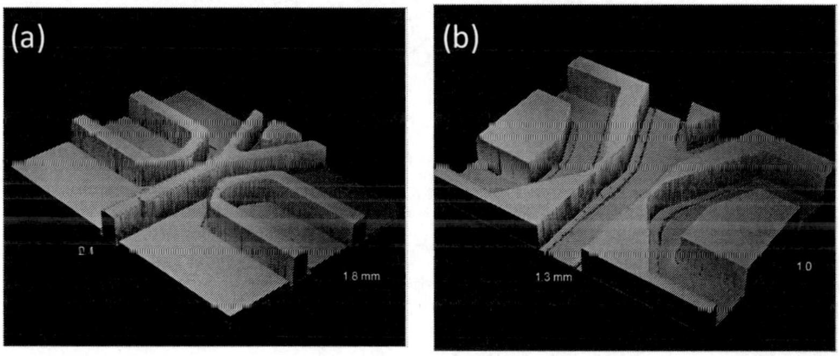

Fig. 2: 3D Optical Image of the (a) PDMS Soft Stamp and (b) Embossed Microchannel in PC

CONCLUSIONS

Microfluidic channels in thick polycarbonate substrates have been fabricated by hot embossing using PDMS stamps. The effects of PDMS stamp precursor, curing agent mix ratio, and post thermal treatment on embossing conditions have been investigated. A soft stamp has been fabricated with 5:1 mix ratio and post-annealed at 150°C resulting in much better embossing properties than un-annealed counterparts.

ACKNOWLEDGEMENTS

This work was performed in part at the Melbourne Centre for Nanofabrication, and supported by Edith Cowan University and the Department of Industry, Innovation, Science, Research and Tertiary Education, Australia.

REFERENCES

[1] Becker, H. and Heim, U., Hot embossing as a method for the fabrication of polymer high aspect ratio structures, Vol. 83, Issues 1–3, pp. 130–135, 2000.

[2] Giboz, U., Copponnex, T. and Mélé, P., Microinjection molding of thermoplastic polymers: A review, *J. Micromech, Microeng.,* Vol. 17, No. 6, R96, 2007.

[3] Rogers, John A., Ralph, G. N., Recent progress in soft lithography, *Materials Today*, Vol. 8, Issue 2, pp. 50–56, 2005.

[4] Goral, V.N., Hsieh, Yi-Cheng., Petzold Odessa N. and Faris, Ronals A., Hot embossing of plastic microfluidic devices using poly (dimethylsiloxane) molds, *J. Micromech. Microeng.,* Vol. 21, pp. 960–967, 2011.

[5] Maurya, D. ., Ng, Wee Yang., Mahabadi, K.A., Liang, Y.N. and Rodríguez, I., Fabrication of lab-on chip platforms by hot embossing and photo patterning, *Biotechnology Journal,* Vol. 2, pp. 1381–1388, 2007.

[6] Duffy, D., McDonald, J., Schueller, O.J.A. and Whitesides, G.M., Rapid prototyping of microfluidic systems in poly(dimethylsiloxane), *Anal. Chem.*, 70, pp. 4974–84, 1998.

[7] Lee, J. ., Park, C. and Whitesides, G. M., Solvent compatibility of poly(dimenthylsiloxane)-based microfluidic devices, *Anal. Chem.*, Vol. 75, pp. 6544–54, 2003.

[8] Herold, K.E. and Rasooly, A., Lab on a Chip Technology: Fabrication and microfluidics, Vol. 1, Horizon Scientific Press, 2009.

INFLUENCE OF MOLAR RATIO ON THE SURFACE CHARACTERISTICS OF FERROUS PHOSPHATE NANOPARTICLE SYNTHESIS

P. RENUKA DEVI AND E. HEMANANTHAN

Biotechnology Centre, Anna University, Regional office, Coimbatore
E-mail: drrenukadevip@gmail.com; hemananth4u@gmail.com

ABSTRACT

In view of the increasing interest in magnetic nanoparticles in the field of health care, a facile synthetic process, which allows controlling the size, properties were developed. Ferrous phosphate nanoparticles were prepared by controlled hydrolysis synthesis under various initial molar ratios without adding any oxidizing and reducing agents in order to clarify effects of the molar ratio on the reaction mechanism. The powdered nanoparticles prepared were characterized by BET Surface area analyzer, Centrifugal particle size analyzer (CPS) and Zeta potential analyzer. At the molar ratio corresponding to the stoichiometric ratio in the synthesized reaction, the particles showed the mean particle size of 85 nm and the mean surface area of 220 m^2/g with the zeta potential value of +6.8 mV, rather other concentrations delivered increased particle size with low surface area. The small positively charged zeta potential value was found to be beneficial in understanding the stability of the $FePO_4$ particles. With these advantageous surface characteristics, the particles were expected to have high intestinal absorption and high bioavailability. It was clarified that the molar ratio of the metal salt used in the synthesis played an important role in determining the surface characteristics of the particles.

INTRODUCTION

Over the last decade, the preparation and characterization of hybrid materials have been extensively studied. These materials offer exceptional opportunities not only to combine the important properties of the organic and inorganic phases but also to create entirely new compositions with truly properties. These characteristics make these materials very promising

V. Rajendran, P. Prabu and K.E. Geckeler (eds.)
Nano Biomaterials, pp. 13–16 (2012).

in different technological areas as adsorbents, sensors, optical, catalysis and electrochemical devices.[1] Additionally this method allows us to prepare materials with different morphological characteristics, by varying the quantity of organics added in the synthesis. Recently, as the applications expand, the demand for iron containing nanoparticles increases. Various methods for preparing iron nanoparticles have been developed.[2–3]

In many hydrothermal methods, a mixed aqueous solution of Fe2+ and Fe3+ is used as the starting solution and an oxidizing or reducing agent is added to the solution. This results in insufficient control of the properties because the reaction rate changes remarkably depending on the molar ratio of Fe2+/Fe3+ and the presence of oxidizing and reducing agents.[4–7] Therefore, it is necessary to clarify the reaction mechanism in order to control the properties although there have been no such studies so far.

The interests in the synthesis of ferrous phosphate nanoparticles have been enormously increasing since its importance in varied applications were indicated by researchers.[8] Ferrous phosphate is a poorly soluble iron compound, with high biological impact and unknown absorption values because of its bulk size. The scientific opinion of the European food society authority indicates the importance of ferrous phosphate compounds as a food additive and its unrevealed nature of absorption and acidic solubility.[9]

This paper reports on the synthesis of $FePO_4$ nanoparticles by a controlled hydrolysis reaction. Furthermore, the dependence of particle diameter, surface characteristics and properties on the molar ratio of ferrous and phosphate ions contained in the metal salt solution is discussed, with the aim of establishing a procedure for the facile formation of ferrous phosphate nanoparticles.

MATERIALS AND METHODS

All chemicals used in the experiments were of analytical reagent grade and were used without further purification.

Synthesis of Ferrous Phosphate Nanoparticles

The ferrous phosphate nanoparticles were synthesised by controlled hydrolysis method. To the aqueous solution of 0.5 M sodium phosphate, 0.5 M of ferrous sulphate solution were added drop wise to obtain the stochiometric concentration (sample A) of the precursors. Besides this, the synthesis procedure was followed, altering the concentration of the Fe salt (0.25 M (B), 0.75 M (C)) to understand the effect of molar ratio and to obtain an appropriate Fe content in the synthesised particles. The solution developed was kept at room temperature under vigorous stirring. The pH of the solution was maintained at 6.2 throughout the process at room temperature. After stirring for 40 minutes, the particles was allowed to settle down, vaccum filtered, washed with ethanol and secondly with deionised water to remove the impurities. The final product was dried at 80°C and subjected for characterization.

Particle Size Distribution Analysis

The disc centrifuge measures the particle size distributions using sedimentation, a well known and reliable method of particle size analysis. Particles settle in a fluid under a gravitational field according to stoke's law. The samples were analysed using CPS disc centrifuge particle size analyser (CPS Instruments, Inc. Florida, USA) operated at a speed of 24,000 rpm for 40 minutes and a sample volume of 350 µl. All the analyses were run against a known calibration standard and the particle size in the solution was observed by size distribution graph.

Surface Area Analysis (BET method)

BET is a technique that is used to measure the surface area and pore volume of a sample. Before subjecting to the surface area analysis (Quantochrome, Nova 12000, USA) samples were degassed with N_2 gas at 150°C prior for the determination of surface area. Degassing was done for 3 hours at N_2 flow rate of 60 cm^3/mm. The analysis was performed under liquid nitrogen.

Zeta potential Analysis

The zeta potential of the nanopowder samples were determined according to laser Doppler electrophoretic mobility theory using the Zetasizer 2000 (Malvern Instruments Ltd, UK) at 25°C. The concentration of the sample was diluted to 0.1% (v/v) by pH 7.0 PBS (0.05 M).

RESULTS AND DISCUSSION

The synthesis of ferrous phosphate nanoparticles were carried out without adding any oxidizing and reducing agents under various initial Fe and P molar ratios in order to clarify effects of the molar ratio on the reaction mechanism. When the molar ratio was 0·5 M corresponding to the stoichiometric ratio in the synthesis reaction of $FePO_4$, the nucleation progressed rapidly in an initial stage of the hydrolysis, resulting in formation of the Fe_3O_4 nanoparticles having a smaller size. On the other hand, when the molar ratio was more and less than 0.5 M, the particle size increased because the $FePO_4$ nanoparticles grew up slowly using surplus $Fe(OH)_2$ under hydrolytic conditions according to the Schikorr reaction.

The size distributions of the particles were observed using the centrifugal particle size analyser. Though the shape and morphology was found to be unique for all three synthesis concentrations, the molar ratio was found to have influenced the size of the particles. The diameters of the particles were shown in Table 1. The size distribution study indicated that altering the concentrations of the metal salt tend to produce changes in the particle size of ferrous phosphate particles. Since, particle size of the compounds alone does not explain the differences in the bioavailability values, the samples were subjected for the analysis of specific surface area.

The surface area is a suitable parameter that can reflect the solubility of the iron powders. The surface area of the samples were analysed using the BET surface area analyser. As a general feature, the specific surface area of the particles obtained by different synthesis experiments ranged from 70 to 220 m^2/g. The sample with low particle size revealed the maximum surface to volume ratio.

Zeta potential is an important and useful indicator of the particle surface charge, which can be used to predict the stability of the synthesised particles. The zeta potential of ferrous phosphate nanoparticles was 6.8, 8.2 and 11.8 mV at pH 7.0 PBS (0.05 M). The particles possessed small positive charges at pH 7.0 indicating that a weak electrostatic repulsive force occurred between the particles. It was found to be beneficial in understanding the stability of the $FePO_4$ nanoparticles.

Table 1: Shows Particle Size Distribution, Specific Surface Area and Zeta Potential Values of the Particles Present in Different Method of Synthesis, Altered by the Molar Ratio of the Metal Salt

Sample	*Particle Size Range (nm)*	*Specific Surface Area (m^2/g)*	*Zeta Potential Value (mV)*
A	70–130	220	+ 6.8
B	110–160	104	+ 8.2
C	180–220	87	+ 13.8

CONCLUSION

The results suggest that the molar ratio is one of the most important operating factors in the control of characteristics of $FePO_4$ nanoparticles. It was clarified that the molar ratio of the metal salt used in the synthesis plays an important role in determining the particle size, surface area and surface charge in the obtained product. Low particle size and high surface area are the advantageous factor of the particles for the bioavailability analysis, obtained by stoichiometric concentration.

REFERENCES

[1] Zhao, Q.T., Hou, L.S. and Huang, R., Synthesis of ZnS nanorods by a surfactant assisted soft chemistry method, *Inorg. Commun.* 6, 971–973, 2003.

[2] Noristi, L., Marchetti, E., Baruzzi, G. and Sgarzi, P., Investigation on the particle growth mechanism in propylene polymerization with MgC12-supported Ziegler Natta catalysts. *J. Polym. Sci. Part A*, 32, 3047–3059, 1994.

[3] Jamjah, R., Zohuri, G.H., Vaezi, J., Ahmadjo, S., Nekomanesh, M. and Pouryari, M., Morphological study of spherical $MgCl_2$.nEtOH supported TiCl4 Ziegler-Natta catalyst for polymerization of ethylene. *J. Appl. Polym. Sci.* 101, 3829–3834, 2006.

[4] JalaliDil, E., Pourmahdian, S., Vatankhah, M. and Taromi, F.A., Effect of dealcoholation of support in MgC12-supported Ziegler-Natta catalysts on catalyst activity and polypropylene powder morphology. *Polym. Bull.* 64, 445–457, 2010.

[5] Mizutani N., Effect of ferrous/ferric ions molar ratio on reaction mechanism for hydrothermal synthesis of magnetite nanoparticles. *Bull. Mater. Sci.*, 31, 5, 713–717, 2008.

[6] Chen, J.Y., Herricks, T. and Xia. Y.N., Polyol synthesis of platinum nanostructures: control of morphology through the manipulation of reduction kinetics, *Angew. Chem.* Int. Ed. 44, 2589–2592, 2005.

[7] Takagi, H., Isoda, T., Kusakabe, K. and Morooka,S., Effects of solvents on the hydrogenation of mono-aromatic compounds using noble-metal catalysts, *Energy Fuels* 13, 1191–1196, 1999.

[8] Siddharth, S. and Debabrata, D., Applying Nanotechnology to human health: Revolution in Biomedical Sciences. *Journal of Nanotechnology,* Vol. 10, pp. 25–28, 2009.

[9] Aguilar, F., Scientific opinion of the panel on Food additives and nutrient sources added to food. *The EFSA journal,* Vol. 951, pp. 1–13, 2009.

A CLINICAL STUDY OF ADVANCED AIDS-KS WITH LIPOSOMAL DOXORUBICIN, PHLEOMYCIN AND VINCRISTINE IN NANO DRUG DELIVERY

POORNACHARI

Vasavi Vidya Trust Group of Institutions, Salem, Tamil Nadu
E-mail: director@vysyamit.in; poornacharimd@gmail.com

ABSTRACT

Liposomes are used for drug delivery due to their unique properties. A liposome encapsulates a region on aqueous solution inside a hydrophobic membrane; dissolved hydrophilic solutes cannot readily pass through the lipids. Hydrophobic chemicals can be dissolved into the membrane, and in this way liposome can carry both hydrophobic molecules and hydrophilic molecules. To deliver the molecules to sites of action, the lipid bilayer can fuse with other bilayers such as the cell membrane, thus delivering the liposome contents. By making liposome in a solution of DNA or drugs (which would normally be unable to diffuse through the membrane) they can be (indiscriminately) delivered to pass the lipid bilayer. There are three types of liposomes—MLV (multilamillar vesicles) SUV (Small Unilamellar Vesicles) and LUV (Large Unilamellar Vesicles). These are used to deliver different types of drugs. Liposomes are used as models for artificial cells. Liposome can also be designed to deliver drugs in other ways. Liposome that contain low (or high) pH can be constructed such that dissolved aqueous drugs will be charged in solution (i.e., the pH is outside the drug's pH range). As the pH naturally neutralizes within the liposome protons can pass through some membranes), the drug will also be neutralized, allowing it to freely pass through a membrane.

Experiments are conducted how different types of lipolomes in different PH levels with some selected drugs like Doxorubicin, Phleomycin and vincristine (50 Mg, sodium actatate) in PH 5.5, 5.0, 4.5, 4.0

V. Rajendran, P. Prabu and K.E. Geckeler (eds.)
Nano Biomaterials, pp. 17–22 (2012).

The release of contents from lipsomes composed Phosphatidylethenolamine and Cholesteryl of hemisuccinate to be a function of PH. 258 patients with advanced AIDS-KS treated with pegylated liposomal doxorubicin (20 Mg/m^2) phleomycin (10 Mg/m^2) and vincristine (1 Mg).

For 14 days 6 cycles. For an overall response rate of 45.9%

Out of 133 patients, 1 achieved complete clinical response.

61 patients achieved partial clinical response.

71 patients had poor response.

Out of 125 patients – received ABV, 31 patients achieved partial clinical response – response rate 24.8%

94% patients had poor response

95% of confidence interval in 1st case – (37% –54%)

95% of confidence interval in 2nd case – (17% –32%)

The difference was stastiseally significant in addition to the objective responses. Prospectively defined clinical benefits and toxicity outcomes also favoured pegylated liposomal doxorubicin and it is more effective and less toxic than the standard combination chemotherapy.

INTRODUCTION

Standard chemotherapy agents have achieved a moderate success in the treatment of Kaposi sarcoma (KS). Kaposi sarcoma is the most common malignancy seen in the setting of HIV infection. Although the percentage of HIV infected patients presenting with K.S as an Aids-defining diagnosis decreased from 30 to 10% in last 10 years. But patients with advanced AIDS with Kaposi sarcoma are in the increase. At the same time treatment with effective anti retro-viral therapy reduces the condition. KS is generally identified as a late manifestation of HIV infection. Advanced AIDS with KS affect quality of life. Tumor with edema cause difficult and pain. Bulky oral dosage leads to speaking trouble. In severe cases, cause respiratory problems, disfigurement and isolation.

Now many research institutes and departments associated with anti HIV drug delivery systems apply nanotechnology to improve AIDS treatment. The basic concept behind the use of nanotechnology based systems for antiretroviral drug delivery is the ability to compartmentalize and modify the qualities and behavior of drugs in the bio medium. Through drugs association with nano structured systems, the release of drugs are determined by the physico- chemical properties of the nano systems and not the properties of drugs. There properties can include the protection from the metabolism, to increase the retention of the drug in the body and the possibility of the specific drug to specific cells or organs. A dosage reduction, more appropriate dosage regimens. lower adverse effects and patients compliance. “There is a possibility of incorporating different anti HIV drugs in the same delivery system which can also contribute to simplification of drug administration schedules”(1). There is improvement of the nanotechnology based studies. Many of them are skill in the pre formation or pre-clinical phases. The potential of nanotechnology based drug delivery systems to improve AIDS treatment is evident.

The goal of the current study is to identify a systematic review with Doxorubicin, Phleomycin and vincristine in isolation and also in combination in nano drug delivery. The most popular surfactant system that incorporates anti retroviral drugs are the liposomes (Ziden *et al.,* 2010, cleylon *et al*., 2009 kent *et al.,* 2008). Nano particles were developed in and around 1970. It was used as carriers for vaccine and anti cancer drugs (2). In order to enhance tumor uptake, the strategy of drug targeting was employed and as a first step, research focused on the development of method to reduce the uptake of the nano particles by the cells of the reticulo-endothelial system (RES) (3). Later the use of nano particles for opthalmic and oral delivery was investigated (4).

OBJECTIVES

1. To study whether liposomes can be used in the treatment of AIDS- KS with drugs.
2. To study at what pH the absorption is peek.
3. To study what will be the ideal ph for combination of drugs.

CLINICAL STUDY AND DRUGS

Doxorubicin

Also known as adriamycin and hydroxydenorubicin. As a drug used in cancer therapy. It is an anthracycline antibiotic, closely related to the natural product dennomycin and like all anthracyclines. It works by intercalating DNA, while most serious adverse effect being life threatening heart damage. It is commonly used in the treatment of a wide range of cancers, including hematological malignancies, many types of carcinoma and soft tissue sarcomas. The drug is administered intravenously, in the form of hydrochloride salt. It is photo sensitive andd often covered by an aluminum bag or from wax peper to prevent light from affecting it. Doxorubicin is available in liposome encapsulated forms as DOXIL, CAELYX and MYOCES.

Phleomycin

It is a glycopeptide antibiotic of the Bleomycin family. Isolated from a mutant strain of S. verticillus. It binds and intercalates DNA thus destroying the integrity of the double helix. It is active against most bacteria, fungi, yeast, plant and animal cells. Phleomycin is used at a concentration of 10 mg/ml.

Formula – C55 H85 N_{20} S_2 CU – HCL.

Vincristine

Vincristine (ONCOVIN) formally known as leurocristine some times abbreviated"VCR" is a Vinca alkaloid. It is a mitotic inhibitor and is used in cancer chemotherapy. Vincristine is created by the coupling of iodole vindoline and catharanthinea in the vinca plant (5).

TREATMENT

Doxorubicin – 20 mg/M^2/day
Phleomycin – 10 mg/M^2/day
Vincristine – 1 mg/day

The related contents from liposomes composed of phosphatidylethnolamine and cholesteryl of hemisuccinate for 258 patients with advance AIDS- KS for 14 days 6 cycles.

Batch–I		*Batch–II*	
Full response	1	Partial response	31
Partial response	61	Poor response	94
Poor response	71		
Total	**133**	**Total**	**125**

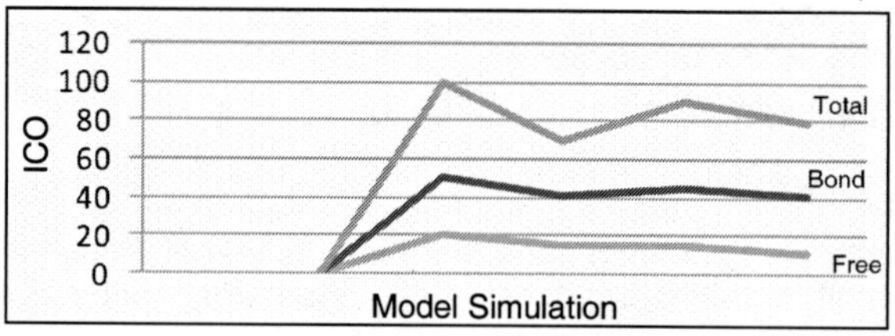

Source: Primary data.

Diagram–I: Intracellular Absorption Using Fluorascence

It is observed that intracellular absorption is slow at acidic PH and increase with alkaline PH.

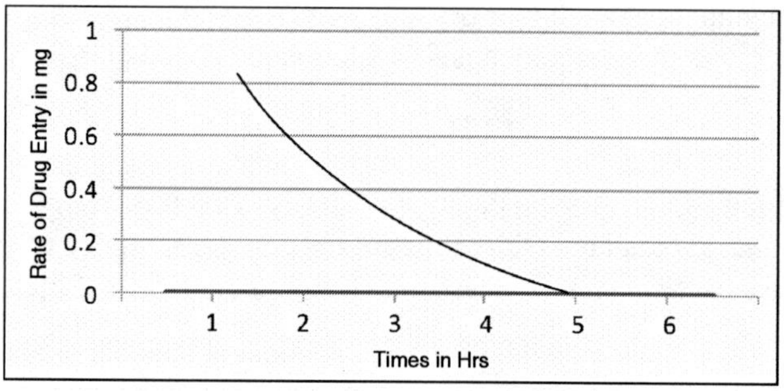

Source: Primary data.

Diagram–II: Time and PH

It is observed at initial rate of drug absorption is more within one hour of administration of drug at a particular PH. As time passes the rate of absorption is in decreasing trend. The above said drug combination, the absorption is high at 6.4 PH and lower at 7.4 PH.

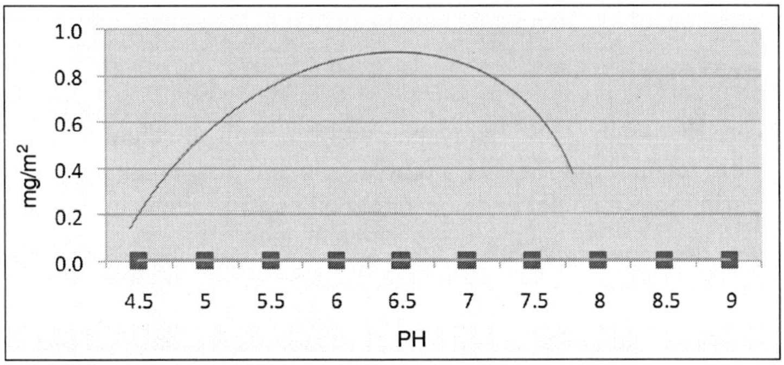

Source: Primary data.

Diagram–III: PH and Intracellular Release of Drugs

It is observed that polymeric vesicles produced by spontaneous self association of liposomes in aqous solutions of PH5 exhibit the PH regulated drug release behavior. Though the electrostatic interaction with ionised drugs molecules can be highly accommodated into either outer or inner surfaces of vesicles when the Ph is adjusted from 5 to 7.4 it depends largely on the structural transition of vesicles in response to PH change. The drug release is accelerated through the disruption of electro static interaction of drugs.

SIDE EFFECTS

Myelosuppression is a frequent problem. Drug concentration may also considered to reduce side effects. Acute effects of nausea, vomiting and arrhythmias. Swelling after 48 hours. Cardio myopathy and death. Peripheralneuropathy, hyponatremia. Constipation and hair loss may also occur.

FINDINGS

By using liposomal agents, the drugs absorption is increased and the retention of drugs in cell membrane is increasing. Absorption increases in acid medium starting from PH 4.5 and maximum at 7.4 and this page is ideal.

CONCLUSION AND RECOMMENDATIONS

The above study is useful in ascertaining the combined drug absorption of liposomal Doxorubicin. Phleomycin and vincristine in nano drug delivery methods under variation of PH levels. Maximum absorption level attained at PH 7.4 and best absorption commences from PH 5. As PH varies the rate of absorption also varies. AIDS- KS remains an incurable tumor . The selection of therapy must be taken into account the potential benefit and adverse effects of treatment. Cosmetically disturbing lesions should be treated early. Radiation therapy is effective local therapy. Systemic chemotherapy used for patients with rapidly progressive or symptomatic and liposome encapsulated drugs are the most effective and less toxic.

ACKNOWLEDGEMENT

I express my sincere thanks to Dr.Sivnarayanan and his team of scientists and technicians of the department of bio technology for their humble help and also thanks to the management of the institute for bearing the financial and non-financial contributions during the study.

REFERENCES

[1] Falvia Chiva carvalso and others. School of pharmaceutical sciences. State university of sao Paulo, UNESP Araraqura. State university of west center, UNICENTRO, Guaraphava, Brazil.

[2] Corvrewr P. Kante, B. Grislain, L. Rolend, M. and Speister P. toxicity of polyalkyl lyamoacrylete nano particles II. Doxorubicin loaded nano particles. *J. Plerm. Sci.* 71-790-792, 1982.

[3] Convreur, arislain, L.leveert, V. Brasseur, F. Guiot, Pand Biernecki, A. Bio degradable polymeric nano paticles as drug carrier for anti timer drugs. In Guiot. P. and convreur. P (EDS) polymeris nano particles and microsphses. CRC press. Bocaration, 1986 PP 27–93.

[4] Lathasetwar. V, Sing .C and levy; R.J. nano paticle drug del-rev, 24, 63–85, 1997.

[5] Pharmacosrogy of vince alkaloids (http://phrrueguerse.info/press/2012/01 Pharmacosrogy of vince alkaloids. Periwinlelel).

Websites

http://www.nacoonline. org/Living – with – HIV AIDS

http://www.cdc. gov/hiv

http://www.who.int./hiv/en

http://www.aidsinfo.nih.gov

http://hivinsite.ucsf.edu.

SILVER HYDROXYAPATITE NANOCOMPOSITE PREPARED BY A NOVEL METHOD FOR C-N BOND COUPLING AND ANTIBACTERIAL ACTIVITY

PERUMAL EMAYAVARAMBAN AND NALLASAMY DHARMARAJ

Department of Chemistry, Bharathiar University, Coimbatore
E-mail: dharmaraj@buc.edu.in

ABSTRACT

Now-a-days, noble metal nanoparticles synthesised by green approach have attracted the attention of several researchers over chemical methods due to the high toxic nature of reducing agents such as $NaBH_4$. The hot water extract of bark of Acacia auriculiformis *was applied to reduce silver (I) ions to metallic silver nanoparticles. A further, silver hydroxyapatite nanocomposite was prepared by adding a solution of calcium nitrate tetra hydrate and di-ammonium hydrogen phosphate as precursors for HAp into AgNPs colloids. The as-synthesised Ag-HAp nanocomposites were characterized by UV-visible spectroscopy, Fourier transforms infrared spectroscopy, powder X-ray diffraction, scanning electron microscopy, transmission electron microscopy and surface area analysis. The synthesised silver nanocomposite was used as heterogeneous catalyst for C-N coupling reactions using imidazole/ benzimidazole and substituted aryl halides. The prepared Ag-HAp was tested for antibacterial activity against some gram positive and gram negative bacteria.*

INTRODUCTION

Calcium hydroxyapatite $Ca_{10}(PO_4)_6(OH)_2$ is particularly interesting because of its chemical similarity to the principal inorganic constituent of bone tissue. Due to its absolute biocompatibility with living tissues, calcium hydroxyapatite ceramics are widely used as biomaterials in medicine.[1] In addition, hydroxyapatite is used as absorbents in chromatography and catalysis many of the organic reactions were carried out as a support material for the conversion of

V. Rajendran, P. Prabu and K.E. Geckeler (eds.)
Nano Biomaterials, pp. 23–28 (2012).

methane, dehydration of alcohols, carbon-carbon bond coupling and others.[2–4] They can also be effectively used for nuclear waste disposal.[5] Hydroxyapatite modified by substitution of other elements to have better catalytic activity, on the either Ca cationic position with Ni, Cu, Zn and Mg or anionic PO_4^{3-} sites with VO_4^{-3}, HPO_4^{-2}, CO_3^{-2}, OH, CO_3^{-2} and F^{-} of the lattices.[6] A thin coating of metals on hydroxyapatite, titanium and CoCrMo alloys, for hip, knee, and dental prostheses, silver atom in hydroxyapatite increase the biocide activity towards pathogens and this material can be used as reconstructing bones and treatment for bone defects. Biosynthesis of silver and gold nanoparticles has been reported earlier using Cinnamomum zeylanicum leaf broth and extract of Azadirachta indica.[7,8] Condensed tannins are polyphenols with multifunctional groups widely distributed in plants. The general characteristics of condensed tannins are that they are able to chelate with many kind of metal ions (Cu^{2+}, Pt^{4+} and Pd^{2+}, etc.).

C-N bond forming reaction is an important one in organic synthesis from lower molecular level to higher proteins. Generally, Ullmann coupling for hetroaromatic with aryl borane, aryl silane and aryl bismuth reagents was used.[9–11] However, these reactions suffered from several drawbacks such as high reaction temperature and extended reaction time. Hence, we reported herein the application of silver-HAp nanocomposite as a green method for N-arylation of benzimidazole.

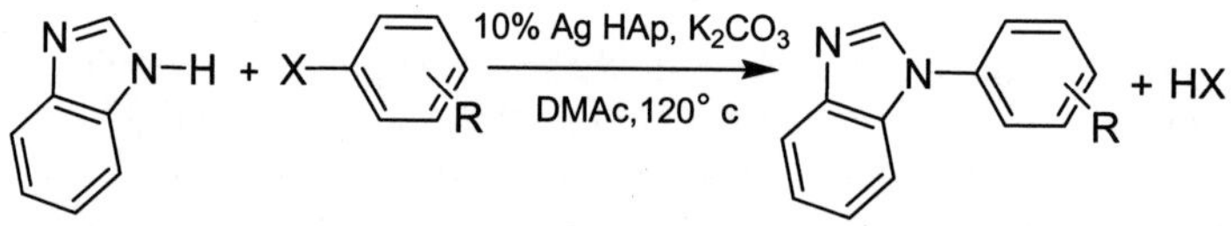

Where X= F, Cl, Br and R= -H, -CN, -NO2, -OCH3 or -COCH3

Scheme–1

EXPERIMENTAL

Preparation of Plant Bark Extracts (PBE)

Acacia auriculiformis plant bark was collected from the Bharathiar university campus at Coimbatore. Samples collected from different trees of barks were dried under sunlight for about 2 days after drying the barks were crushed in to fine powders and was used for throughout the our investigation. About 5 g of powered bark was added in to a 50 mL of double distilled water and was heated up to 10 minutes in a water bath filter the extract by whatman filter paper residue was thrown off and the filtrate was preserved at 4°C for further study.

Catalyst Preparation

Three different concentrations of silver nanoparticles loaded hydroxyapatite nano composites were prepared by a green rout. For all the three composites, 5%Ag HAp, 10%Ag HAp and 15%Ag HAp, silver nitrate was used as precursor for silver. For the optimization of quantity of plant bark extract (PBE) to reduce silver ions into silver metal nanoparticles, 200 μL, 400 μL, 600 μL, 800 μL, 1 mL and 2 mL of the extract was added to 1mM of silver nitrate. In the first step, 1mL of PBE was added in to 1mM of $AgNO_3$ solution (100 mL) and continuously stirred for 30 minutes during which the silver nitrate changes from colorless to yellow color this indicates that silver (I) ions is reduced to Ag NPs. Second step add 0.01M, 0.005 M and 0.003 M of $Ca(NO_3)_2.4H_2O$ (20 mL)in to silver nanoparticles solution and a drop wise addition of 0.01 M, 0.005 M 0.003 M of di-ammonium hydrogen phosphate (80 mL) was done. The black/brown colored precipitate were forming continuously stirred for about 3 hours and then samples were freeze dried and kept under air tight vacuum desiccator.

Procedure for the N-Arylation Reaction

About 50 mg of 10%Ag HAp was added to a mixture of benzimidazole (1 mM), 4-Cl-benzonitrile (1.2 mM) and K_2CO_3(2 mM) in 4 mL of N,N-dimethyl acetamide (DMAc) and refluxed at 120° C (Scheme 1). After completion of the reaction monitored by thin layered chromatography (TLC), the reaction mixture was centrifuged to separate the catalyst and washed several times with ethyl acetate. The filtrate was quenched with aqueous sodium hydrogen carbonate and the product were extract with ethyl acetate. The combined organic layer were dried over anhydrous sodium sulfate and filtered. The filtrate was concentrated and the residue was purified by column chromatography on silica gel (petroleum ether and ethyl acetate).

RESULT AND DISCUSSION

From the UV-visible measurements in the range 300 to 800 nm for samples 5% Ag HAp, 10% Ag HAp and 15% Ag HAp, the surface Plasmon band appeared at 381nm, 400 nm and 427 nm respectively this indicates that silver NPs are in different sizes whereas hypsochromic shifts (blue shift) represents particles are smaller in size and bathochromic sifts (red shift) particles are larger in size. In absence of silver nanoparticles, hydroxyapatite did not exhibit any absorption bands in the absorption spectrum as presented in Figure 2.

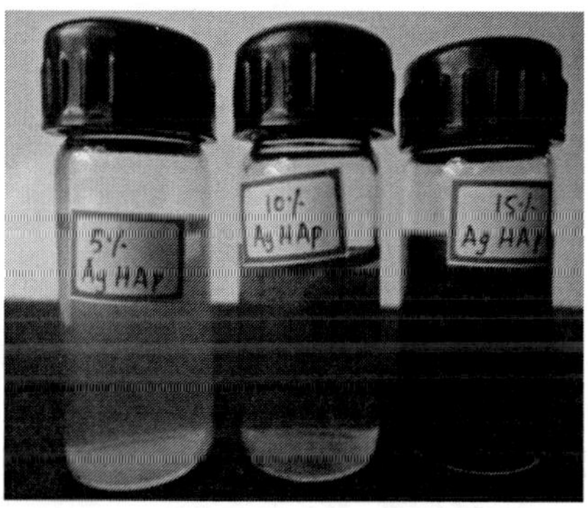

Fig. 1: Different Percentage Loaded Ag NPs in HAp Nanocomposites

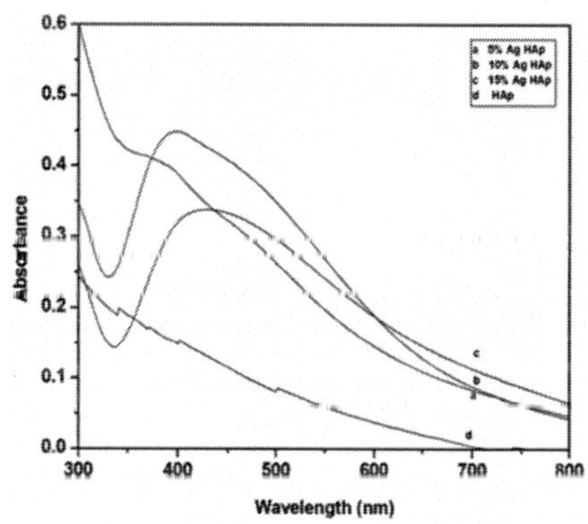

Fig. 2: Absortion Spectra of Ag NPs in HAp Nanocomposites

In FT-IR spectrum of all three composite shows that a broad band observed around 3432 cm^{-1} and another weak band around 1412 cm^{-1} were assigned to the stretching and bending vibrations of hydroxyl group respectively. Further, the appearance of three bands at 1032 cm^{-1}, 602 cm^{-1} and 562 cm^{-1} were characteristics of P-O bonds in $(PO_4)^{3-}$ group. There is no change in the stretching and bending vibrations modes of phosphate and hydroxyl groups in HAp owing to presence of silver nanoparticles. In powder XRD patterns, well crystalline HAp registered at 2θ values 26.1, 29.0, 32.3, 47, 49.7 and 53.4. It was found that metallic silver phase at 2θ value 38.2, 44.0, 64.4 and 77.5 observation from the XRD patterns, the hydroxyapatite peaks decreases with the increase in silver nanoparticles percentage, were as Ag NPs are in cubic phase (JCPDS No. 87–0720). From the SEM analysis, surface of the silver hydroxyapatite nanocomposite were found to be porous like structure was shown in the Figure 5. Similarly from the TEM micrograph showed in the Figure 6, flake like structure of silver hydroxyapatite nanocomposites was observed. The silver particles size was conformed to be less than 5 nm and selected area diffraction pattern was inserted in Figure 5b. It was clearly shown that polycrystalline nature of the material. The Brunauer-Emmett-Teller (BET) surface area of silver hydroxyapatite of three different loading percentage 5% Ag HAp, 10% Ag HAp, 15% Ag HAp with corresponding values are 88.93 sq./gm, 99.60 sq./gm, 60.36 sq./gm respectively.

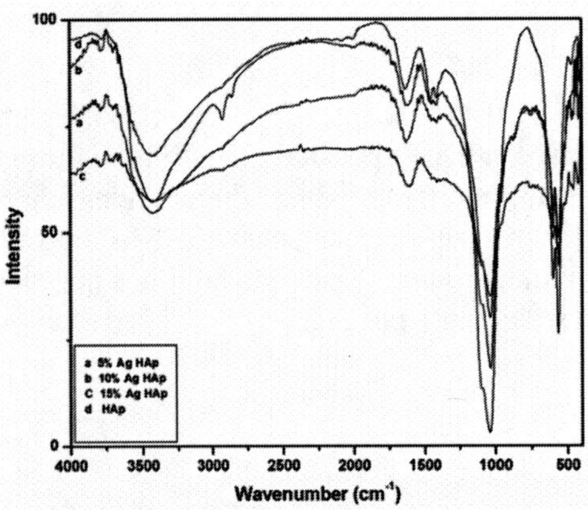

Fig. 3: FT-IR Spectra of Silver Nanoparticles Loaded Hydroxyapatite Nanocomposites

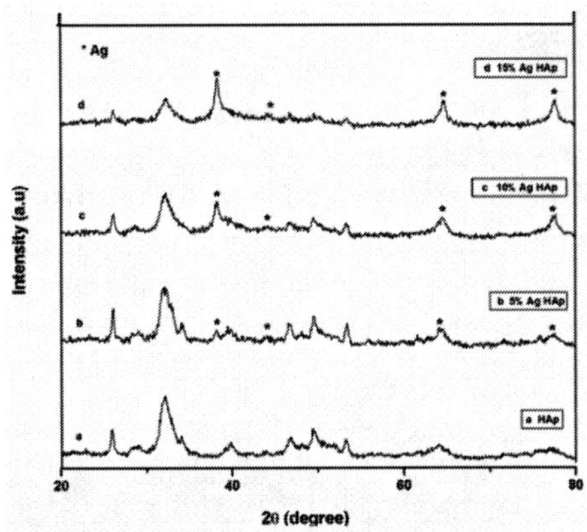

Fig. 4: XRD Patterns of Silver Loaded Hydroxyapatite Nanocomposites

PRODUCT ANALYSIS

4-Benzoimidazole-1-yl-benzonitrile was characterized by FT-IR, proton and carbon 13 NMR techniques.

FT-IR (KBr, υ, cm^{-1}): 2223 (C≡N), 1590 (C=N)

^{1}H NMR ($CDCl_3$): δ = 7.42 (m, 2H, C5', C6'-H), 7.61 (m, 2H, C3, C5-H), 7.72 (dd, *J*=4.5 Hz, 2H, C2, C6-H), 7.93 (m, 2H, C4', C7'-H), 8.19 (s, 1H, C2'-H).

^{13}C NMR ($CDCl_3$): δ = 110.40, 111.94, 117.72 (**C**≡N), 120.71, 124.13 (C3 and C5), 124.80 (C5' and C6'), 132.63, 134.26 (C2 and C6), 139.82, 141.48, 143.10.

Isolated yield: 81%

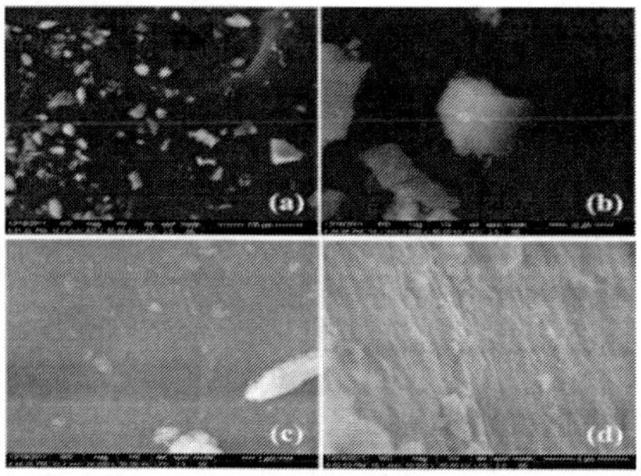

Fig. 5: SEM Images of 10% Ag HAp Nanocomposites Low Magnification to High Magnification (a), (b), (c) and (d)

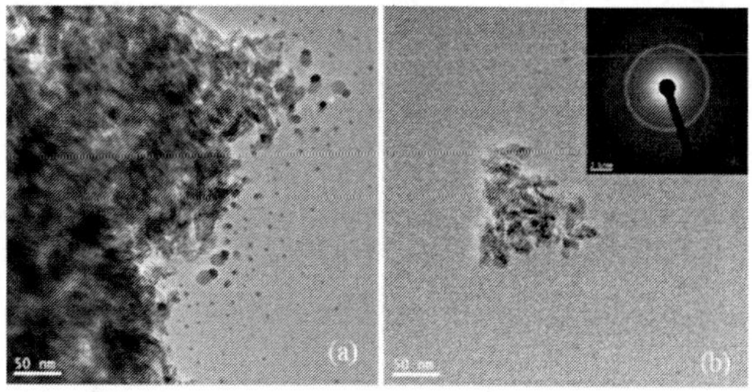

Fig. 6: (a) and (b) are TEM Image of 10%Ag HAp and Corresponding SEAD Patterns are Inserted in (b)

ANTIMICROBIAL ACTIVITY

Bactericidal effect of silver hydroxyapatite was studied against the four bacterial species Psedudomonas fluorescence, Bacillus subtilis, Eschericia coli and Staphylococcus aureus. The experiments clearly indicates that the tested sample with silver nanoparticles concentration 5%, 10% and 15% showed inhibitory effect against all four bacterial species. 15% silver hydroxyapatite showed the higher inhibition against all the four bacterial species. The HAP sample without silver nanoparticles did not show any antibacterial activity. The bar diagram (Figure 7) shows the inhibitory effect of sample with 15 % Ag hydroxyapatite nanocomposites.

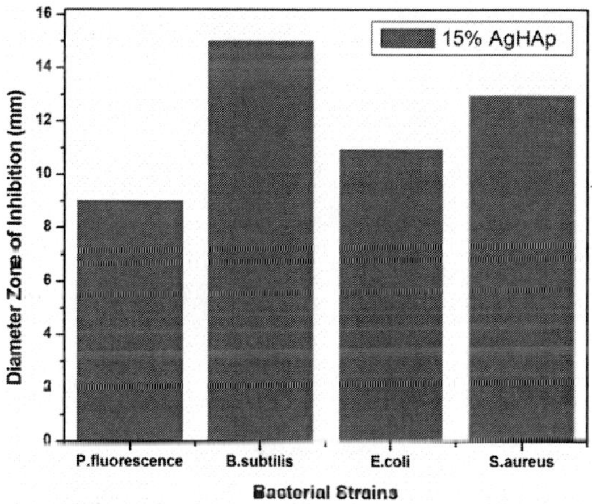

Fig. 7: Bar Diagram for the Antimicrobial Activity

CONCLUSION

The work describes a green chemical route to synthesize silver hydroxyapatite nanocomposites. The catalyst showed better activity for N-arylation reactions compared to Ag-HAp nanocomposites prepared via chemical method. Antibacterial activity tested against gram positive and gram negative bacteria's showed higher activity mainly with 15% Ag-HAp nanocomposites.

ACKNOWLEDGEMENTS

The authors wish to thank the University Grants Commission (UGC), New Delhi for the financial support in the form of a major research project (UGC F. No. 37-98/2009(SR).

REFERENCES

[1] Ardanova, L.I., Get'man, E.I., Loboda, S.N., Prisedsky, V.V., Tkachenko, T.V., Marchenko, V.I., Antonovich, V.P., Chivireva, N.A., Chebishev, K.A. and Lyashenko, A.S., *Inorg. Chem.* 49 (2010) 10687–10693.

[2] Sugiyama, S., Moffat, J.B., *Catal. Lett.* 81 (2002), 77–81.

[3] Yamaguchi, K., Mori, K., Mizugaki, T., Ebitani, K. and Kaneda, K., *J. Am. Chem. Soc.,* 122 (2000), 7144–7145.

[4] Hara, T., Kanai, S., Mori, K., Mizugaki, T., Ebitani, K., Jitsukawa, K. and Kaneda, K.J., *Org. Chem*., 71 (2006), 7455–7462.

[5] Ewing, R.C., *Prog. Nucl. Energy,* 49 (2007), 635–643.

[6] Cazalbou, S., Combes, C., Eichert, D. and Rey, C.J., *Mater. Chem*. 14 (2004), 2148–2153.

[7] Smitha, S.L., Philip, D. and Gopchandran, K.G., *Spectrochim. Acta A*, 74 (2009), 735–739.

[8] Khan, Z., Hussain, J.I. and Hashmi, A.A., *Colloids and Surf. B.*, 95 (2012), 229–234.

[9] Lopez-Alvarado, P., Avendano, C., Menendaz, J.C., *J. Org. Chem*., 61 (1966), 5865–5870.

[10] Elliott, G.I. and Konopelski, J.P., *Org. Lett.,* 2 (2000), 3055–3057.

[11] Lam, P.Y.S., Deudon, S., Averill, K.M., Li, R., He, M.Y., Deshong, P. and Clark, C.G., *J. Am. Chem. Soc*., 112 (2000), 7600–7601.

GREEN SYNTHESIS OF COPPER OXIDE NANOPARTICLES AND THEIR ANTIMICROBIAL AND DNA CLEAVAGE STUDIES

T. ABIRAMAN, S. MALATHI, M.D. BALAKUMARAN[1], P.T. KALAICHELVAN[1] AND S. BALASUBRAMANIAN

Department of Inorganic Chemistry, University of Madras, Chennai
[1]CAS in Botany, University of Madras, Chennai
E-mail: bala2010@yahoo.com

ABSTRACT

Chitosan stabilized cubic and flake-like shaped copper oxide nanoparticles are synthesized by a simple and green chemical reduction route. The synthesis is performed by the reduction of copper (II) salt with various reducing agents like glucose, urea and ascorbic acid in aqueous solution under ambient condition in the presence of chitosan as capping agent. The advantages of this synthetic method include the use of green reducing agents, environmentally benign solvents and no inert atmosphere is required. The resulting copper oxide nanoparticles are characterized by UV-Visible spectroscopy, XPS and FESEM studies. The results demonstrate that control of pH is important for the formation of the copper oxide nanoparticles. The resulting copper oxide nanoparticles exhibit better antibacterial activity on Staphylococcus aureus (gram positive bacteria) when compared to Escherichia coli (gram negative bacteria). Besides, the copper oxide nanoparticles exhibit DNA cleavage, increased reactive oxygen species levels and induction of apoptosis. The chemicals involved in the synthesis of nanoparticles are commonly available, cheap, and non-toxic.

INTRODUCTION

Nanotechnology is one of the growing technologies and their products are quite useful in all fields, because of their small size (10^{-9} nm) and large surface area. In recent years the

V. Rajendran, P. Prabu and K.E. Geckeler (eds.)
Nano Biomaterials, pp. 29–34 (2012).

transition metal and metal oxide nanoparticles, due to their special properties and also small dimensions, find important applications in anti microbial studies, DNA cleavage, gas sensor, waste water treatment, hydrogen generation, heterogeneous catalysis,[1] superconductors, optical switches, lithium ion electrode materials, field-emission emitters and photo catalysis. The copper oxide nanoparticles have good antibacterial activity and antimicrobial formulations comprising nanoparticles could be used as an effective bactericidal agent [2–6] against *Escherichia coli* (*E. coli*) and *Staphylococcus* strains. Similarly, copper oxide nanoparticles exhibit DNA cleavage property. Copper oxide nanoparticles have been synthesized by green chemical reduction method, using glucose and urea as reducing agents and water as solvent. The anti-microbial and DNA cleavage studies of copper oxide nanoparticles are reported in the present investigation.

EXPERIMENTAL

Synthesis of Copper Oxide Nanoparticles

Copper (II) sulfate pentahydrate (2.0025×10^{-3} M), (500 mg) was dissolved in 10 ml of distilled water. The capping agent chitosan (50 mg), (1%), dissolved in 5 ml of distilled water was added to the copper sulfate solution and the resulting heterogeneous solution was vigorously stirred to form a homogeneous solution and it was heated to 60–70°C with continuous stirring. D-Glucose (0.0277 M), (5 mg) and urea (8.3250×10^{-4} M), (50 mg) were dissolved in 50 ml of distilled water and their pH was maintained at 8–9 using 0.05 M sodium hydroxide. The glucose/urea solution was slowly added drop wise to the above hot homogeneous solution. The blue homogeneous solution changes to yellow suspension after 15 minutes due to the reduction of copper salt. The resulting yellow suspension was carefully transferred to a centrifuge tube, and the solid residue was recovered by centrifugation (7000 rpm for 30 minutes), which was repeatedly washed with distilled water and dried under vacuum.

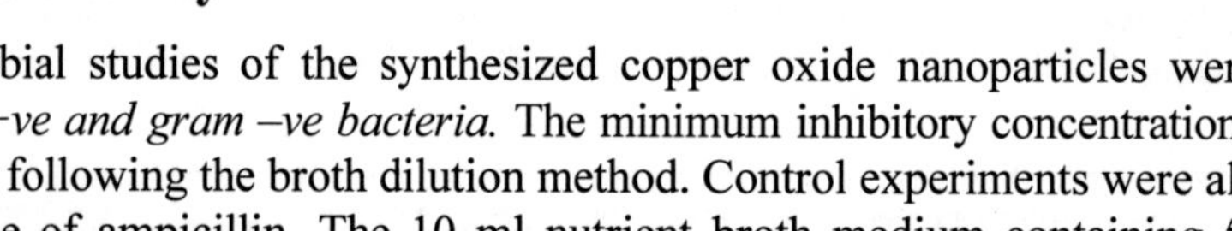

Schematic Illustration of Copper Oxide Nanoparticles

Antibacterial Activity

The antimicrobial studies of the synthesized copper oxide nanoparticles were carried out against *gram +ve and gram –ve bacteria.* The minimum inhibitory concentrations (MIC) were determined by following the broth dilution method. Control experiments were also carried out in the presence of ampicillin. The 10 ml nutrient broth medium containing 50–100 μg/ml copper oxide nanoparticles were prepared. Each set was inoculated aseptically with 10 ml of respective bacterial suspension (approximately 108 CFU/ml). The inoculated sets were incubated at 37°C for 24 h. The experiments were carried out in triplicates. The viable colonies were counted by naked eye.

RESULTS AND DISCUSSION

The copper oxide nanoparticles were synthesized with chitosan as a capping agent. Besides providing long-term stability to the nanoparticles by preventing particle agglomeration, polymer capping agent such as chitosan makes the particles dispersible in aqueous solution. It has been found that several factors including NaOH concentration, reaction time and temperature affect the composition size, morphology, and the degree of agglomeration of the resulting copper oxide nanoparticles.

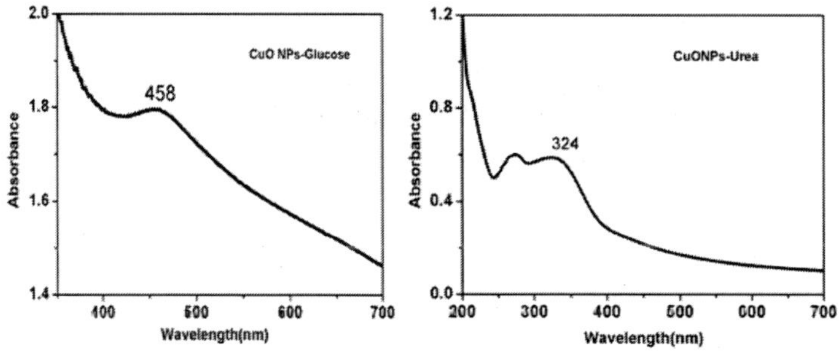

Fig. 1: UV-Visible Spectrum of Copper Oxide Nanoparticles

Figure 1 displays the uv-visible spectrum of CuO NPs formed when alkaline glucose was used as a reducing agent in the presence of chitosan. The characteristic plasmon absorption band for copper oxide nanoparticles was observed at 458 nm. However, the plasmon absorbance band for CuO NPs was observed at 354 nm when urea was used as a reducing agent. Swarankav *et al.* have reported the synthesis of CuO NPs by irradiation method using CTAB as a reducing agent. In order to examine the influence of various reducing agents on the characteristics of copper oxide nanoparticles, glucose, ascorbic acid and urea were employed in the present investigation. The interaction with these reducing agents occurs under various conditions and hence the physicochemical properties of CuO NPs obtained vary widely (Table 1).

Fig. 2: FESEM Images of Copper Oxide Nanoparticles

Table 1: Reaction Conditions and Spectral Data of Copper Oxide Nanoparticles

Reducing Agents	*pH (0.05 M NaOH)*	*Temperature (°C)*	*Time (min)*	*NPs (λ_{max})*	*Color*
D-Glucose	8–9	70–80	15	CuO (458)	Yellow
Urea	8–9	RT	5	CuO (354)	Brown

The FESEM image of CuO NPs synthesized via glucose as a reducing agent is reproduced (Figure 2A). The crystalline nanoparticles are cubic in shape and their size is 100 nm. The FESEM image of CuO NPs synthesized with urea as a reducing agent shows flake like structure (Figure 2B). The temperature of the reaction influences the morphology of CuO NPs. The cubic shape is formed at 70–80°C while the flake like structure is obtained at room temperature. The interiors of the hexahedron structures cannot be seen in the FESEM image, since all nano cubics are well constructed without having broken facets, even though some nanocubes are in contact with others and show twisted distorted facets. The nanostructure observed in the present case is essentially nanocubes.

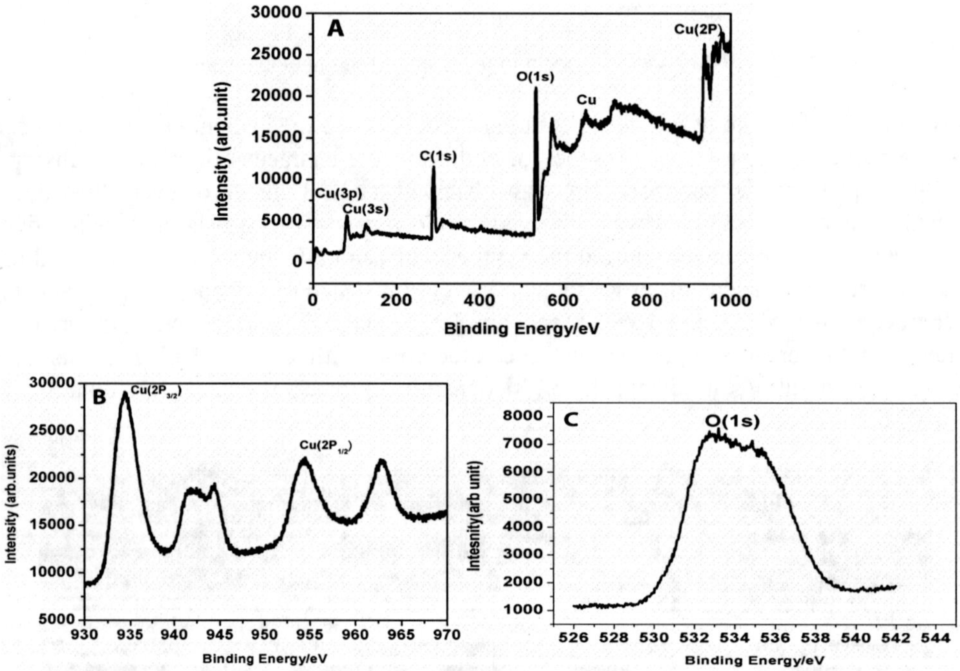

Fig. 3: (A) X-Ray Photoelectron Spectrum of CuO (flakes) Nanoparticles, (B) Cu (2P), (C) O (1s)

The XPS surface analysis of the copper oxide nanoparticles indicates their surface chemical composition. An elemental XPS analysis was carried out on the glucose reduced copper oxide nanoparticles. The XPS spectrum of the CuO nanoparticles (flakes) is shown in Figure 3A. The most interesting features were observed in the high resolution Cu $2p_{3/2}$ spectrum of the NPs (Figure 3 B&C). The Cu $2p_{3/2}$ NPs spectrum shows the characteristic main peak at 934 eV and shakeup satellites at higher binding energies.

The shakeup satellites are characteristic of materials having a d electronic configuration in the ground state, i.e. Cu^{2+}. The two peaks at 933.1 eV and 953.3 eV are assigned to Cu ($2p_{3/2}$) and Cu ($2p_{1/2}$), respectively which indicates the presence of paramagnetic Cu^{2+}. The C1s peak is observed at 288.8 eV, while the broad peak at 534eV is attributed to O (1s).

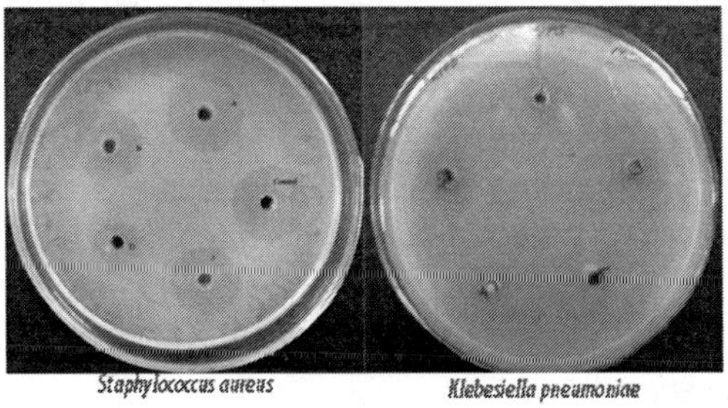

Fig. 4: Antimicrobial Studies of CuONPs Using Gram +ve and Gram –ve Bacteria

The sedimentation of the bacteria as observed in the presence of broth medium did not exhibit any change in the absence of nanoparticles. Three gram +ve strains (*Staphylococcus aureus, Staphylococcus epidermis, Bacillus subtilis*) and three gram –ve strains (*Klebsiella pneumaniae, Escherichia coli, Proteus vulgaris*) were used for the study. It has been found that the nanoparticles exhibit growth inhibition to only gram +ve *Staphylococcus aureus* organism (Figure 4A).

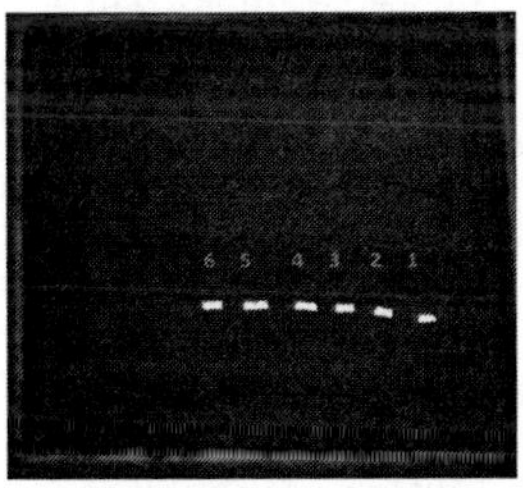

Fig. 5: DNA Cleavage Studies on CuONPs: 1. Control, 2. $CuSO_4$, 3. CuONPs (100 um), 4. CuONPs (150 um), 5. CuONPs (200 um), 6. CuONPs (250 um)

The effect of CuO NPs on bacterial genomic DNA isolated from Escherichia coli was tested by treating the DNA with gradually increasing the concentration of CuO NPs (10 100 μM) for 3h at 37°C in PBS at pH 7.4. After the incubation, the DNA was analyzed by agarose gel electrophoresis. It was observed that the CuO NPs induced DNA degradation increases with increase in the concentration of CuO NPs.

CONCLUSION

Copper oxide nanoparticles were synthesized by green chemical reduction route using glucose and urea as the reducing agents and chitosan as a capping agent. The size and shape of the CuO NPs was controlled by the nature of green reducing agents and pH. Copper oxide nanoparticles show a plasmon peak at 458 nm and 324 nm respectively. The FESEM reveals that the average size of cubic CuO NPs is ~70 nm and flake like structure is <100 nm. The XPS spectrum of CuO NPs shows the characteristic peaks at 933eV and 534eV attributed to Cu (2P3/2) and O (1s) respectively. The chemicals involved in the synthesis of nanoparticles are commonly available, cheap and non-toxic. Antimicrobial activities of CuO NPs were carried out with gram +ve and gram –ve bacteria. The results indicate that nanoparticles exhibit growth inhibition to only gram +ve organism. The present method suggests the possibility of using this nano material in water purification, air filtration, air quality management and antibacterial packaging. Furthermore, the DNA degradation potential can be utilized in designing better and more active cancer drugs by chemically modifying the CuO NPs with a number of macromolecules.

ACKNOWLEDGEMENTS

We are thankful to NCNST, University of Madras for FESEM and XPS studies.

REFERENCES

[1] White, B., Yin, M., Hall, A., Le, D., Stolbov, S., Rahman T., Turro, N. and Brien, S.O., Complete CO Oxidation over Nanoparticles supported on silica gel. *Nano Lett.*, 2006, 6, 2095.

[2] Stoimenov, P.K., Klinger, R.L., Marchin, G.L. and Klabunde, K.J., Metal oxide nanoparticles as bactericidal agents. *Langmuir,* 2002; 18(17): 6679–6686.

[3] Ravikumar, S., Gokulakrishnan, R., Selvanathan, K. and Selvam, S., Antibacterial activity of metal oxide nanoparticles against ophthalmic pathogens. *Int J Pharm Res Dev,* 2011; 3(5): 122–127.

[4] Rajendran, R., Balakumar, C., Hasabo, A.M.A., Jayakumar, S., Vaideki, K. and Rajesh, E.M., Use of zinc oxide nanoparticles for the production of antimicrobial textiles. *Int J Eng Sci Technol,* 2010; 2(1): 202–208.

[5] Padmavathy, N. and Vijayaraghavan, R., Enhanced bioactivity of ZnO nanoparticles-an antimicrobial study. *Sci. Technol Adv Mater,* 2008, 9: 7.

[6] Hosseinkhani, P., Zand, A.M., Imani, S., Rezayi, M. and Zarchi, S.R., Determining the antibacterial effect of ZnO nanoparticle against the pathogenic bacterium, Shigella dysenteriae (type 1). *Int J Nano Dimens,* 2011; 1(4): 279–285.

SYNTHESIS, CHARACTERIZATION AND ELECTROCHEMISTRY OF RUTHENIUM TERPYRIDINE COMPLEX ANCHORED TO GOLD NANOPARTICLE AND ITS ANTICANCER ACTIVITY

T. Ezhilarasu, S. Malathi and S. Balasubramanian

Department of Inorganic Chemistry, University of Madras, Maraimalai Campus, Chennai
E-mail: bala2010@yahoo.com

ABSTRACT

Supramolecular chemistry has become one of the most interesting areas in modern chemistry. 2, 2': 6',2"-terpyridines have been of great interest in the recent years, principally because of their ability to chelate transition metals. The reduction of tetrachloroaurate with isoniazid (INH) in the presence of 4'-(4-mercaptomethylphenyl)-2, 2': 6', 2"- terpyridine resulted in terpyridine functionalized AuNPs (TPy-Au). Metal complexation of TPy-Au with ruthenium trichloride, $RuCl_3.3H_2O$, gave a new AuNPs functionalized with ruthenium-terpyridine complex (Ru-TPy-Au). The resulting Ru-TPy-Au nanoparticles were characterized by UV-Visible, FTIR, NMR and Mass spectral studies as well as by cyclic voltammetric and HRTEM studies. The Ru-TPy-Au compound was subjected to invitro anticancer activity against skin cancer cell lines.

INTRODUCTION

Many metal complexes have been shown to possess bioactivity and several drugs based on metal complexes have been developed. These include Pt, Au, Ru and Bi compounds used in the treatment of certain types of cancer, arthritis and stomach ailments.[1] 2 ,2':6', 2"-terpyridine (tpy) derivatives are quite attractive candidates for electronic conductors between the functional groups and the substrates due to their favourable π conjugation, but also versatile ligands suitable for formation of complexes with transition metal ions.[2] Particular interest focuses on 4'-substituted tpy ligands in which various substitutents with different electronic or

V. Rajendran, P. Prabu and K.E. Geckeler (eds.)
Nano Biomaterials, pp. 35–40 (2012).

steric characteristics can be directly introduced by the Krohnke reaction or improved Krohnke reaction. These tpy derivatives are further converted into thiol-functionalized tpys, which can subsequently be introduced onto gold surfaces. Hence it is important to develop convenient and feasible approaches to access thiol-functionalized tpy compounds.

EXPERIMENTAL

Synthesis of $[Ru(mpttpy)_2](PF_6)_2$

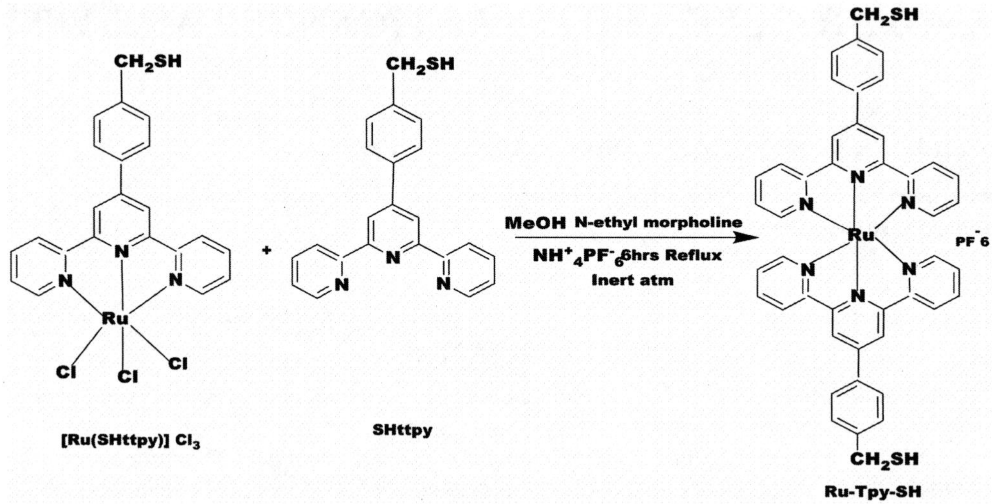

Terpyridine was synthesized from 2-acetyl pyridine and mercaptomethylphenyl terpyridine was synthesized using thiourea.[6] The ligand mptpy (64 mg, 0.18 mmol) was added to $[Ru(mptpy)]Cl_3$ (100 mg, 0.18 mmol) in methanol along with N-ethyl morpholine (5 drops). The mixture was refluxed for 6h under an inert atmosphere. The resulting deep red solution was filtered and a three-fold excess of methanolic ammonium hexafluorphosphate was added to the filtrate, and the red precipitate formed was filtered, recrystallized from acetone-acetonitrile solution and dried under vacuum.[4]

Synthesis of [Ru-tpy-Au]

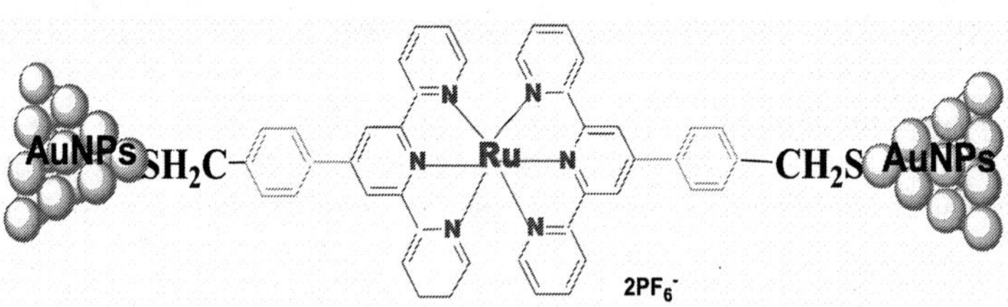

Ru-TPy-Au

The ligand-exchange reaction of Ru–Tpy–Au with a starch capped AuNPs was carried out as follows. A mixture of $[Ru–(TpySH)_2]^{2+}$ (50 mg) and AuNPs (5ml) taken in MeCN(10 ml) was stirred at room temperature for 48 h. Ethanol (200 ml) was added to the mixture and the resulting precipitate was collected by filtration and washed several times with EtOH. The nanoparticles were dissolved in MeCN precipitated with EtOH, and then the particles were isolated by filtration.[7]

RESULTS AND DISCUSSION

Several methods are available for the synthesis of terpyridine ring systems. The PEG300 was chosen as a reaction medium as reported by smith *et al.*[3] 4′-(4-Methylphenyl)-2, 2′:6′, 2″-terpyridine(ttpy) was synthesized by the reaction of 2-acetylpyridine with 4-methyl benzaldehyde in the molar ratio 2:1 at 0°C. 4′-(4-Bromomethylphenyl)-2, 2′:6′, 2″-terpyridine (Brttpy) was synthesized according to the literature method.[5] Brttpy was converted into 4′-(4-mercaptomethylphenyl)-2, 2′:6′, 2″-terpyridine (mptpy) using thiourea according to the literature method.[6] The terpyridine ligands were characterized by UV-Visible, IR, ^{1}H-NMR, ^{13}C-NMR and mass spectral studies. In the ^{1}HNMR spectra of ttpy and mptpy lignads, the signals for aromatic protons of the terpyridine rings were observed in the region δ 7.29–8.70. The methyl protons appear as a singlet in the region of δ2.46 and it undergoes a downfield shift in mptpy which is observed at δ 3.6, whereas the SH protons appear at δ 1.8 Figure 1. The positive ion FAB mass spectra of Ruthenium(II) terpyridine complexes provide compelling evidence for the formation of the complexes. For example the symmetrical complex $[Ru(mptpy)_2](PF_6)_2$ shows a molecular ion peak at m/z = 1101. The overall fragmentation pattern in the FAB mass spectrum of the complex strongly supports the proposed formulation of the complex.[4]

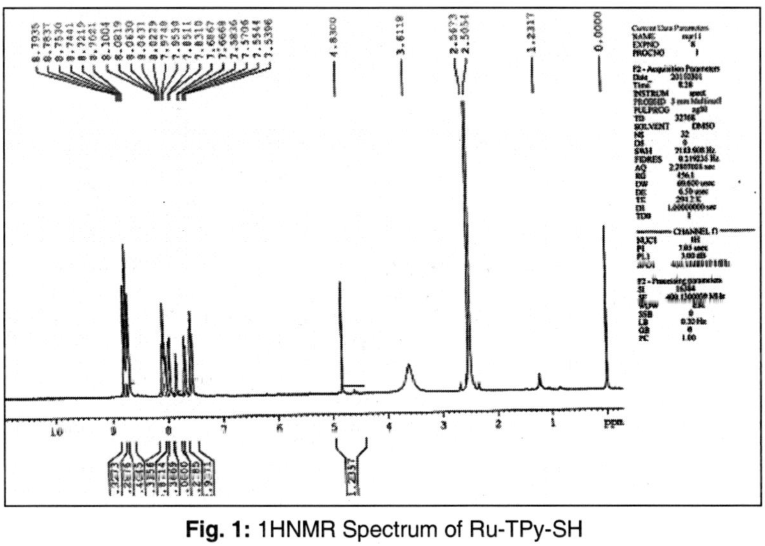

Fig. 1: 1HNMR Spectrum of Ru-TPy-SH

The electronic spectra of the ruthenium complexes were recorded in acetonitrile medium and the relevant data are summarized in Table 1. The UV-Visible spectrum of Ru(mptpy)$(PF_6)_2$ exhibits the typical metal to ligand charge transfer transition of the Ru(II)-terpyridine system at λ_{max} = 492 nm and the bands at 381 nm and 280 nm are due to ligand centered transitions. The electronic spectrum of starch capped gold nanoparticles exhibit a peak at 530 nm indicating the formation of gold nanoparticles. In the case of ruthenium complex anchored to AuNPs, the electronic spectrum shows a red shift due to the presence of ruthenium complex (570 nm) (Figure 2).[4]

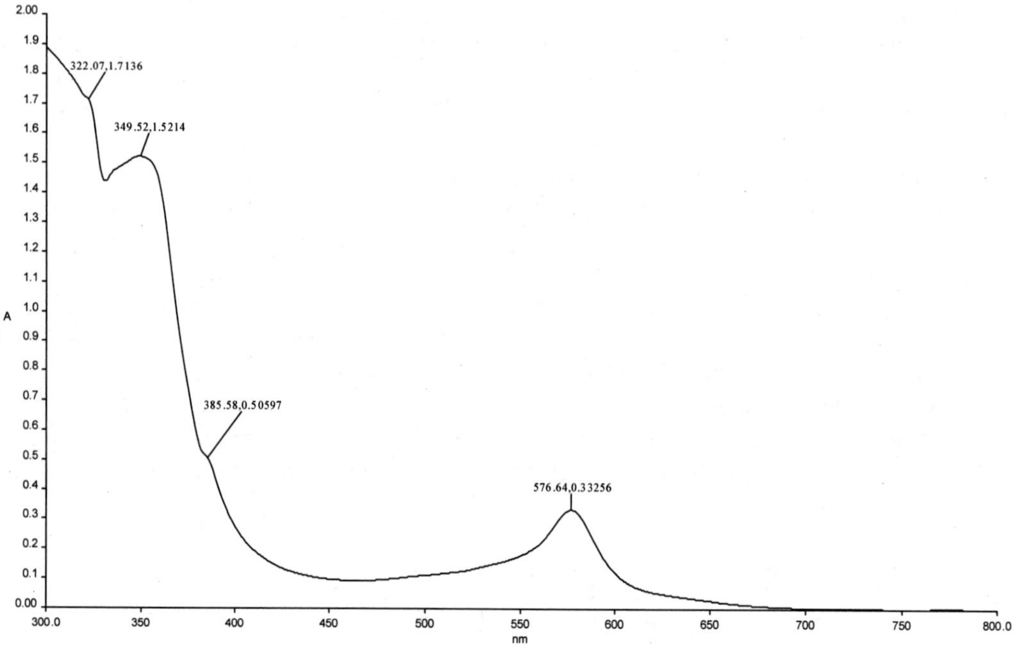

Fig. 2: UV-Visible spectrum of Ru-TPy-Au

The FTIR spectrum of ttpy shows the CH_3 stretching frequency at 1390 cm^{-1}. CH_2Br stretching frequency in Brttpy at 1260 cm^{-1} and the sharp SH stretching frequency at 2555 cm^{-1}. The HRTEM image of gold functionalized ruthenium mercapto terpyridine shows spherical nanoparticles which shows a particles are less than 50 nm (Figure 3).[7]

The electrochemical properties of the monomer ruthenium complexes were examined in acetonitrile medium. The cyclic voltammogram of the [Ru(tpy)]$(PF_6)_2$ complex exhibits a reversible oxidation peak at –0.779V which can be attributed to the Ru^{II}/Ru^{III} process. The first scan of the CV of Ru–Tpy–Au in 0.1 M TBAP–acetonitrile at a GC electrode exhibits a single irreversible oxidation peak at E_{pa} = +1.13 V which seems to overlap with the oxidation peaks of the Ru^{II}/Ru^{III} couple. The multiple scans show an increase in the redox-peak current.

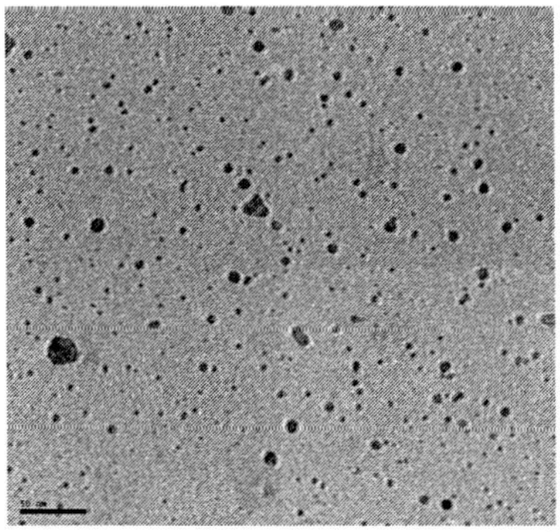

Fig. 3: HRTEM Images Ru-TPy-Au

After the electrode was rinsed copiously with solvent and dipped in fresh 0.1 M TBAP–MeCN solution. The CV of the Ru–Tpy–Au modified GC electrode shows two irreversible peaks due to terpyridine at 1.5 and 1.79 V and one reversible redox peak due to the Ru^{II}/Ru^{III} couple at +0.87 V (ΔEp =136 mV). Thus metal nanoparticles bearing a Ru-terpyridine complex are readily immobilized on metal electrode surfaces (Figure 4).

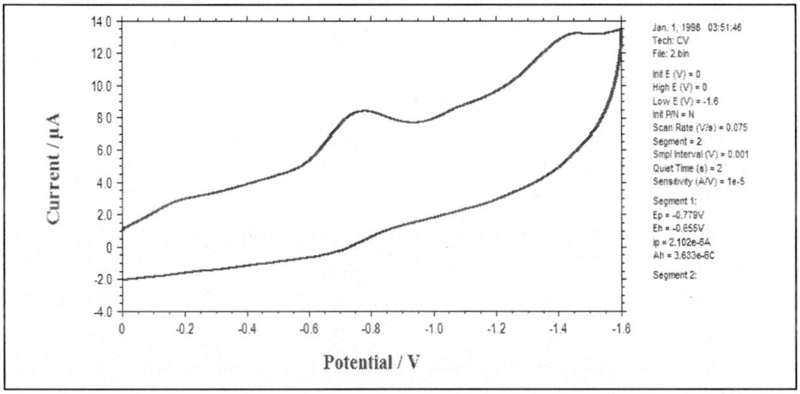

Fig. 4: Cyclic Votammogram of Ruthenium Complex

Table 1: Spectral Data for Ligand and Ruthenium Complexes

Compound	$\lambda_{max}(nm)$	$\upsilon(cm^{-1})$	δ *(ppm)*
Ligands (a)ttpy (b)Brttpy (c)SHttpy	 279 280 276, 380	 – C=C–str 1581 – CH_2Br-str 1259 – CH_2SH-str 2555	 2.59(CH_3) 4.59(CH_2Br) 1.18(SH),3.64(CH_2SH)
Ru-TPy-SH	492, 316, 284	CH_2SH str 2560	1.18(SH), 3.64(CH_2SH)
Ru-TPy-SH-Au	576, 385, 349	– CH_2SH-str 2562	–

The anticancer activity of the ligand SHttpy, Ru(SHttpy) and Au functionalized Ru complex towards A431 cell line was examined by MTT assay. The gold functionalized ruthenium complex shows significant anticancer activity when compared to SHttpy and the unbound ruthenium complex. The mercapto methyl phenyl moiety and gold nanoparticles enhances the activity of gold functionalized ruthenium complex.

CONCLUSION

The synthesis of a Ru(II) complex with the 4′-(4-mercaptophenyl)-2,2′:6′,2″-terpyridine ligand has been achieved by a multiple step procedure. The complex was characterized by ^{1}H NMR, FTIR, UV-VIS, HRTEM spectral studies and CV studies. The UV-Visible spectrum clearly exhibits the typical metal-ligand charge transfer (MLCT) band of the Ru-Tpy-SH complex at 492 nm as well as the π–π^* transition of the terpyridine at $\lambda = 316$ nm. The Au functionalized ruthenium complex shows λ_{max} at 576 nm indicating the formation of gold is modified the ruthenium complex. The results show more cytotoxicity towards the gold functionalized ruthenium complex when compared to the ligands.

REFERENCES

[1] Alessio, E., *et al.*, Novel Diastereomers with Opposite Chirality at Ruthenium Formed by N7, α-PO_4 Chelation of 5'-dGMP to the Antimetastatic Agent trans-$RuCl_2(DMSO)_4$: NMR and CD Evidence., *J. Am. Chem. Soc.*, 1989, 111, 7068.

[2] Patrick Sullivan, B., Calvert, J.M. and Meyer, T.J., Cis-Trans Isomerism in (tpy) (PPh_3) $RuCl_2$. Comparisons between the Chemical and Physical Properties of eight Cis-Trans Isomeric Pair, *Inorg. Chem.,* 1980, 19, 1404–1407

[3] Smith, C.B., Raston, C.L. and Sobolev, A.N., Poly(ethyleneglycol) (PEG): A versatile reaction medium in gaining access to 4□-(pyridyl)-terpyridine, *Green Chem.*, 2005, 7, 650.

[4] Anthonysamy, A., Balasubramanian, S., Shanmugaiah, V. and Mathivanan, N., Synthesis, characterization and electrochemistry of 4′-functionalized 2, 2′:6′,2″- terpyridine ruthenium (II) complexes and their biological activity. *Dalton. Trans.*, 2008, 2136–2143.

[5] Tang, B., *et al.*, A Near-Infrared Neutral pH Fluorescent Probe for Monitoring Minor pH Change: Imaging in Living HepG2 and HL-7702 Cells, *J. Am. Chem. Soc.*, 2009, 131, 3016–3023.

[6] Zhang, X., Li, D. and Zhou, X.-P., From large 3D assembly to highly dispersed spherical assembly: Weak and strong coordination mediated self-aggregation of Au colloids, *New J. Chem.*, 2006, 30, 706–711.

[7] Ito, M., Tsukatani, T. and Fujihara, H., Preparation and characterization of gold nanoparticles with a ruthenium terpyridyl complex, and electropolymerization of their pyrrole-modified metal nanocomposites, *J. Mater. Chem.,* 2005, 15, 960–964.

PHARMACOKINETICS AND BIODISTRIBUTION OF NEGATIVELY CHARGED PECTIN NANOPARTICLES ENCAPSULATING PACLITAXEL

ANITA K. VERMA AND A. KUMAR

Department of Zoology, K.M. College, University of Delhi, Delhi
E-mail: akamra23@hotmail.com; akverma@kmc.du.ac.in

ABSTRACT

Relevance of Work: *Pectin, a naturally occurring biopolymer has been found to have increasing applications in the pharmaceutical and biotechnology industry. Sugars with their three dimensional structures are important for many biological functions. We report preparation of negatively charged pectin nanoparticles encapsulating Paclitaxel, a broad spectrum anticancer drug for possible therapeutic applications*

Important Results: *The mean diameter of the nanoparticles was ~300–350 nm with an encapsulation efficiency (E.E) ~17%. The TEM studies indicated that the particles were spherical in shape and their sizes are in unison with the DLS size spectra. The surface charge of pectin polymer was –5 mV and of nanoparticles were –32 mV. The enhanced surface charge shows greater stability. The electrophoretic mobility of nanoparticles was also high ~3.5 to 1.5 μmcm/Vs that confirmed the nano-size of particles. Cytotoxicity of the encapsulated drug was reduced when incubated with cultured Hep G2 hepato-carcinoma cell lines as compared to Paclitaxel per se. The pharmacokinetics and biodistribution studies on Balb/c mice indicated that the nanoparticles had prolonged plasma retention of the drug with major accumulation in liver tissue after an i.v. tail vein injection of 20 mg/kg drug.*

Conclusion: *The in-vitro studies clearly indicated that the efficacy of the drug was not compromised by encapsulation, making it a good candidate to deliver bio-pharmaceuticals. Nanoparticles produced free radicals in the free cell system and this*

V. Rajendran, P. Prabu and K.E. Geckeler (eds.)
Nano Biomaterials, pp. 41–46 (2012).

ability caused oxidative stress, which may give rise to inflammation, cell destruction, and genotoxicity. Thus, the results obtained in this study holds great promise for pectin nanoparticles to be exploited for passive delivery of paclitaxel to tumor tissues, in particular, liver cancers.

INTRODUCTION

Pectin has gained wide acceptance as matrix materials because of their inherent biocompatibility, poly-functionality, biodegradable nature, flexible structural networks, non-toxicity and low production costs.[1] The physiological properties, functionality and application of pectin is dependent on the degree of methoxylation.[2–3] Pectin is used in food industry as a thickening agent, a gelling agent and a colloidal stabilizer, a carrier and coating material in pharmaceutical sciences.[4] Pectin is highly hydrophilic in nature, and hydrophilic matrices are generally used for the oral drug delivery and for the preparation of modified release formulation.[5] However, the major challenge of using pectin for the development of modified drug formulation was to overcome its solubility in aqueous medium which can contribute to the undesirable premature and local release of the active principle from the polysaccharide matrix.[6–7]

We have previously reported that coordination bond between the Platinum(II) atom and the carboxylic group in the galacturonate of pectin causes spontaneous folding to form nano-conjugates,[8] as well as formation of pectin nanoparticles prepared by inducing the gelification and cross-linking of pectin solution with calcium chloride and L-asparagine at low pH.[9] At higher pH values, the polycarboxylate groups that are present can be cross-linked by the addition of calcium ions which leads to macromolecular aggregates.[10] It has been widely reported that thiolation of a polymer such as chitosan [11] poly(acrylic acid),[12] and alginates,[13] enhances its cohesive properties leading to oxidative cross-linking of thiol groups.[14] In continuation, our present investigation focuses on the *in-vitro*, cytotoxicity, and *in-vivo* pharmaco-kinetics and biodistribution of the negatively charged Pectin nanoparticles.

EXPERIMENTAL

In-Vitro Cytotoxicity

HepG2, hepatic carcinoma cell line was procured from American Type Culture Collection, (Rockville, MD), and maintained at 37°C in 5% CO_2 at 95% relative humidity. Pectin nanoparticles [PPN] for cytotoxicity as per our previously published protocol [20] using a standard MTT assay in a 24 hr, 48 hr and 72 hr assay. MTT assay is a non-radioactive assay done routinely to assess the viability of the cell culture. Briefly, 5×10^3 cells/well were incubated in RPMI-1640 supplemented with 10% FCS with Paclitaxel [Pax] *per se* and Paclitaxel encapsulated Formazan crystals formed by the cellular reduction of MTT were dissolved in 150 μl of DMSO and plates were read on an ELISA-reader using 570 nm filter. The relative cell viability (%) related to control wells containing cells without nano-particles was calculated by:

$$[A]_{control} - [A]_{test}/[A]_{control} \times 100$$

Where, $[A]_{test}$ is absorbance of the test sample and $[A]_{control}$ is the absorbance of the control sample.

Animal Studies: Pharmacokinetics and Biodistribution

Studies were conducted under standardized procedures on *in bred*, male 6–8 week old Balb/c (18–20 g) mice.[15] Destructive sampling method was adopted for acquisition of biological samples (blood and tissues) for analysis. The animals were anesthetized with ether inhalation and paclitaxel *per se* and re-dispersed PPN containing drug at a dose of 20 mg/kg body weight, and PBS as control were then administered intravenously (*i.v.*). Blood was drawn from tail vein at pre-determined intervals at 5, 10, 15, 30, 60, 90, 120 min. Mice were sacrificed by cervical dislocation and dissected to obtain various tissues (kidney, liver, lung and spleen), each tissue was weighed and stored vial at –20°C for further analysis. Frozen tissue samples were thawed for 45 min and homogenized.

Determination of Paclitaxel

The drug was extracted from biological samples with tert-butyl methyl ether.[16] The sample was mixed with the organic reagent at a 1:1 ratio on a vortex mixture. After centrifugation, the organic layer was transferred to a clean tube and evaporated. The residue was re-dissolved in ethanol and analyzed by HPLC.

RESULTS AND DISCUSSION

Pectin nanoparticles were prepared by ionotropic gelation of degraded pectin chains with calcium ions.[9] The objective of this modification was to reduce the polarity of pectin by reducing the number of free hydroxyl groups. Pectin nanoparticles thus formed, assumed an oval shape with clear edges as can be clearly observed in TEM pictures.

The mean diameter (Mean R_H) of the samples was in the range 350 nm–500 nm with a polydispersity index of 0.3 (Figure 1). A ξ value of –32 mV at physiological pH was sufficient to keep the negatively charged particles under repelling effect thus bringing stability to the nanoparticle dispersion. Chains were crosslinked with Ca^{++} and folded into particulate shape in the presence of L-Asn, and the intrinsic viscosity [η] that indicates the hydrodynamic volume of a polymer molecule was measured and found to be reduced more than 6 times.[9] A reduced [η] for nanoparticles as compared with that of pectin chains indicated that they can be administered at a much higher concentration *in vivo.* The nanoparticle solution was lyophilized to a white amorphous powder for further experiments.

We compared the cytotoxic effect of Pax *per se*, PPN and pectin chains on Hep G2, hepatic carcinoma cell line. Dose dependent cytotoxicity was observed, wherein ~21.7 ± 3.2% cytotoxicity was observed by PPN (Figure 2), but Pax *per se* showed ~55.6 ± 3.5% cytotoxicity in a72 hr assay. The slow release of the drug can be attributed to the cytotoxicity observed as a function of time. Pectin chains showed negligible toxicity even at a concentration of 5 mg/ml.

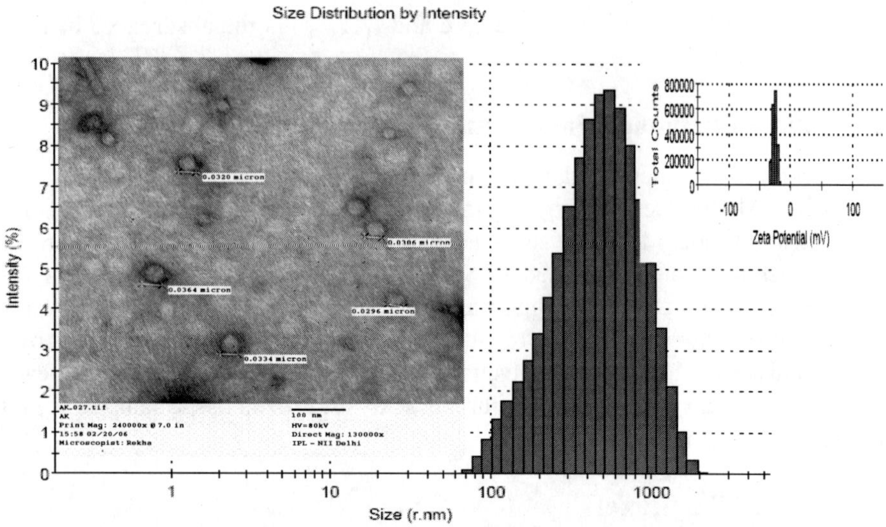

Fig. 1: A Representative Graph of the Size Distribution Spectra (~350 nm with PDI= 0.30) and Zeta Potential (–26 mv). Inset TEM Image of Pectin Nanoparticles

The plasma concentration-time curve of Pax and PPN at a dose of 20 mg/kg body weight of Balb/c mice has been shown in Figure 3. A non-compartmental analysis of the pharmacokinetics graph indicates a biphasic pattern with initial rapid phase followed by a slower decay of Pax with the level falling beyond detection limit by 48 hours. It was noticed that the blood profile remains much higher for PPN as compared to Pax *per se*.

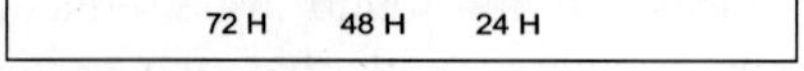

Fig. 2: Dose Dependant Cytotoxicity of Pax and PPN on HepG2 [Conc. 0.02 μM]. Pure Pectin (5 mg/ml)

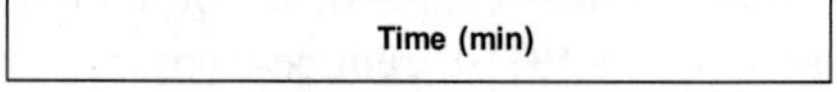

Fig. 3: Pharmacokinetic Profile of Pax Post *i.v.* inj.[20 mg/kg PPN] in Balb/c

Tissue biodistribution of Pax and PPN was also evaluated after an *i.v.* dose of 20 mg/kg PPN. Pax accumulates rapidly in various tissues with a maximum concentration found in the kidney. However, after 48 hrs, Pax accumulated in the lymphatic tissues, spleen and liver. More than 90% of Pax binds rapidly and extensively to plasma proteins.[20] On the contrary, the biodistribution was not altered as explicitly as with PPN at an early point of analysis, i.e. after 2 hrs, accumulation in kidney was markedly less. 48hrs post injection, tissue specificity changes (Figure 4) from the difference in the rank order of concentration, i.e. liver > kidney > lung > spleen for the PPN and spleen > liver > kidney ≥ lung for Pax *per se*. Higher accumulation of Pax in spleen maybe attributed to splenic filtration.

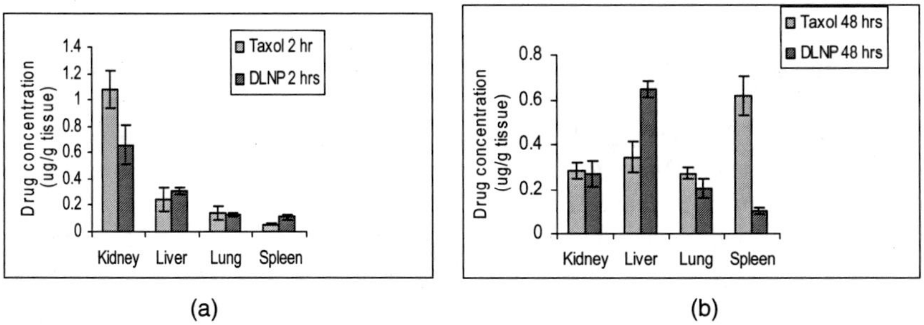

Fig. 4: Biodistribution Profile of Paclitaxel and PPN after (a) 2 hours and (b) 48 Hours of an *i.v.* Infusion of 20 mg/kg in Balb/c mice. Mean ± S.D. (n=4)

CONCLUSION

The biomedical potential of the negatively charged pectin nanoparticles have been assessed. Paclitaxel was encapsulated in the aqueous nanoparticle dispersion with high efficiency. The *in vivo* studies reflected that the nanoparticles significantly delayed plasma clearance with detection of Pax possible upto 48 hrs. It was evident from the biodistribution studies that enhanced PPN accumulation was observed in liver but a greater accumulation of pax *per se* in kidney. The reported pectin nanoparticles forms a good candidate for passive delivery of paclitaxel to liver tissue and warrants further investigation to ascertain its efficacy to regress tumor implants in mice.

ACKNOWLEDGEMENTS

This work was supported by a DBT grant for Nano Science and NanoTechnology Research Initiative GOI, (BT/PR/9895/NNT/28/55/2007).

REFERENCES

[1] May, C.D., Industrial pectins: sources, production and applications. *Carbohydrate Polym*, 12, 79–99, 1990.

[2] Thibault, J.F., Renard, C.M.G.C., Axelos, M.A.V., Roger, P. and Crépau, M.J., Studies of the length of homogalacturonic regions in pectins by acid hydrolysis, *Carbohydrate Res.*, 238, 287–339, 1993.

[3] Garnier, C., Axelos, M.A.V. and Thibault, J.F., Phase diagrams of pectin-calcium systems: Influence of pH, ionic strength and temperature on the gelation of pectins with different degrees of methylation, *Carbohydrate Res.*, 240, 219–232, 1993.

[4] Endress, H.U., Nonfood uses of pectin, in the chemistry and technology of pectin, Walter, R.H., Taylor, S., Eds., pp. 251–68, USA, Academic, 1991.
[5] Sinha, V.R., Kumria, R., Microbiologically triggered drug delivery to the colon, *Eur. J. Pharm. Sci.,* 18, 3–18, 2003.
[6] Sande, S.A., Pectin-based oral drug delivery to the colon, *Expert Opin. Drug Deliv*., 2, 441–450, 2005.
[7] Baojian, W., Daoyin, D., Yi, L. and Wei, W., Biphasic release of indomethacin from HPMC/ Pectin/Calcium matrix tablet: II. Influencing Variables, stability and Pharmacokinetics, *Eur. J. Pharm. Biopharm,* 69, 294–302, 2008.
[8] Verma, A.K. and Sachin, K., Novel Hydrophilic Drug Polymer Nano-Conjugates of Cisplatin Showing Long Blood Retention Profile—Its Release Kinetics, Cellular Uptake and Bio-Distribution, *Current Drug Delivery,* 5, 120–126, 2008.
[9] Verma, A.K., Chanchal, A. and Kumar, A, Potential of Negatively charged Pectin nanoparticles encapsulating Paclitaxel: Preparation and Characterization, *Proceedings of the IEEE xplore,* 1–8, 2011.
[10] Kohn, K., Ion binding on polyuronates—alginate and pectin, *Pure Appl. Chem.,* 42, 371–397, 1975.
[11] Kast, C.E., Bernkop-Schnürch, A., Thiolated polymers—thiomers: development and *in vitro* evaluation of chitosan–thioglycolic acid conjugates, *Biomaterials,* 22, 2345–52, 2001.
[12] Marschütz, M.K., Bernkop-Schnürch, A., Thiolated polymers: self-crosslinking properties of thiolated 450 kDa poly(acrylic acid) and their influence on mucoadhesion, *Eur. J. Pharm. Sci*., 15, 387–94, 2002.
[13] Bernkop-Schnürch, A., Kast, C.E. and Richter, M.F., Improvement in the mucoadhesive properties of alginate by the covalent attachment of cysteine, *J. Control Release,* 71, 277–85, 2001.
[14] Oosterveld, A., Beldman, G. and Voragen, A.G.J., Oxidative cross-linking of pectic polysaccharides from sugar beet pulp, *Carbohydrate Res*., 328, 199–207, 2000.
[15] Verma, A.K., Chanchal, A. and Chuttani, K, Augmentation of anti-tumour activity of Cisplatin by Pectin nano-conjugates in B-16 mouse model—Pharmacokinetics and in-vivo biodistribution of radio-labeled, hydrophillic Nano-conjugates, *IJNT,* Vol. 9, Nos. 10/11/12, 872–885, 2012.
[16] Verma, A.K., Sachin, K., Saxena, A. and Bohidar, H.B., Release Kinetics from bio-polymeric nanoparticles encapsulating protein synthesis inhibitor-Cycloheximide, for possible therapeutic applications, *Curr Pharm Biotech*, 6, 121–130, 2005.
[17] Xu, Z., Gu, W., Huang, J., Sui, H., Zhou, Z. *et al.*, In Vitro and In Vivo Evaluation of Actively Targetable Nanoparticles for Paclitaxel Delivery, *Int. J. Pharm*., 288(2), 361–368, 2005.
[18] Powell, D.A., Morris, E.R., Gidley, M.J. and Rees, D.A., Conformations and interactions of Pectins II. Influence of residue sequences on chain association in calcium pectinate gels, *J. Mol. Biol.*, 155, 517–531, 1982.
[19] Grant, G.T., Morris, E.R., Rees, D.A., Smith, P.J.C. and Thom, D., Biological interactions between polysaccharides and divalent cations: the egg-box model, *FEBS letters*, 32, 195–198, 1973.
[20] Bodmeier, R. and Paeratakul, O., A novel multiple-unit sustained release indomethacin hydroxylpropyl methylcellulose delivery system prepared by ionotropic gelation of sodium alginate at elevated temperatures, *Carb. Polym.,* 16, 399–408, 1973.

DEVELOPMENT AND EVALUATION OF PROPELLANT FREE QUICK BREAK FOAM FORMING CLOBETASOL PROPIONATE NANOEMULSION FOR EFFICIENT TREATMENT OF PSORIASIS

N. Subramanian, P. Chandra Sekar, S. Abimanyu, S.P. Sharavanan, M. Nandagopal, D. Vijula Rani and N. Jamuna Rani

Laboratory for Lipid Based Systems, Department of Pharmaceutical Technology, Anna University of Technology Tiruchirappalli, Tiruchirappalli
E-mail: natesansubbu@gmail.com

ABSTRACT

Clobetasol propionate is a highly lipophilic drug and showed good potential for treatment of psoriasis as topical application. The aim of the present study is to design, develop and characterize the propellant free quick breaking foam forming Clobetasol propionate nanoemulsions (Captex 500P/Solutol HS 15/Tween 80/Water) for the effective treatment of psoriasis. Formulated foam forming nanoemulsions were characterized for its globule size, zeta potential, cloud point, viscosity, pH, transparency, surface morphology, foamability, foam stability and in-vitro drug release. Average globule size and zeta potential of the developed foam forming nanoemulsion formulations was found to be in the range of 10 to 12 nm and –3.09 to –4.0 mV respectively. The cloud point of formulations was above 63° C. The viscosity and pH of the nanoemulsions were varied from 3 to 30 Pa s and 6.3 to 6.5, respectively. Surface morphology of the formulated nanoemulsion was found to smooth and spherical in shape. Formulation F3 and F4 showed the good formation of foam and exhibited quick foam breaking properties. The highest percentage of drug release observed with nanoemulsion formulation F4. Hence we conclude that the formulated quick breaking foam forming nanoemulsion (F4) may be efficient foam forming topical delivery system for Clobetasol propionate and it may have a great potential in the treatment of psoriasis.

V. Rajendran, P. Prabu and K.E. Geckeler (eds.)
Nano Biomaterials, pp. 47–52 (2012).

INTRODUCTION

Clobetasol propionate, a highly potent topical corticosteroid, has been used to treat several dermatological conditions, including scalp and body psoriasis.[1] Conventionally, it is available in the form of alcoholic solutions, lotions or creams for the treatment of psoriasis. The longer exposure of more viscous products (creams and lotions) produces irritation and less viscous solutions/products leads to high fluidity and difficult to control the loss of medicament.[2] Since, the conventional Clobetasol propionate formulations showed high degree of ineffectiveness and patient non-compliance. Clobetasol propionate in aerosol form containing propellant is not freely available in the Indian market and has low level of patient acceptability since the propellant based aerosols are produced irritation in certain patients and their high cost.[3]

Quick breaking foam formulation is an emulsion, comprises one or more active substances, surfactants, aqueous/non-aqueous liquids and propellants. It produces thermally unstable foam and collapses in a relatively short time upon exposure to skin temperature.[4] The quick breaking foam vehicles are able to offers minimal residue and ease of application, rapid localize action in large area and enhanced permeation.[5] But, propellant based quick breaking foam forming system produces irritation in fewer patients and the high cost of the systems made less affordable. Hence, the aim of the present study is to design, develop and characterize the propellant free quick breaking foam forming Clobetasol propionate nanoemulsions (NE) for the effective treatment of psoriasis.

MATERIALS AND METHODS

Materials

Clobetasol propionate was obtained from Apex Laboratories, Chennai. Captex 200, Captex 355, Captex 500P, Captex 810D, Capmul MCM, Capmul MCM C10 was gifted by Abitech corporation, U.S.A. Solutol HS 15 was gifted by BASF, Mumbai. Vitamin E TPGS was obtained from Cognis, Mumbai. Tween 20 and Tween 80 were purchased from S.D fine chemicals, Mumbai. All other reagents used were of analytical grade.

Determination of Solubility

Solubility of Clobetasol propionate in various oils were determined by dissolving the excess amount of drug in oil, centrifuged and the absorbance of the supernatant oil was measured in UV–Spectrophotometer at 242 nm.

Preparation of Clobetasol Propionate Foam Forming Nanoemulsions

Fixed concentration of Clobetasol Propionate was dissolved in oil (Captex 500P). The oil was mixed with varying concentrations of surfactants (solutol HS 15 and Tween 80) to get clear solution. Distilled water was added to the above solution and mixed using cyclomixer until get transparent nanoemulsion. The composition of the nanoemulsions is given in Table 1.

Table 1: Composition of Clobetasol Propionate Foam Forming Nanoemulsions

Component (%w/v)	*Formulation Code (%)*			
	F1	*F2*	*F3*	*F4*
Clobetasol propionate	0.1	0.1	0.1	0.01
Captex 500P	5	7.5	10	12.5
Solutol HS 15	10	12.5	15	17.5
Tween 80	5	7.5	10	12.5
Distilled water	80	72.5	65	57.5

CHARACTERIZATION

Foam forming nanoemulsion was evaluated for its average globule size and zetapotential by Zetasizer Nano ZS (Malvern Instruments, UK). Transparency of formulations was evaluated by spectrophotometrically at 400 nm. Viscosity and pH of the formulations was measured using Ostwald viscometer and pH meter, respectively. The surface morphology of foam forming nanoemulsion was studied by transmission electron microscopy (TEM) operating at 200kv (Philips EM-430 TEM, USA). Foam stability and foamability of the nanoemulsion formulations were assessed by visual observation of the change in foam volume and foam structure at different time intervals and the changes were also recorded by using digital camera (Cannon, USA). The foam volume was plotted as a function of time to obtain foam stability.[5] Cloud point of the formulations was visually observed by placing them in waterbath and the temperature was gradually increased. The temperature at which the appearance of cloudiness was noted. Drug content of nanoemulsion formulations was evaluated spectrophotpmetrically at 242 nm. *In-vitro* drug release was studied by dialysis bag method and the amount of drug released was measured spectrophotpmetrically at 242 nm.

RESULTS AND DISCUSSION

Solubility and Formulation of Clobetasol Propionate Foam Forming Nanoemulsions

Solubility of clobetasol propionate in various oils was determined and the results are represented in Table 2. The maximum amount of drug (62.77 mg/ml) was dissolved in captex 500P and it was selected for the development of nanoemulsions. Nanoemulsion formulations were made by mixing of varying the concentration of oil, surfactants and water and the formulated naneomusions were transparent as seen in Figure 1. Biocompatible Solutol HS 15 was combined with high HLB value surfactant Tween 80 due to their high miscibility with the oil phase and ability to reduce the globule size.

Table 2: Solubility of Clobetasol Propionate in Various Oils

Name of the Oil	*Solubility (mg/ml)*
Capmul MCM	27.14 ± 2.93
Capmul PG-8	47.27 ± 4.48
Captex 810 D	56.26 ± 5.07
Captex 500 P	62.77 ± 5.48
Captex 355	22.25 ± 2.15
Capmul MCM C8	20.29 ± 1.76
Captex 200	12.68 ± 0.97
Capmul MCM C10	13.26 ± 1.18
Coconut oil	01.72 ± 0.36

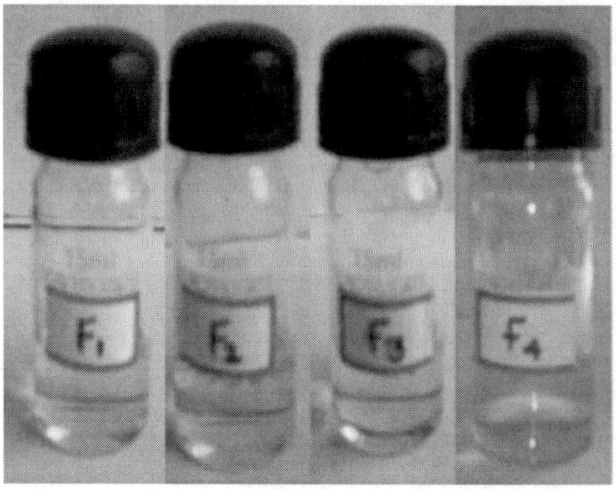

Fig. 1: Clobetasol Propionate Foam Forming Nanoemulsion Formulations

Characterization

The mean size of the globules was varied between 10 to 12 nm which indicates that the presence of globules in the formulation were narrow in range (Table 3 and Figure 2). Significant change in droplet size was not observed when storing the formulation for 3 months. The zetapotential of the nanoemulsions were lies in the range of –3 to –4 mV (Table 3 and Figure 3). The surfactants present in the formulations are non-ionic nature which attributes the charges of the formulations closure to zero. The highest transparency was found with formulation F1 and a low transparency value was found with F4 (Table 3). The lowest transparency obtained with formulation F4 might be due to the presence of higher concentration of surfactants and the pale yellow colour of the Tween 80. But the higher concentration of surfactants and oil offers greater wetability and solubilization of the drug. All the nanoemulsions formulations were showed the pH nearer to neutral pH (Table 3). The viscosity of the nanoemulsion was found to increase from 3.02–30 Pa.s (Table 3) with increasing concentration of oil and surfactant. It confirms the suitability of the formulation for topical administration to the skin.

Cloud point is an essential factor for the formation of a stable nano/micro emulsion. The developed formulations exhibited a slight cloudiness above 63°C (Table 3). At high temperature, dehydration of polyethylene oxide moiety of the non-ionic surfactant leads to phase separation of emulsions. Since, both drug solubilization and formulation stability will decline with this phase separation, the cloud point of the formulation should be over 37°C. This study suggests that, our formulated foam forming nanoemulsions are stable below the temperature of 60°C.

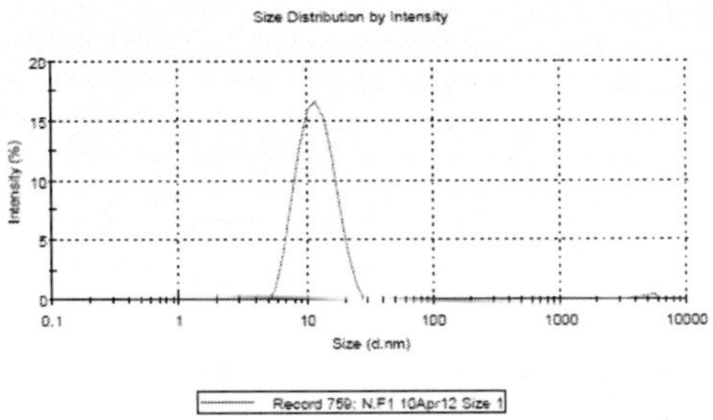

Fig. 2: Globule Size of Formulation (F2)

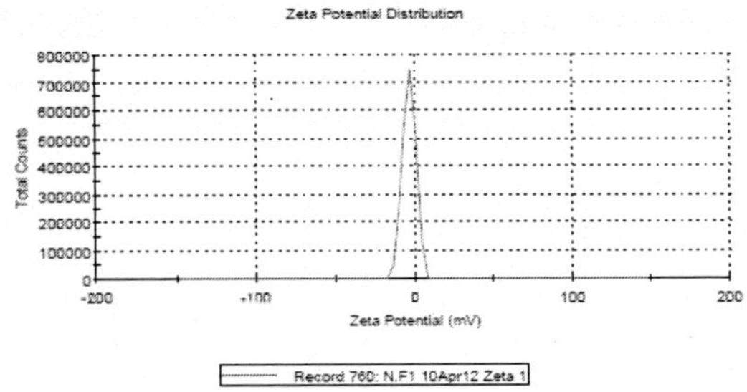

Fig. 3: Zetapotential of Formulation (F2)

Table 3: Characterization of Foam Forming Nanoemulsions

Formulation	*Globule Size (nm)*	*Zeta Potential (mV)*	*Transparency (%)*	*pH*	*Viscosity (Pa s)*	*Cloud Point (°C)*
F1	10.12 ± 0.33	–3.09 ± 0.04	51.77 ± 2.02	6.52 ± 0.32	3.027 ± 0.10	63 ± 2
F2	11.29 ± 0.42	–3.38 ± 0.06	38.56 ± 4.16	6.37 ± 0.14	4.513 ± 0.08	64 ± 4
F3	12.05 ± 0.28	–3.91 ± 0.03	23.18 ± 3.11	6.41 ± 0.16	14.834 ± 0.06	65 ± 3
F4	10.80 ± 0.56	–4.0 ± 0.02	20.75 ± 3.00	6.44 ± 0.11	30.007 ± 0.08	70 ± 2

TEM image of the foam forming nanoemulsion showed the presence of small spherical globules with smooth surface (Figure 4). Foamability and foam stability of the nanoemulsions were evaluated and the data is shown in Table 4. Formulation F1 and F2 were showed higher foamability and stability. Formulation F3 and F4 showed the good formation of foam and exhibit quick breaking properties (Figure 5). Foamability of the formulation was increased with increasing concentration of water. The foam stability was decrease with increasing concentration of oil. The quick breaking foam of the developed products was applied on the skin which quickly breaks and converted into liquid form.

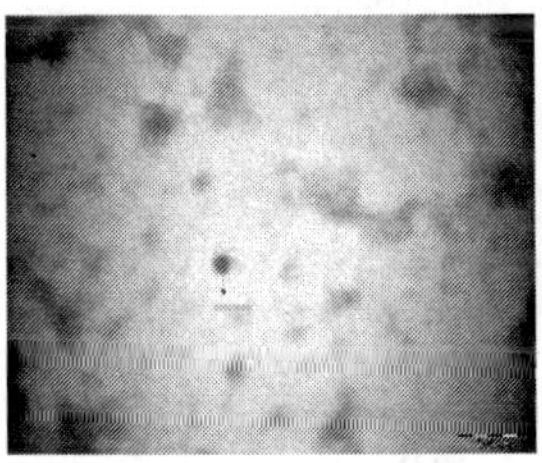

Fig. 4: TEM Image of Foam Forming NE

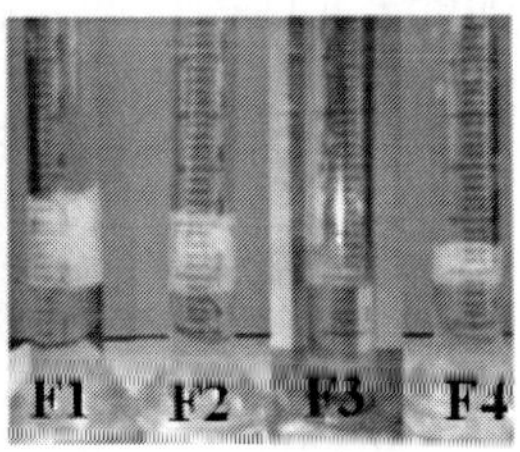

Fig. 5: Foam Stability Test

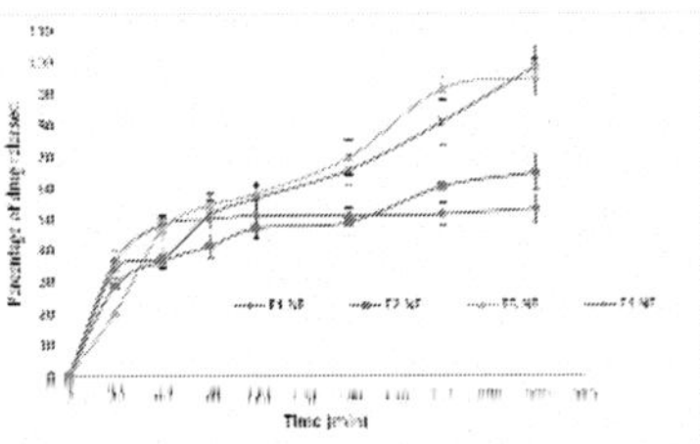

Fig. 6: *In-vitro* Drug Release of Foam Forming Nanoemulsions

The profile of the drug release from foam forming nanoemulsion is shown in Figure 6. The formulation F3 and F4 showed highest drug release of 94 and 98%, respectively with in 300 min. It was apparent from the *in-vitro* release profile of Clobetasol propionate that initial rapid release was followed with the gradual drug release. The highest percentage of drug release observed with nanoemulsion formulations might be caused by the greater amount of incorporated surfactants, which results in greater wetting and solubilization of the drug.

Table 4: Foamability and Foam Stability of Foam Forming Nanoemulsions

Formulation	*Breaking Volume (mins)*				
	0	*5*	*10*	*15*	*20*
F1	3.8 ± 0.10	3.4 ± 0.22	3.2 ± 0.16	3.2 ± 0.20	3.2 ± 0.34
F2	3.0 ± 0.22	2.8 ± 0.34	2.6 ± 0.24	2.6 ± 0.12	2.6 ± 0.22
F3	2.8 ± 0.15	2.6 ± 0.26	2.4 ± 0.20	2.4 ± 0.34	2.4 ± 0.28
F4	2.5 ± 0.20	2.3 ± 0.10	2.3 ± 0.32	2.3 ± 0.26	2.3 ± 0.16

CONCLUSION

The Captex P 500 possessing high solubility was used as oil phase and combination of Solutol HS 15, Tween 80 was used as surfactant along with water for the development of quick break foam forming nanoemulsions. The developed foam forming nanoemulsions showed uniform small spherical droplets with smooth surface within the size range of 10 to 12 nm. The cloud point was above 63°C. Formulation F3 and F4 showed the good formation of foam and exhibit quick breaking properties. The highest percentage of drug release observed with nanoemulsion formulations. Hence we conclude that the formulated quick breaking foam forming nanoemulsion (F4) may be efficient foam forming topical delivery system for Clobetasol propionate and it may have a great potential in the treatment of psoriasis.

ACKNOWLEDGEMENTS

The authors are very much thankful to BASF, Mumbai, Cognis, Mumbai and Abitech Corporation, U.S.A, for providing surfactants samples to carry out this research work.

REFERENCES

[1] Breneman, D., Fleischer, A.B., Kaplan, D., Lebwohl, M., Miller, B. *et al.*, Clobetasol propionate 0.05% lotion in the treatment of moderate to severe atopic dermatitis: a randomized evaluation versus clobetasol propionate emollient cream. *J Drugs Dermatol,* 4(3), 330–6, 2005.

[2] Koo, J.Y.M., New Developments in Topical Sequential Therapy for Psoriasis. *Skin Therapy Letter,* 10(9), 1–4, 2005.

[3] Woodford, R. and Barry, B.W., Bioavailability and activity of topical corticosteroids from a novel drug delivery system, the aerosol quick-break foam. *J Pharm Sci.*, 66(1), 99–103, 1977.

[4] Shrewsbeury, Robert, Applied Pharmaceutics in Contemporary Compounding, 2nd ed., Morton Publishing Company, Allied Health Titles, USA. 2008.

[5] Arzhavitina, A. and Steckel, H., Foams for pharmaceutical and cosmetic application. *Int J Pharm.*, 394(1–2), 1–17. 2010.

DEVELOPMENT OF ELECTROSPUN POLY (PROPYLENE CARBONATE) AND POLYCAPROLACTONE SCAFFOLDS FOR TISSUE ENGINEERING

NAVEEN NAGIAH, SIVAKUMAR SINGARAVELU, LAKSHMI MADHAVI[1], R. ANITHA[2], TIRUCHIRAPALLI SIVAGNANAM UMA AND TIRUPATTUR SRINIVASAN NATARAJAN[2]

Bioproducts Lab, Central Leather Research Institute, Adyar, Chennai
[1]Cavinkare Research Centre, Ekkatuthangal, Chennai
[2]Conducting Polymer Lab, Department of Physics, Indian Institute of Technology Madras, Chennai
E-mail: suma67@gmail.com: tsniit@gmail.com

ABSTRACT

Electrospinning is a simple and versatile technique to produce ultrathin fibers for myriad applications. Recent advances in this decade include exploiting the potential of these fibers for tissue engineering applications. Blended electrospun fibers were obtained of submicron dimensions by blending polycaprolactone with poly(propylene carbonate) using 2, 2, 2 trifluoroethanol as solvent. The resultant fibers were characterized using Scanning electron microscopy (SEM), Differential scanning calorimetry (DSC) and Attenuated total Reflection-Fourier Transformed Infrared spectroscopy (ATR-FTIR). The fibrous scaffold supports growth and proliferation of human dermal fibroblasts and keratinocytes thus proving its reliability in using it as a potential scaffold for skin tissue engineering applications.

INTRODUCTION

Electrospinning is an alternative method to fabricate scaffolds, owing to its increased surface areas, ability to mimic the extra cellular matrix and is versatile in creating a wide range of

V. Rajendran, P. Prabu and K.E. Geckeler (eds.)
Nano Biomaterials, pp. 53–58 (2012).

scaffolds.[1,2] Furthermore, electrospinning aids in mimicking the structure and biological function of native extracellular matrix (ECM) proteins, which provide mechanical support and regulate cell activities First synthesized by Inoue and coworkers in 1969, poly(propylene carbonate), is an aliphatic polycarbonate composed of carbon dioxide and propylene epoxide.[3,4] Poly (propylene carbonate) has competent mechanical and biodegradable properties and a potential for wide range of biomedical applications.[5,6] Polycaprolactone (PCL) is a bioresorbable polymer and is significantly less expensive and is readily available in large quantities.[7] In our present study, blended fibrous mats of poly(propylene carbonate) and polycaprolactone were evaluated for its mettle as a scaffold for skin tissue engineering.

MATERIALS AND METHODS

Materials

Polycaprolactone, 2, 2, 2, triflouroehanol, poly (propylene carbonate) of average molecular weight M_n~50,000 Dulbecco's Modified Eagle's medium (DMEM), Fetal calf Serum (FCS), keratinocytes growth medium and the supplementary antibiotics for tissue culture were purchased from Sigma Aldrich, Bangalore, India. Rest of the chemicals and culture wares were purchased from Sigma Aldrich unless specified otherwise.

Electrospinning of Polymer Fibers

Polymer solutions of 10 wt% concentration were prepared by dissolving 1 g of polycaprolactone and poly (propylene carbonate) in 10 mL of 2, 2, 2, triflouroehanol. The dissolved solution obtained after stirring for 6 to 8 hours were mixed to obtain three blended solutions of polycaprolactone and poly (propylene carbonate) in the ratio of 50:50. The polymer solution was then loaded in 5 mL syringes with 24 G needle connected to the positive terminal of a high voltage DC supply (ZEONICS, Bangalore, India). The polymer solutions were extruded at 0.6 mL/h using a computer controlled syringe pump and was subjected to an electric potential of 0.8 kV/cm. The fibers were deposited onto a grounded aluminum substrate wound around a rotary drum placed at a distance of 15 cm perpendicular to the needle. The details of the apparatus used is published elsewhere.[8] The obtained samples were stored at room temperature until further use.

Scanning Electron Microscopy

The aluminum substrates coated with the electrospun fibers were mounted on brass stubs and observed under a scanning electron microscope (F E I Quanta FEG 200 - HRSEM) operating at an accelerating voltage of 5–20 kV. Fibers were electrospun with different concentrations and by varying different process parameters like flow rate, etc. The diameters of about 50 different fibers were measured in each of the above case using the UTHSCA Image tool to obtain their average diameter.

ATR-FTIR Spectroscopy

Attenuated total reflection-Fourier Transform Infrared spectroscopic measurements were carried out on peeled fibrous membranes using ABB 3000 spectrometer with Grams as the operating software. Typically, 21 scans were signal-averaged to reduce spectral noise. The spectrum of the samples was recorded from 600 to 4000 cm^{-1}.

Differential Scanning Calorimetry

Differential Scanning Calorimetry (DSC) analysis of the fiber was performed from 0°C to 200°C at 10°C min^{-1} using universal V4.4A TA instruments. The instrument was calibrated using an indium standard and the calorimeter cell was flushed with liquid nitrogen at 50 mL min^{-1} prior to the experiment.

Tensile Strength Measurement

Two samples of electrospun nanofibrous membranes of dimension 100 × 5 mm^2, one dry and the other soaked in phosphate buffer saline of pH 7.4 for 24 hours were tested for their mechanical strength. Load-elongation measurement was carried out at a crosshead speed of 5 mm/min, 25°C temperature and RH of 65% respectively. Percentage of elongation at break (%) was measured using a universal testing machine (INSTRON model 1405) according to Vogel at an extension rate of 5 mm/min.

Biocompatibility of the Electrospun Fibers

Skin biopsies collected from patients undergoing cosmetic surgery were brought in sterile condition to the tissue culture laboratory (for dealing with human samples, the study was conducted in accordance to the guidelines of the ethical committee of Central Leather Research Institute [C.L.R.I], Chennai). The tissue was washed with Phosphate Buffered Saline (PBS) with 10x antibiotics thrice and then twice with PBS-1x antibiotics. The samples were placed in 0.3% trypsin at 4°C overnight. The next day epidermal and dermal layers were separated and the dermal portion was cut into small pieces, with collagenase for digestion and then transferred to the flask containing DMEM supplemented with 10% FCS and epidermal layers were treated with trypsin and the released keratinocytes were grown in keratinocyte growth medium, 4mM L-glutamine, streptomycin (100 μg/mL), penicillin (100 U/mL), gentamycin (30 μg/mL), fungizone (2 μg/mL) at 37°C, in a humidified 5% CO_2/95% air atmosphere. After allowing the cultures to grow to confluence, experiments were carried out using early passages of cells (2–4 passage).

A monolayer of the human fibroblast cells and keratinocytes were grown on fibrous membranes placed in a 12-well culture plate (Corning, NY) and maintained in DMEM with 10% fetal calf serum supplemented with penicillin (120 units per mL), streptomycin (75 $mgmL^{-1}$), gentamycin (160 $mgmL^{-1}$), and amphotericin B (3 $mgmL^{-1}$) at 37°C, humidified with 5% CO_2. Cells cultured in a blank well were used as a control. After 48 h, the culture medium was replaced with a serum-free medium containing thioazolyl blue (MTT) and incubated for 4 h at 37°C in a humidified atmosphere of 5% CO_2. The medium was aspirated and the formazan needles were dissolved in 500 mL of dimethyl sulfoxide (DMSO). The absorbances of the solutions were read at 570 nm and the results were expressed as mean of three experiments, calculated (using student's 't' test) with $p < 0.05$.

RESULTS AND DISCUSSION

The SEM micrograph of electrospun fibers is depicted in Figure 1(a). The average diameter of the fibers was found to be 310 ± 60 nm. Figure 1(b) depicts the ATR-FTIR of electrospun blended polycaprolactone and poly (propylene carbonate) fibers.

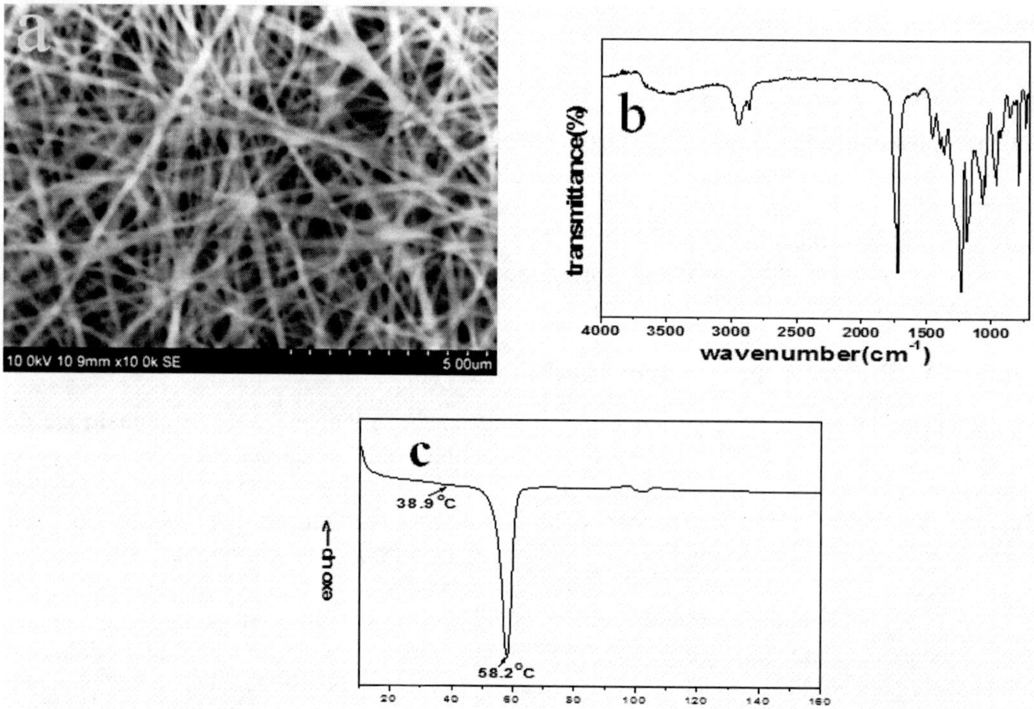

Fig. 1: (a) SEM Micrograph of Blended Electrospun Fibers of Polycaprolactone and Poly (propylene carbonate), (b) ATR-FTIR of Blended Electrospun Fibers of Polycaprolactone and Poly (propylene carbonate), (c) DSC of Blended Electrospun Fibers of Polycaprolactone and Poly (propylene carbonate)

Figure 1b represents the ATR-FTIR spectrum of the electrospun blended fibrous mat. The peaks at 1742, 1452, 1223 and 1081 cm^{-1} correspond to C=O stretching of amorphous carbonyl group of and poly(propylene carbonate). Asymmetric CH2 stretching, COC, C-O stretching and Carbonyl stretching at 2947, 2866, 1169, 1157, 1720 of polycaprolactone respectively. The peak observed at 1169 cm^{-1} corresponds to the C–O–C stretching.[9, 10]

Figure 1(c) represents the DSC of blended fibers of polycaprolactone and poly (propylene carbonate). The glass transition temperature of poly (propylene carbonate) is visible at 58.2°C and polycaprolactone show at 38.9°C both mass with each other. DSC and ATR-FTIR confirms the interaction between the polymers as the peaks mass with one another in both DSC and ATR-FTIR was observed. The tensile properties of the electrospun fibers are tabulated in Table 1. The tensile properties of the fibrous at was found to be competent for skin tissue engineering applications. MTT assay of electrospun fibers with human dermal fibroblasts and keratinocytes is depicted in Figure 2. The developed scaffold aided in the adhesion and proliferation of both human dermal fibroblasts and keratinocytes proving its applicability as a potential scaffold for skin regeneration.

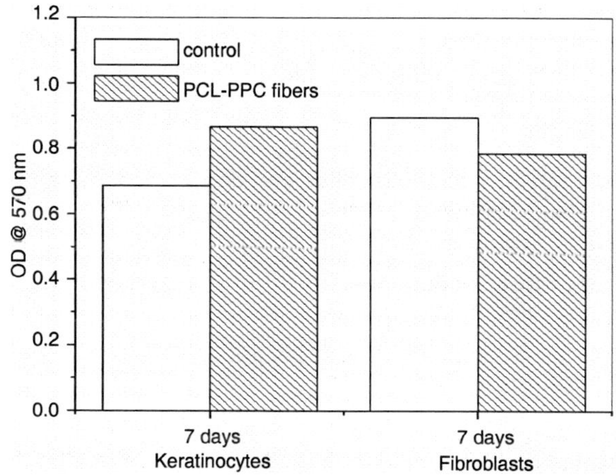

Fig. 2: MTT Assay of Blended Electrospun Fibers of Polycaprolactone and Poly (propylene carbonate)

Table 1: Tensile Properties of Blended Electrospun Fibers of Polycaprolactone and Poly (propylene carbonate)

Maximum Load (N)	*Tensile Strength (MPa)*	*Elongation At Break (%)*	*Tensile Strain (%)*	*Extension at Maximum Load (mm)*	*Young's Modulus (MPa)*
0.40	0.40	4.83	4.83	0.97	18.23

CONCLUSION

Fibers of poly (propylene carbonate) and polycaprolactone were electrospun using 2, 2, 2, triflouroethanol as solvent. The fibrous mats exhibited competent thermal properties and ensured the complete evaporation of solvent during electrospinning. The mats supported the growth of human dermal fibroblast and keratinocytes proving its potential as a scaffold in skin tissue engineering.

REFERENCES

[1] Zahedi, P., Rezaeian, I., Seyedomid, R.S., Siadat, R., Jafari, S.H. and Supaphol, P.A., review on wound dressings with an emphasis on electrospun nanofibrous polymeric bandage, *Polym. Adv. Technol.*, 21, 77–95, 2010.

[2] Shanshan, X., Junxing, L., Aihua, He., Wenwen, L., Xingyu, J., Jianfen, Z., Charles, C.H., Benjamin, S.H., Benjamin, C. and Dufei, F., Chemical crosslinking and biophysical properties of electrospun hyaluronic acid based ultra-thin fibrous membranes, *Polymer,* 50, 3762–3769, 2009.

[3] Li, J., Lai, M.F. and Liu, J.J., Effect of poly (propylene carbonate) on the crystallization and melting behavior of poly (β-hydroxybutyrate-co-β-hydroxyvalerate), *J. Appl. Polym. Sci.,* 92, 2514–2521, 2004.

[4] Wang, S.J., Du, L.C., Zhao, X.S., Meng, Y.S. and Tjong, S.C., Synthesis and characterization of alternating copolymer from carbon dioxide and propylene oxide, *J. Appl. Polym. Sci.,* 85, 2327–2334, 2002.

[5] Binyuan, L., Xiaojiang, Z., Xianhong, W. and Fosong, W., Thermal degradation kinetics of poly (propylene carbonate) obtained from the copolymerization of carbon dioxide and propylene oxide, *J. Appl. Polym. Sci.,* 90, 947–953, 2003.

[6] Du, L.C., Meng, Y.Z., Wang, S.J. and Tjong, S.C., Synthesis and degradation behavior of poly(propylene carbonate) derived from carbon dioxide and propylene oxide, *J. Appl. Polym. Sci.*, 92, 1840–1846, 2004.

[7] Venugopal, J.R., Zhang, Y. and Ramakrishna, S., *In vitro* culture of human dermal fibroblasts on electrospun polycaprolactone collagen nanofibrous Membrane, *Artif. Organs*., 30, 440–446, 2006.

[8] Naveen,N., Ramadhar, K., Balaji, S., Uma, T.S., Natarajan, T.S. and Praveen, K.S., Synthesis of nonwoven nanofibers by electrospinning—a promising biomaterial for tissue engineering and drug delivery, *Adv. Eng. Mat.*, 12, B380–387, 2010.

[9] Shengjie, W., Yuhui, H., Bin, L., Guo, L., Guangmin, C. and Liban, C. Structure and properties of poly (propylene carbonate), *Int. J. Polym. Anal. Ch.*, 3, 131–143, 1997.

[10] Tamara Elzein, Mohamad Nasser-Eddine, Christelle Delaite, Sophie Bistac, and Philippe Dumas FTIR study of polycaprolactone chain organization at interfaces *J. Colloid. Interf. Sci.*, 273 381–387, 2004.

SYNTHESIS OF BLENDED FIBERS OF POLY (3-HYDROXYBUTYRIC ACID) AND POLY (PROPYLENE CARBONATE) SCAFFOLDS FOR TISSUE ENGINEERING

NAVEEN NAGIAH, GIRIPRASATH RAMANATHAN, LAKSHMI MADHAVI[1], R. ANITHA[1], TIRUCHIRAPALLI SIVAGNANAM UMA AND TIRUPATTUR SRINIVASAN NATARAJAN[2]

Bioproducts Lab, Central Leather Research Institute, Adyar, Chennai
[1]Cavinkare Research Centre, Ekkatuthangal, Chennai
[2]Conducting Polymer Lab, Department of Physics, Indian Institute of Technology Madras, Chennai
E-mail: suma67@gmail.com; tsniit@gmail.com

ABSTRACT

Electrospinning is one of the cheapest and up-scalable techniques to mimic the structure of Extracelluar Matrix (ECM) of tissue and organs. Fibers of submicron and nanoscale dimensions were obtained by electrospining blended polymer solutions of poly (3-hydroxybutyric acid) and poly (propylene carbonate) using 1, 1, 1, 3, 3, 3 hexafluoro-2-propanaol and 2, 2, 2 trifluoroethanol as solvent. The resultant fibers were characterized using Scanning Electron Microscopy (SEM), Differential Scanning Calorimetry (DSC) and Attenuated Total Reflection-Fourier Transformed Infrared spectroscopy (ATR-FTIR). The fibrous scaffold had competent tensile properties and supports growth and proliferation of human dermal fibroblasts and keratinocytes with normal morphology thus proving its reliability in using it as a potential scaffold for skin regeneration.

INTRODUCTION

Poly (3-hydroxybutyric acid) is natural polyester stored by wide range of microorganisms.[1] Poly (3-hydroxybutyric acid) can be used as suitable replacement for petrochemical plastics

V. Rajendran, P. Prabu and K.E. Geckeler (eds.)
Nano Biomaterials, pp. 59–64 (2012).

(polyethylene, polypropylene etc.), as they completely degrade to carbon dioxide and water through natural microbiological mineralization.[2] The challenging combination of biomedical and biodegradable properties of PHB made it a potential tool in design of novel medical devices and tissue engineering.[3, 4] 3-hydroxybutyric acid is a common metabolite in all higher living organisms, which is evidence for the non-toxicity of poly (3-hydroxybutyric acid).[5]

Poly (propylene carbonate) is amorphous with a glass transition temperature in the range of our body temperature (35–40°C) making it neither a typical engineering material nor a rubber (modifier).[6] This property along with the biodegradable property of poly (propylene carbonate) has been evaluated as engineered constructs for tissue engineering applications.[7, 8] Poly (propylene carbonate) and poly (3-hydroxybutyric acid) were chosen for this study because of its cheap production from wastes.[9] Hydrophilic polymers usually has a lower half life *in vivo* and are majorly used for acute wounds while chronic wounds needs a higher half life *in vivo* due to which hydrophobic polymers or their blends have been used in this study. Essentially, a tissue engineering scaffold should mimic the structure and biological function of native extracellular matrix (ECM) proteins, which provide mechanical support and regulate cell activities.[10] The structural requirements include small fiber diameters, high-matrix porosity, large void volume, large surface-to-volume ratio, mechanical strength and a three dimensional structure to allow cell infiltration.[11]

Much attention has been paid to electrospinning, a remarkably simple and powerful technique for the generation of polymeric fibers in the sub-micrometer scale, ranging from about 50 nm to several micrometers.[12] The present study evaluates the potential of blended poly (propylene carbonate) and poly (3-hydroxybutyric acid) scaffold for skin regeneration.

MATERIALS AND METHODS

Materials

Poly (3-hydroxybutyric acid), 1, 1, 1, 3, 3, 3 hexafluoro-2-propanaol, poly (propylene carbonate) of average molecular weight M_n~50,000 Dulbecco's Modified Eagle's medium (DMEM), Fetal calf Serum (FCS), keratinocytes growth medium and the supplementary antibiotics for tissue culture were purchased from Sigma Aldrich, Bangalore, India. Rest of the chemicals and culture wares were purchased from Sigma Aldrich unless specified otherwise.

Electrospinning of Polymer Fibers

Polymer solutions of 6 wt% concentration were prepared by dissolving 0.6 g of poly (3-hydroxybutyric acid) and poly (propylene carbonate) in 10 mL of 1, 1, 1, 3, 3, 3 hexafluoro-2-propanaol. The dissolved solution obtained after stirring for 6 to 8 hours were mixed to obtain three blended solutions of poly (3-hydroxybutyric acid) and poly (propylene carbonate) in the ratio of 50:50. The polymer solution was then loaded in 5 mL syringes with 24 G needle connected to the positive terminal of a high voltage DC supply (ZEONICS, Bangalore, India). The polymer solutions were extruded at 0.75 mL/h using a computer controlled syringe pump and was subjected to an electric potential of 1 kV/cm. The fibers were deposited onto a grounded aluminum substrate wound around a rotary drum placed at a distance of 17 cm perpendicular to the needle. The details of the apparatus used is published elsewhere.[14] The obtained samples were stored at room temperature until further use.

Scanning Electron Microscopy

The aluminum substrates coated with the electrospun fibers were mounted on brass stubs and observed under a scanning electron microscope (FEI Quanta FEG 200 - HRSEM) operating at an accelerating voltage of 5–20 kV. Fibers were electrospun with different concentrations and by varying different process parameters like flow rate, etc. The diameters of about 50 different fibers were measured in each of the above case using the UTHSCA Image tool to obtain their average diameter.

ATR-FTIR Spectroscopy

Attenuated total reflection-Fourier Transform Infrared spectroscopic measurements were carried out on peeled fibrous membranes using ABB 3000 spectrometer with Grams as the operating software. Typically, 21 scans were signal-averaged to reduce spectral noise. The spectrum of the samples was recorded from 600 to 4000 cm^{-1}.

Differential Scanning Calorimetry

Differential Scanning Calorimetry (DSC) analysis of the fiber was performed from 0ºC to 200ºC at 10ºC min^{-1} using universal V4.4A TA instruments. The instrument was calibrated using an indium standard and the calorimeter cell was flushed with liquid nitrogen at 50 mL min^{-1} prior to the experiment.

Tensile Strength Measurement

Two samples of electrospun nanofibrous membranes of dimension 100×5 mm^2, one dry and the other soaked in phosphate buffer saline of pH 7.4 for 24 hours were tested for their mechanical strength. Load-elongation measurement was carried out at a crosshead speed of 5 mm/min, 25°C temperature and RH of 65% respectively. Percentage of elongation at break (%) was measured using a universal testing machine (INSTRON model 1405) according to Vogel at an extension rate of 5 mm/min.

Biocompatibility of the Electrospun Fibers

Skin biopsies collected from patients undergoing cosmetic surgery were brought in sterile condition to the tissue culture laboratory (for dealing with human samples, the study was conducted in accordance to the guidelines of the ethical committee of Central Leather Research Institute (CLRI), Chennai). The tissue was washed with Phosphate Buffered Saline (PBS) with 10x antibiotics thrice and then twice with PBS-1x antibiotics. The samples were placed in 0.3% trypsin at 4°C overnight. The next day epidermal and dermal layers were separated and the dermal portion was cut into small pieces, with collagenase for digestion and then transferred to the flask containing DMEM supplemented with 10% FCS and epidermal layers were treated with trypsin and the released keratinocytes were grown in keratinocyte growth medium, 4 mM L-glutamine, streptomycin (100 µg/mL), penicillin (100 U/ mL), gentamycin (30 µg/mL), fungizone (2 µg/mL) at 37°C, in a humidified 5% CO_2/95% air atmosphere. After allowing the cultures to grow to confluence, experiments were carried out using early passages of cells (2–4 passage).

A monolayer of the human fibroblast cells and keratinocytes were grown on fibrous membranes placed in a 12-well culture plate (Corning, NY) and maintained in DMEM with 10% fetal calf serum supplemented with penicillin (120 units per mL), streptomycin (75 $mgmL^{-1}$), gentamycin (160 $mgmL^{-1}$), and amphotericin B (3 $mgmL^{-1}$) at 37ºC, humidified with 5% CO_2. Cells cultured in a blank well were used as a control. After 48 h, the culture medium was replaced with a serum-free medium containing thioazolyl blue (MTT) and incubated for 4 h at 37ºC in a humidified atmosphere of 5% CO_2. The medium was aspirated and the formazan needles were dissolved in 500 mL of dimethyl sulfoxide (DMSO). The absorbances of the solutions were read at 570 nm and the results were expressed as mean of three experiments, calculated (using student's 't' test) with $p < 0.05$.

RESULTS AND DISCUSSION

Figure 1(a) depicts the SEM image of blended fibers of poly (3-hydroxybutyric acid) and poly (propylene carbonate). Coalescence of fibers at junctions is clearly visible which would contribute to an additional enhancement in tensile strength of fibers.[13]

Figure 1(b) represents the ATR-FTIR spectrum of the electrospun blended fibrous mat. The peaks at 1742, 1452, 1223 and 1081 cm^{-1} correspond to C=O stretching of amorphous carbonyl group, Asymmetric CH_3, CH_3 vibration and symmetric C–O stretching of poly(3-hydroxybutyric acid) and poly(propylene carbonate) respectively. The peak observed at 1169 cm^{-1} corresponds to the C–O–C vibration.[14, 15]

Figure 1(c) and 1(d) represent the DSC of blended fibers of poly (3-hydroxybutyric acid) and poly (propylene carbonate). The glass transition temperature of poly (propylene carbonate) is visible at 38.9°C. The sharp endothermic peak at 170.8°C represents the melting peak of poly (3-hydroxybutyric acid). DSC and ATR-FTIR confirms the non-interaction between the polymers as separate peaks in both DSC and ATR-FTIR was observed. The tensile properties of the electrospun fibers are tabulated in Table 1. The tensile properties of the fibrous at was

Table 1: Tensile Properties of Blended Electrospun Poly (3-hydroxybutyric acid) and Poly (propylene carbonate) Fibers

Maximum Load (N)	*Tensile Strength (MPa)*	*Elongation At Break (%)*	*Tensile Strain (%)*	*Extension at Maximum Load (mm)*	*Young's Modulus (MPa)*
2.81	5.63	49.08	59.42	9.82	94.42

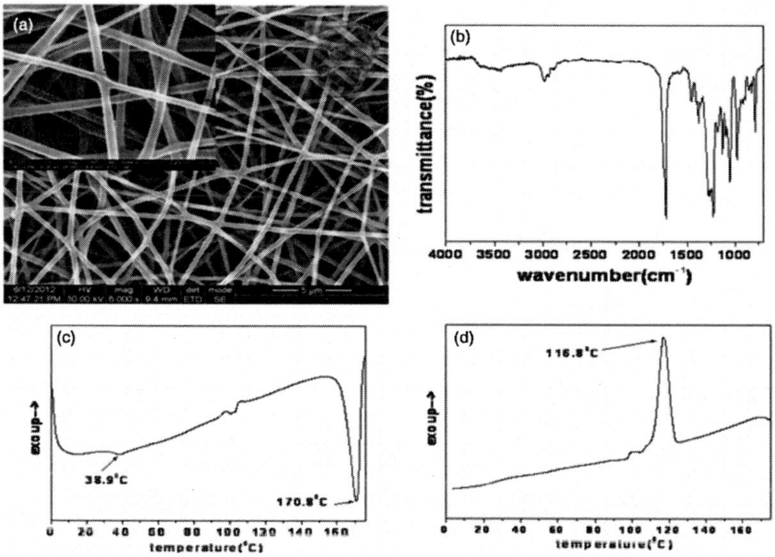

Fig. 1: (a) SEM Micrograph of Blended Electrospun Poly (3-hydroxybutyric acid) and Poly (propylene carbonate) Fibers, (b) ATR-FTIR of Blended Electrospun Poly (3-hydroxybutyric acid) and Poly (propylene carbonate) Fibers, (c) DSC (heating), and (d) DSC (cooling) of Blended Electrospun Poly (3-hydroxybutyric acid) and Poly (propylene carbonate) Fibers

found to be competent for skin tisuue engineering applications. MTT assay of electrospun fibers with human dermal fibroblasts and keratinocytes is depicted in Figure 2. The developed scaffold aided in the adhesion and proliferation of both human dermal fibroblasts and keratinocytes proving its applicability as a potential scaffold for skin regeneration.

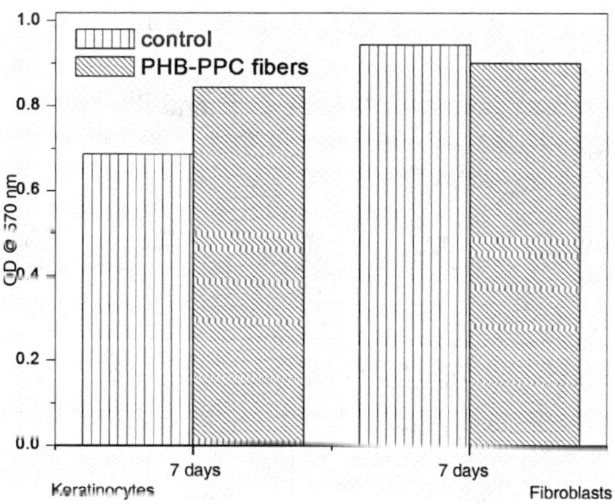

Fig. 2: MTT Assay of Blended Electrospun Poly (3-hydroxybutyric acid) and Poly (propylene carbonate) Fibers

REFERENCES

[1] Madison, L.L. and Huisman, G.W. "Metabolic engineering of poly (3-hydroxyalkanoates): from DNA to plastic", *Microbiol. Mol. Biol. Rev.*, 63, 21–53, 1999.

[2] Sudesh, K., Abe, H. and Doi, Y. "Synthesis, structure and properties of polyhydroxyalkanoates: Biological polyesters", *Prog. Polym. Sci.,* 25, 1503–1555, 2000.

[3] Frier, T., Kunze, C., Nischan, C., Kramer, S., Sternberg, K., Sab, M., Hopt, U.T. and Schmitz, K.H. "*In vitro* and *in vivo* degradation studies for development of a biodegradable patch based on poly (3-hydroxybutyrate)", *Biomaterials*, 23, 2649–2657, 2002.

[4] Theron, S.A., Zussman, E. and Yarin, A.L. "Experimental investigation of the governing parameters in the electrospinning of polymer solutions", *Polymer*, 45, 2017–2030, 2004.

[5] Williamson, D.H., Patrica, L. and Krebs, H.A. "The redox state of free nicotinamide-adenine dinucleotide in the cytoplasm and mitochondria of rat liver", *J. Biochem.,* 103, 90–96, 1967.

[6] Gerrit, A.L. "Poly (Propylene Carbonate) old copolymers of propylene oxide and carbon dioxide with new interests: catalysis and material properties", *Polym. Rev.,* 48, 192–219, 2008.

[7] Alexander, W., Kroger, M., Doring, M., Niederer, K., Pindel, E. and Chronakis, I.S. "Electrospun aliphatic polycarbonates as tailored scaffold materials", *Biomaterials*, 28, 2211–2219, 2007.

[8] Jun, Z., Hongxu, Q., Hongjun, W., Ping, H., Lailiang, O., Shuhua, G., Jing, L., Yongzhe, C., Yaoting, Y. and Deling, K. "Engineering of vascular grafts with genetically modified bone marrow mesenchymal stem cells on poly (propylene carbonate) graft", *Artif. Organs*, 30, 898–905, 2006.

[9] Du, L.C., Meng, Y.Z., Wang, S.J. and Tjong, S.C. "Synthesis and degradation behavior of poly(propylene carbonate) derived from carbon dioxide and propylene oxide", *J. Appl. Polym. Sci.*, 92, 1840–1846, 2004.

[10] Park, K.E., Kang, H.K., Lee, S.J., Min, B.M. and Park, W.H. "Biomimetic nanofibrous scaffolds: preparation and characterization of PGA/chitin blend nanofibers", *Biomacromolecules*, 7, 635–643, 2006.

[11] Cheryl, L.C., Yang, W., Farach Carson, M.C. and Rabolt, J.F. "Coating electrospun collagen and gelatin fibers with perlecan domain for increased growth factor binding", *Biomacromolecules.*, 8, 1116–1123, 2007.

[12] Huang, Z.M., Zhang, Y.Z., Venugopal, J., Lim, C.T. and Ramakrishna, S. "Characterization of the surface biocompatibility of the electrospun PCL-collagen nanofibers using fibroblasts", *Biomacromolecules*, 6, 2583–2589, 2005.

[13] Ishii, D., Lee, D.W., Kasuya, K.I. and Iwata, T. "Fine structure and enzymatic degradation of poly(R)-3-hydroxybutyrate and stereocomplexed poly (lactide) nanofibers", *J. Biotechnol.*, 132, 318–324, 2007.

[14] Shengjie, W., Yuhui, H., Bin, L., Guo, L., Guangmin, C. and Liban, C. "Structure and properties of poly (propylene carbonate)", *Int. J. Polym. Anal. Ch.,* 3, 131–143, 1997.

[15] Mirko, N., Gerhilt, S., Andreas, J., Frank, S., Dieter, P. and Carsten, W. "Low pressure plasma treatment of poly (3-hydroxybutyrate): Toward tailored polymer surfaces for tissue engineering scaffolds", *J. Biomed. Mater. Res.*, 59, 632–638, 2002.

DESIGN AND DEVELOPMENT OF ARTEMISININ AND DEXAMETHASONE TOPICAL NANODISPERSION FOR THE EFFECTIVE TREATMENT OF AGE RELATED MACULAR DEGENERATION

SUBRAMANIAN NATESAN, CHANDRA SEKAR PONNUSAMY, ABIMANYU SUGUMARAN AND VENKATESHWARAN KRISHNASWAMI

Laboratory for lipid based systems, Department of Pharmaceutical Technology Anna University of Technology Tiruchirappalli, Tiruchirappalli, Tamil Nadu
E-mail: natesansubbu@gmail.com

ABSTRACT

The ocular bioavailability of the drugs that applied topically was enhanced by minimizing the precorneal loss factors and increasing the corneal residence time. The particulate drug delivery systems, nanodispersions, has been showed enhanced ocular residence time and controlled and prolonged drug delivery to the disease site. On instillation, the nanodispersions might form a thin transparent layer on cornea, which results in increased contact and residence time and reduced precorneal loss. Age related macular degeneration (AMD) is a disease which affects the macula due to the formation of new blood vessels in presences of VEGF, resulted in blindness. Artemisinin is a poorly water soluble natural antimalarial agent and recently it has been reported to have the potential to stop the growth of the excessive blood vessels and since it can be used as an anti-angiogenic agent. Dexamethasone is a Corticosteroids, are able to inhibit the Vascular Endothelial Growth Factor (VEGF). Hence, in the present study we developed and characterized nanodispersion containing Anti-angiogenic agent (artemisinin) and corticosteroid (Dexamethasone) for the topical administration to the eye and for the improved treatment of AMD. Various concentrations of polymer (<10%) and polymeric surfactant (<5%) were used to

V. Rajendran, P. Prabu and K.E. Geckeler (eds.)
Nano Biomaterials, pp. 65–70 (2012).

formulate the drug loaded nanodispersion. Drug-excipients compatibility was evaluated by FTIR. The nanodsiperson was characterized for its particle size, zeta potential, cloud point, transparency, surface morphology and evaluated for drug content, drug release and toxicity. FT-IR spectra showed there is no remarkable interaction between drug and excipients. The mean particle size and zeta potential is in the range of 12 to 26 nm and –6 mV to 10 mV, respectively. The surface morphology of the nanodispersion was smooth and spherical in shape. In-vitro drug release result showed that 90 to 101%w/v of artemisinin and 55 to 103% w/v of dexamethasone were released from the nanodispersions. The blank formulation with high concentration of polymeric surfactant showed acceptable level of haemolysis and DNA damage. Hence the topical administration of formulated artemisinin and dexamethasone nanodispersion may have the great potential for the treatment of AMD.

INTRODUCTION

One of the main problems in ophthalmic topical drug delivery is the rapid elimination of conventional liquid eye drops from the eye and this issue still remains unsolved. Only the very small amount of about l–5% of the dosage actually penetrates through the cornea and is able to reach intraocular tissues.[1] Nanodispersions are sub-micron colloidal dispersions of drug particles in an outer liquid phase. The nanodispersion contain polymer are known to provide higher bioadhesion and corneal penetration. Age related macular degeneration (AMD) is a disease associated with aging that gradually destroys sharp and central vision. The wet AMD occurs due to the formation of new blood vessels at large and which is responsible for vision loss. Artemisinin is the anti-malarial agent obtained from *Artemisia annua L* and recently it has been reported that artemisinin posses anti-proliferation, anti-angiogenesis and anti-apoptosis in addition to their anti-malarial properties. Currently certain corticosteroid such as dexamethasone is suggested for the treatment of the wet form of AMD due to their anti-VEGF property. Anti-angiogenic, Anti-VEGF, corticosteroids and photodynamic therapies are utilized individually or in combination to treat the wet AMD.[2] Hence, in the present study we developed and characterized nanodispersion containing Anti-angiogenic agent (artemisinin) and corticosteroid (Dexamethasone) for the topical administration to the eye and for the improved treatment of AMD.

Preparation of Artemisinin and Dexamethasone Nanodispersion

Artemisinin and dexamethasone were dissolved along with the excipients, polymer PVP K90 and polymeric surfactant Poloxamer 407 (2:1), in ethanol, and dried to get drug loaded polymeric particles. The polymeric particles were dispersed in phosphate buffer pH 7.4 by gentle vortex to form a stable nanodispersion. The composition of drug, polymer and polymeric surfactant used to prepare various nanodispersion (ND) formulations are shown in Table 1.

Table 1: Compositions of Artemisinin and Dexamethasone Nanodispersion

S.No.	*Ingredients*	*Composition (%)*		
		F1	*F2*	*F3*
1.	Polyvinyl pyrolidone K90	5	8	10
2.	Poloxamer 407	2.5	4	5
3.	Artemisinin	0.025	0.025	0.025
4.	Dexamethasone	0.025	0.025	0.025

CHARACTERIZATION OF NANODISPERSION

Drug and excipients compatibility was determined by using Fourier transform infrared spectroscopy (FTIR). Transparency of the formulated nanodispersions was evaluated at 400 nm by UV Spectrophotometer using distilled water as a blank. The average particle size, size distribution and zeta potential of the nanodispersions were determined by bluewave particle size analyser and Zetasizer Nano ZS (Malvern Instruments, UK), respectively. The surface morphology of nanodispersions was examined by transmission electron microscopy (TEM) operating at 200 kv (Philips EM-430 TEM, USA). The cloud point of the formulation was evaluated by placing in a water bath and the temperature was gradually increased. The decrease in % transmittance due to the formation of cloudiness in the formulations was measured spectrophoto metrically at 400 nm.

Evaluation of Nanodispersion

Artemisinin and dexamethasone content of nanodispersion were estimated by RP-HPLC technique after suitable dilution of formulation with phosphate buffer pH 7.4 in a system consists of LC- 20AD pump, photodiode array UV detector, LC solution data acquisition software (Shimadzu, Japan) and Phenomenex-C18 (5 µm, 4.6 mm × 250 mm) analytical column. The mobile phase consisting of mixture of 45% Acetonitrile and 55% methanol-water (2:3) was used in the flow rate of 1.0 mL/min and the chromatographic detection was carried out at 219 nm.

The release of Artemisinin and dexamethasone from nanodispersions were evaluated for 8 hours using a dialysis system, consisting of dialysis membrane (Himedia, India) loaded with 1 mL of nanodispersion and suspended in 50 mL of dissolution medium (phosphate buffer solution pH 7.4) maintained at 37 ± 1°C. The dissolution medium was stirred at 600 RPM using magnetic stirrer. Aliquots of 2 mL were withdrawn at predetermine intervals and replaced by an equal volume of the fresh dissolution medium. Then the aliquots were analyzed in RP-HPLC at 219 nm.

Toxicity Studies

In-vitro haemolytic potential of the nanodispersion without drug was studied using the method described by Amin and Dannenfelser.[3] Percentage haemolysis of the nanodispersion was calculated using the following formula.

$$\% \text{ Haemolysis} = \frac{(\text{Negative control} - \text{Blank nanodispersion})}{(\text{Negative} - \text{Positive control})} \times 100$$

The toxic effect of the nano carrier on the genetic materials of the cells was evaluated by the alkaline comet assay as described by Rajaguru *et al.*[4] Overall 50 cells each from 2 slides of same test sample was examined for the length of DNA mass (head + tail) and width of comet head using the software, Comet score (v1.5, Tritek corporation) and the mean ratio of length to width was calculated.

RESULTS AND DISCUSSION

Preparation of Artemisinin and Dexamethasone Loaded Nanodispersion

Different concentration of PVP K90 and Poloxamer 407 in the ratio of 2:1 was used to formulate the artemisinin and dexamethasone loaded nanodispersion. The formulated nanodispersion was clear and transparent. The amount of PVP K 90 and Poloxamer 407 below the formulation F1 leads to the formation of drug crystals on storage and above the amount of formulation F3 is not acceptable for ocular administration.

Characterization of Nanodispersion

FT-IR spectra of artemisinin, dexamethasone, PVP K90, poloxamer 407, and dried nanoparticles were showed no interaction between drug with excipients. The result confirms the compatibility of the formulated nanodispersion. The transparency and cloud point of the formulated nanodispersions is shown in Table 2. The transparency of the formulation decreases from 99% to 96% upon the increases in PVP K 90 and Poloxamer 407 concentration. The cloud point of the formulated nanodispersion was decrease with the increased concentration of polymeric surfactant.

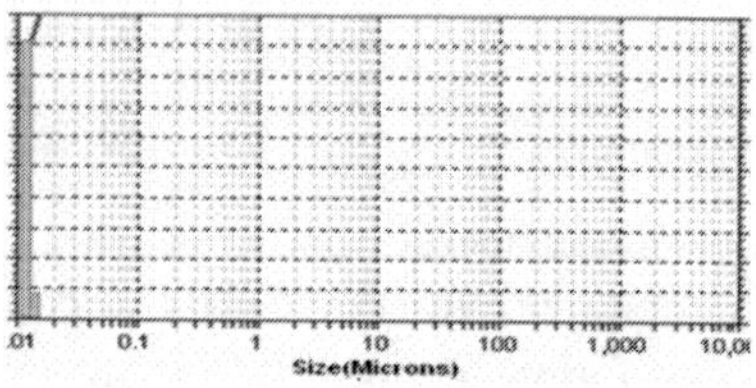

Fig. 1: Particle Size of the Nanodispersion (F3)

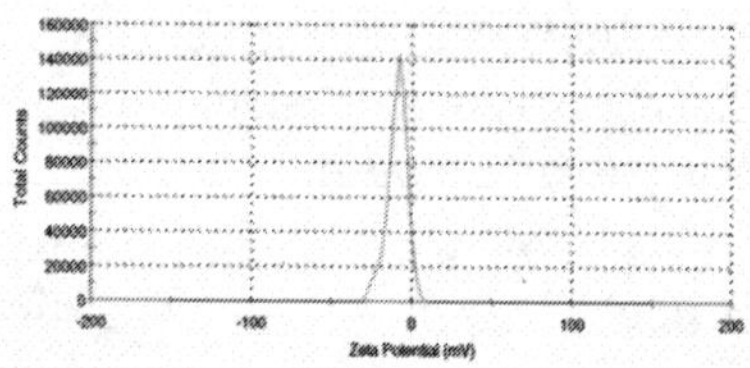

Fig. 2: Zeta Potential of the Nanodispersion (F3)

The mean particles size and zetapotential of the nanodispersions is shown in Table 2 and Figure 1 and it was found to be in the range of 12 to 26 nm and –6mV to 10 mV, respectively. The polydispersity index of all the formulations ranges from 0.01 to 0.02.

Surface morphology of nanodispersions was characterized by TEM analysis. The photomicrograph (Figure 3) showed that the dispersed particles are in spherical shape with smooth surface and distributed uniformly in nano size range. The amount of artemisinin and dexamethasone in nanodispersion were found to be 99–102%w/v and 99.9–103%w/v respectively.

Table 2: Characterization of Nanodispersion

Formulationcode	*Particle Size (nm)*	*PDI*	*Zeta Potential (mV)*	*Transparency (%)*	*Cloud Point (°C)*
F1	12 ± 1.8	0.01 ± 0.001	–6.0 ± 1.3	99 ± 1.2	70 ± 2
F2	15 ± 2.2	0.02 ± 0.001	–9.2 ± 1.4	98 ± 1.5	69 ± 3
F3	26 ± 1.6	0.02 ± 0.002	–10.0 ± 2.8	96 ± 1.8	68 ± 3

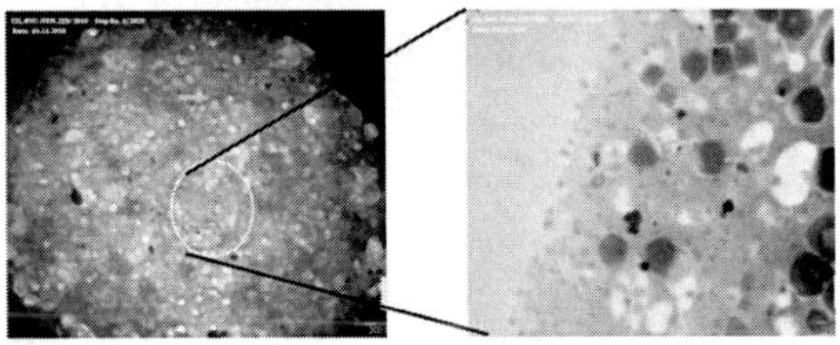

Fig. 3: TEM Image of Artemisinin Dexamethasone Loaded Nanodispersion

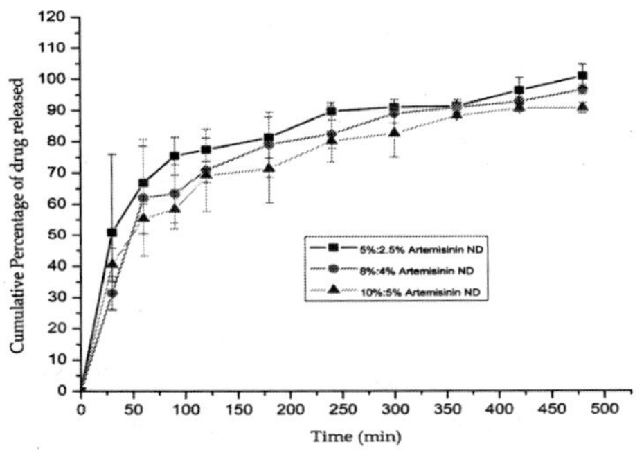

Fig. 4: *In-Vitro* Release of Artemisinin from Nanodispersions

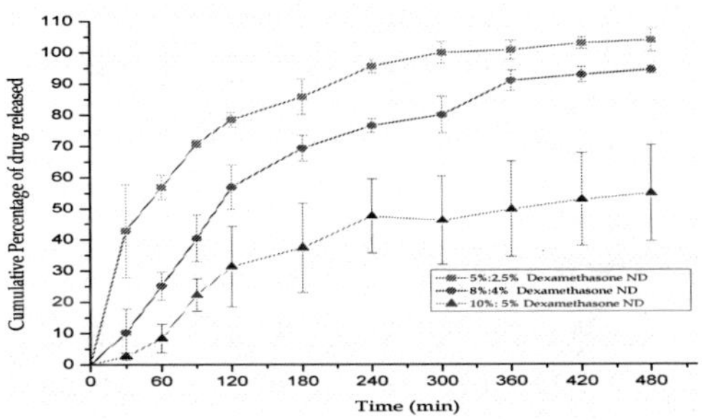

Fig. 5: *In-vitro* Release of Dexamethasone from Nanodispersions

In-vitro drug release from nanodispersions was studied by using dialysis membrane bag method. The results showed that 90 to 101%w/v of artemisinin and 55 to 103% w/v of dexamethasone were released from the nanodispersions and the release patterns are shown in Figures 4 and 5. The results confirm that the polymer and surfactant concentration are indirectly proportional to drug release. High concentration of polymer and polymeric surfactant reduced the drug release were as low concentration of polymer and surfactant faster the *in-vitro* drug release.

The toxicity of the blank formulation containing higher concentration of polymer and surfactant was evaluated by haemolytic studies for % of haemolysis and alkaline comet assay for DNA damage. The data were shown in Figure 6. The results showed that the high concentration of polymer and polymeric surfactant used blank formulations showed below the acceptable limit of percentage haemolysis 12% and do not have any remarkable DNA damages.

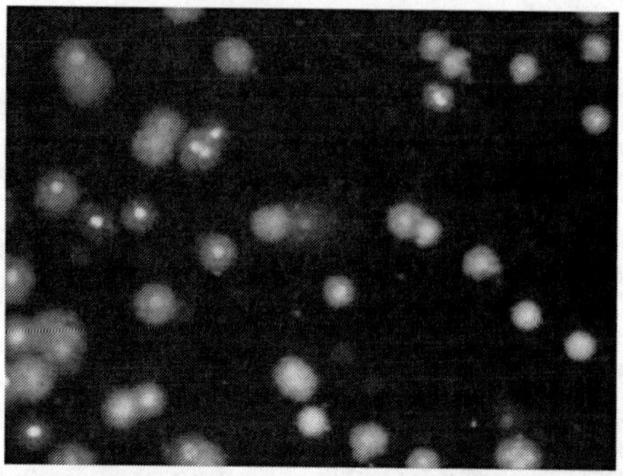

Fig. 6: Alkaline Comet Assay of Blank ND and Fluorescent Image of Lymphocytes

CONCLUSION

Artemisinin and dexamethasone loaded nanodispersion (F3) was formulated with the droplet size of 26 nm. It posses 96% transparency and cloud point at 68°C. The surface morphology of the nanodispersion was smooth and spherical in shape. Nearly 90% of the artemisinin and 55% of the dexamethasone was released from the formulation F3 within 8 hours. The blank formulation consisting of high concentration of polymer and polymeric surfactant showed acceptable level of haemolysis and DNA damage. Hence the topical administration of formulated artemisinin and dexamethasone nanodispersion (F3) may have the great potential for the treatment of AMD.

ACKNOWLEDGEMENTS

The authors are very much thankful to BASF, Mumbai, for providing polymer and polymeric surfactant to carry out this research work. One of the authors Mr. P. Chandra sekar is very thankful to University Grants Commission for providing the fellowship.

REFERENCES

[1] Geroski, D.H. and Edelhauser, H.F., Drug Delivery for Posterior Segment Eye Disease, *Inve Opht and Vis Sci.*, 41, 961–4, 2000.

[2] Augustin, A.J., Puls, S. and Offermann, I., Triple therapy for choroidal neovascularization due to age-related macular degeneration: verteporfin PDT, bevacizumab, and dexamethasone, *Retina*, 27(2), 133–40, 2007.

[3] Amin, K. and Dannenfelser, R.M., In vitro hemolysis: guidance for the pharmaceutical scientist, *J Pharm Sci,* 95(6), 1173–6, 2006.

[4] Rajaguru, P., Vidya, L., Baskarasethupathi, B., Kumar, P.A., Palanivel, M., *et al.*, Genotoxicity evaluation of polluted ground water in human peripheral blood lymphocytes using the cometassay, *Mutation Research/Genetic Toxicology and Environmental Mutagenesis*, 1–2, 27, 29–37, 2002.

FORMULATION DEVELOPMENT AND PHARMACEUTICAL EVALUATION OF LORNOXICAM MAGNETIC NANOEMULSION

S. LATHA, P. SELVAMANI, C. PRABU AND M. ANIS KUMAR[1]

Department of Pharmaceutical Technology, Anna University of Technology, Tiruchirappalli, Tamil Nadu
[1]Department of Biotechnology, Pavendar Bharathidasan College of Engineering and Technology, Tiruchirappalli, Tamil Nadu
E-mail: lathasuba2010@gmail.com

ABSTRACT

Lornoxicam is a potential NSAID belonging to the oxicam class widely used as analgesic, anti-inflammatory and anti-arthritic agent, with plasma half-life of 3–5 hours. In this study, lornoxicam magnetic nanoemulsion (LMNE) was developed by co-precipitation method under alkaline condition and spontaneous emulsification method. The nanoemulsion were evaluated concerning Droplet size, Zeta potential, Encapsulation efficiency, Drug content association to oily phase, Magnetic susceptibility, and TEM. The possible use in targeted and sustained drug delivery of lornoxicam magnetic nanoemulsion was studied in rheumatoid arthritis induced animal model. In-vitro release profile assessed by dialysis membrane shows a release kinetics square root time dependent with 91.5% of lornoxicam having been released within 36 hours. The developed lornoxicam magnetic nanoemulsion might be useful to provide sustained drug release at the targeted site.

INTRODUCTION

Targeted drug delivery is the most important goal of pharmaceutical research and development. In this drug targeting is defined in the broadest sense, that is, to optimize a drug's therapeutic

V. Rajendran, P. Prabu and K.E. Geckeler (eds.)
Nano Biomaterials, pp. 71 76 (2012).

index by strictly localizing its pharmacological activity to the site or organ of action.[1] And the basic challenge in drug delivery is the transfer of drug agents to the targeted site at the appropriate time.[2] Magnetic field-induced, targeted drug delivery is currently attracting significant interest,[3,4] due to their potential application on drug delivery, hyperthermia of tumors magnetic bioseparations, and magnetic resonance imaging.[5] Anti-inflammatory and analgesic properties of lornoxicam have greater potency and it also prevents gastrointestinal side effects as compared to other NSAIDs.[1] Despite all efforts towards its control and prevention, rheumatoid arthritis still remains a global health problem affecting all ages. Although a decline in its prevalence has been witnessed in some developed countries, the level of this decline has been exaggerated. Thus we focus on the various aspects of magnetic nanoemulsion and their potential as a promising strategy in targeted drug delivery for rheumatoid arthritis.

EXPERIMENTAL

Synthesis of Magnetic Fluid (Ferrofluid)

A solution of $FeCl_3.6H_2O$ and $FeCl_2.4H_2O$ mixed in a molar ratio of 2:1. Ammonia aqueous solution (25%) of 15 ml was then quickly charged into the solution with vigorous stirring, under nitrogen atmosphere followed by adding excess ammonia aqueous solution into the mixture slowly while stirring against air until the pH value of the solution reached 11. A black precipitate was obtained. Oleic acid (5% v/v) was added and intensely stirred at 60°C for 30 min. Then the mixture was neutralized while cooling and then the magnetic fluid was prepared.

$$2FeCl_3 + FeCl_2 + 8NH_3 + 4H_2O \rightarrow Fe_3O_4 + 8NH_4Cl \quad \ldots (1)$$

Formulation of Lornoxicam Magnetic Nanoemulsion

Oil phase span 80, tween 80, olive oil were dissolved in 28 ml of ethyl alcohol and the magnetic fluid is added in the organic phase were mixed in the cyclomixer. In a different beaker Lornoxicam drug was mixed in 56 ml of distilled water. The spontaneous emulsification process occurred immediately when the oil phase was slowly incorporated into the water phase using a syringe and slight magnetic stirring. At the end of the oil-phase incorporation, the organic co solvent was removed by reduced pressure at 50°C, using an evaporator. Then remaining amount of distilled water was added slowly to get the final preparation 100% w/w.

RESULTS AND DISCUSSION

Droplet Size Analysis (Particle Size Distribution)

For Lornoxicam magnetic nanoemulsion the droplet size was important physicochemical parameters. The average size of the emulsions was in the range of 120–243 nm is shown in Table 1 which are acceptable for parental emulsion.

Table 1: Particle Size Analysis, Polydispersity Index (PI), Encapsulation Efficiency and Drug Loading (%) of LMNE

Formulation	*Particle Size (nm)*	*PI*	*Encapsulation Efficiency (%)*	*Drug Loading (%)*
LMNE1	120	0.216	54.8	19.74
LMNE2	190	0.249	62.5	16.95
LMNE3	198	0.267	73.4	15.33
LMNE4	200	0.293	79.0	13.21
LMNE5	202	0.315	83.5	11.4
LMNE6	243	0.345	85.2	9.6

Drug Encapsulation Efficiency and Drug Loading

Increase of initial surfactant concentration augment the Lornoxicam loading and similar results were also obtained when encapsulation efficiency of lornoxicam in magnetic nanoemulsion decreased with increasing drug loading concentration. Percentage drug loading and encapsulation efficiency were in Table1.

Determination of Zeta Potential

The zeta potential is a stability indicative parameter in colloidal systems like Lornoxicam magnetic nanoemulsion shows average ZP values around –15.53 to –17.10 mV. As we observed surfactants such as Tween-80, Span-80 with its negatively charged can able to produce a satisfactory negative surface potential for all formulations. The particle surface charge values were determined for all formulations. The results showed no significant difference in various formulations.

Transition Electron Microscope (TEM)

Figures 1 and 2 reveal that the emulsion droplet were almost spherical in shape and has an amorphous core. The results corroborate with the droplet size analysis, showing that droplets are present in the nanometer range, varying from approximately 100 to 250 nm.

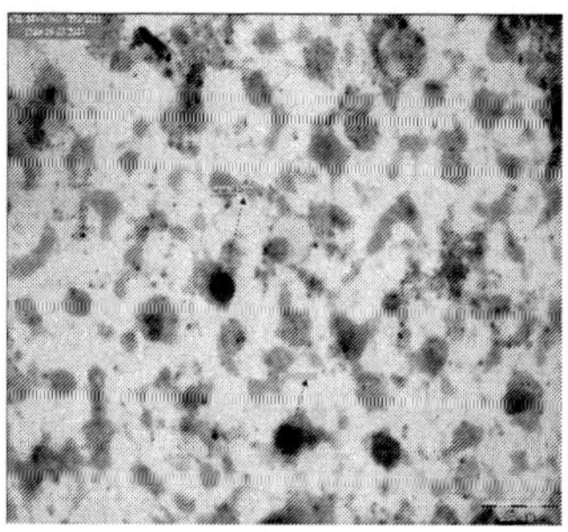

Fig. 1: TEM of LMNE

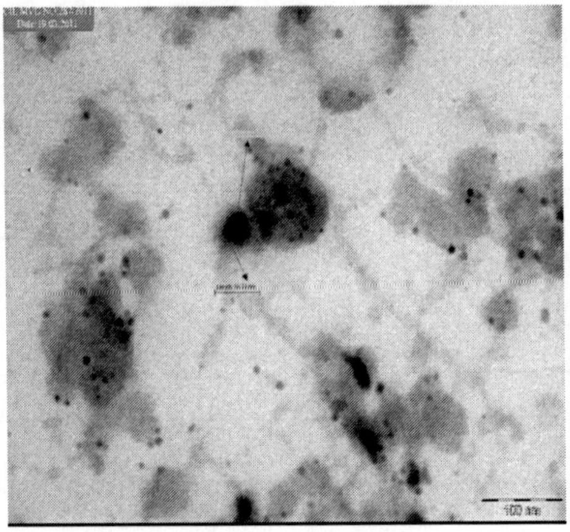

Fig. 2: TEM of LMNE

Fourier Transform Infrared Spectroscopy (FTIR)

The intermolecular interaction of nanoemulsion was characterized by FTIR. It was observed that the spectrum of Lornoxicam magnetic nanoemulsion, highlighted in the wave number range from 1000 to 3500 cm^{-1}. The peak of 3429.82 cm^{-1} became wider and flatter; indicating that hydrogen bonding was enhanced. Due to the electrostatic interaction between octadecyl groups of Tween and Span 80 in magnetic nanoemulsion. FTIR spectra revealed that there was no such interaction between the drug and the surfactants used for nanoparticles formulation.

In-vitro Release Kinetics

Lornoxicam release from the formulations F1, F2, F3, F4, and F5 are 72.54, 84.15, 82.54, 81.24, and 91.50 in 36 hrs respectively. Formulations F2, F3 and F5 released highest percentage of drug compare to the formulations F1 and F4. Through the data obtained shows that that the *In-vitro* release kinetics follows the rate of release is in-dependent of concentration.

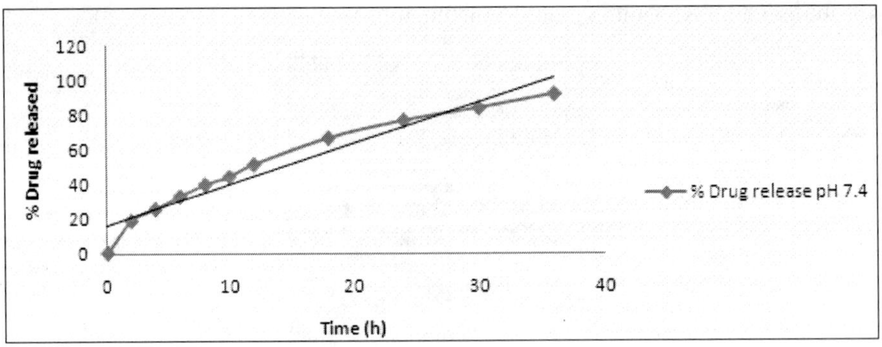

Fig. 3: *In-vitro* Release Kinetics for Formulation F5

In-vivo studies X-ray photograph of rat hind paw upon treatment with Lornoxicam magnetic nanoemulsion

Figure 4 shows deformity, bone erosion and inflammation on 1st day and its shown been reduced by administration of Lornoxicam magnetic nanoemulsion were seen on 14th and 28th day through x-ray analysis.

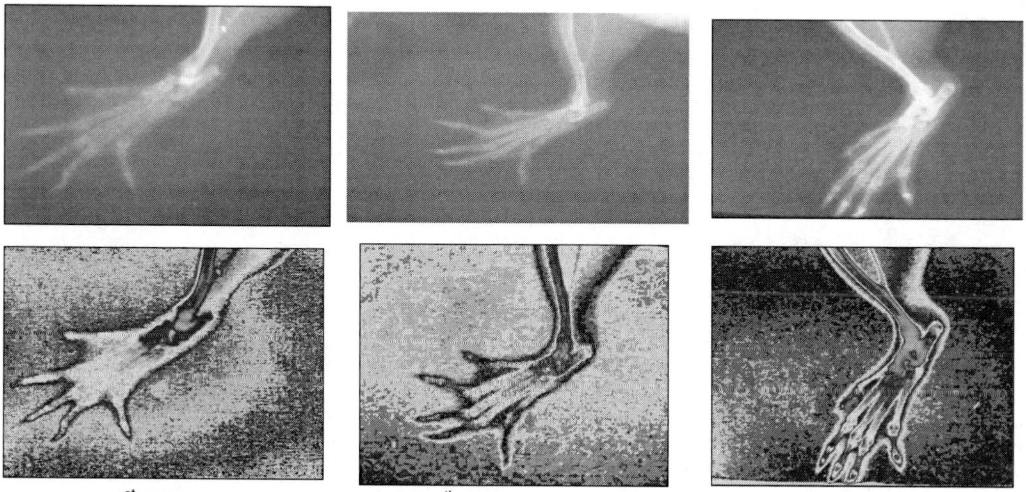

1st DAY 14th DAY 28th DAY

Fig. 4

CONCLUSION

Lornoxicam magnetic nanoemulsion (O/W) was prepared by using a mixture of non-ionic surfactants (Tween 80 and Span 80). By selecting spontaneous emulsification as an appropriate method, nanoemulsion obtained in narrow droplet size distribution in the range of 200 to 250 nm. Ostwald ripening destabilization was observed resulting in a slight increase in the droplet size and a decrease in viscosity. From the determination of zeta potential the stability of nanoemulsion formulation established. *In-vitro* release studies concluded that the formulated magnetic nano emulsion fit with Zero order kinetics, and shows release upto 91.5% of drug 36 hours. *In-vivo* studies shows that reduced bone erosion and inflammation, and the present study confirmed that the nanoemulsion formulation could be used as a possible alternative to traditional oral formulations of Lornoxicam with the advantages of improved targeted and sustained drug delivery.

REFERENCES

[1] Membranes and Barriers: Targeted Drug Delivery, Monograph Series, 154, NIDA, USA, 1995.

[2] Orive, G., *et al.* Drug Delivery in Biotechnology: Present and Future. *Curr Opin Biotech.*, 14, 659, 2003.

[3] Jeong, Y.H., *et al.* Biodegradable Block Copolymers as Injectable Drug-Delivery Systems. *Nature*, 388, 860, 1997.

[4] Souza, A.J., *et al.,* Polyvinylpyrrolidone—Drug Conjugate: Synthesis and Release Mechanism. *Jour Cont Rel.*, 94, 91, 2004.

[5] Pankhurst, Q.A., *et al.,* Application of Magnetic Nanoparticles in Biomedicine. *Jour Phys D: Appl Phys.,* 36, 167, 2003.

PREPARATION AND EVALUATION OF PREDNISOLONE MAGNETIC NANOSUSPENSION FOR POSSIBLE USE IN RHEUMATOID ARTHRITIS THERAPY

S. LATHA, P. SELVAMANI, T. SILAMBARASI AND M. ANIS KUMAR[1]

Department of Pharmaceutical Technology, Anna University, BIT Campus, Tiruchirappalli, Tamil Nadu
[1]Department of Biotechnology, Pavendar Bharathidasan College of Engineering and Technology, Tiruchirappalli, Tamil Nadu
E-mail: lathasuba@yahoo.co.in

ABSTRACT

The main objective of our study is to prepare and evaluate prednisolone magnetic nanosuspension by solvent evaporation method coupled with ultrasonication for therapy of rheumatoid arthritis in order to have enhanced bioavailability at the site of action, rapid onset of action with less side effects. Two formulations were developed by varying the composition of the recipe which contains polyethylene glycol 4000, tween 80, hydroxypropyl methyl cellulose, magnetite, sodium citrate, citric acid, mannitol and methylparaben by ultrasonication method. The prepared prednisolone magnetic nanosuspension were characterized by different techniques such as FT-IR, particle size, polydispersity index, zetapotential conductivity and in-vitro release kinetics. The results obtained upon evaluation of the developed formulation were promising and will certainly make new dimension and milestone in the targeted drug delivery for anti-arthritic therapy.

INTRODUCTION

Prednisolone is a synthetic anti-inflammatory glucocorticoid derived from cortisone. It is biologically inert and converted to prednisolone in the liver. Prednisolone is useful for the

V. Rajendran, P. Prabu and K.E. Geckeler (eds.)
Nano Biomaterials, pp. 77–82 (2012).

treatment of a wide range of inflammatory and auto-immune conditions such as asthma, uveitis, *Pyodermagangrenosum*, rheumatoid arthritis, ulcerative colitis.[1,2] Prednisolone has poor aqueous solubility and in order to improve its solubility and its bioavailability. Magnetic drug delivery system using magnetic nanoparticle carriers targeted by an external magnetic field is a promising alternative to avoid the problems associated with conventional chemotherapy.[3] Rheumatoid arthritis (RA) is a chronic inflammatory autoimmune disease, which is characterized by a series of pathological processes of the joints, such as leukocyte infiltration, pannus formation and extensive destruction of the articular cartilage and bone.[4]

EXPERIMENTAL

Preparation of Magnetic Nanosuspension

In order to produce the magnetic nanosuspension, Prednisolone powder (1% w/v) was dispersed in an aqueous solution containing polysorbate 80 (0.8%) and different ratio of ie polyethyleneglycol 4000, suspending agent (0.2% of hydroxypropyl methyl cellulose), magnetite and glycerol (2.5 ml) was added and stirred under magnetic stirring for 10 minutes. The other ingredients like mannitol, methylparaben, propylparaben sodium citrate and citric acid was added and then made up to the volume and mixed under cyclomixer for 10 minutes. Formulated suspension was kept for particle size reduction under ultrasonicator (7:3 On: Off) cycle for 60min at 20°C. The formulated prednisolone suspension was mixed once again under cyclomixer for 15 minutes.

RESULTS AND DISCUSSION

FT-IR Spectrophotometry

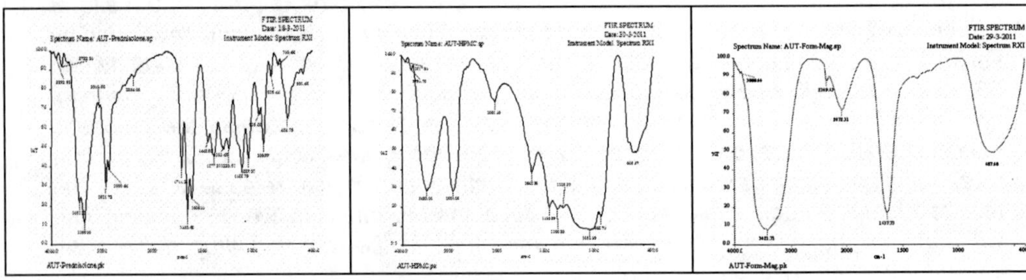

Fig. 1: FT-IR Spectrum of Drug, Suspending Agent and E Magnetite

Table 1: Characteristics of Prednisolone Magnetic Nanosuspension

Formulation Code	*Particle Size (nm)*	*Poly Dispersity Index*	*Zeta Potential (mV)*	*Conductivity (S/cm)*	*Magnetic Susceptibility*
F1	244.5	0.393	–5.87	0.453	33×10^{-6}
F2	281.1	0.314	–9.12	0.332	29×10^{-6}

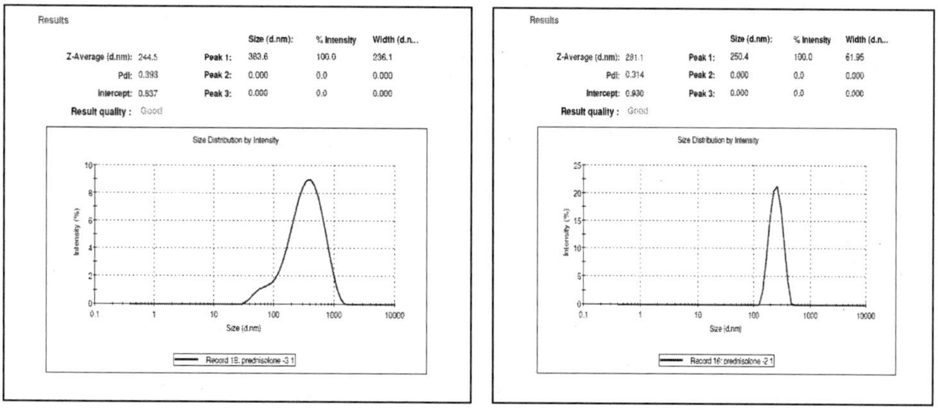

Fig. 2: Particle Size Analysis of Prednisolone Magnetic Nanosuspension

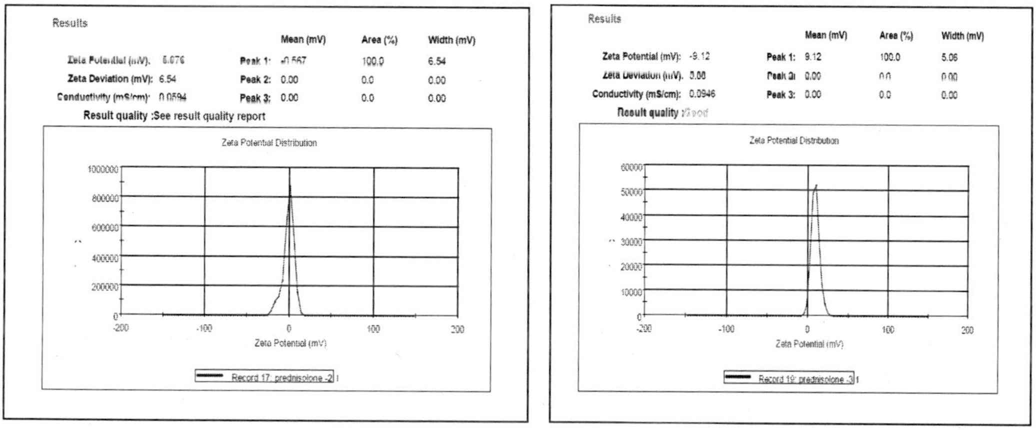

Fig. 3: Zetapotential Distributionofprednisolone Magnetic Nanosuspension

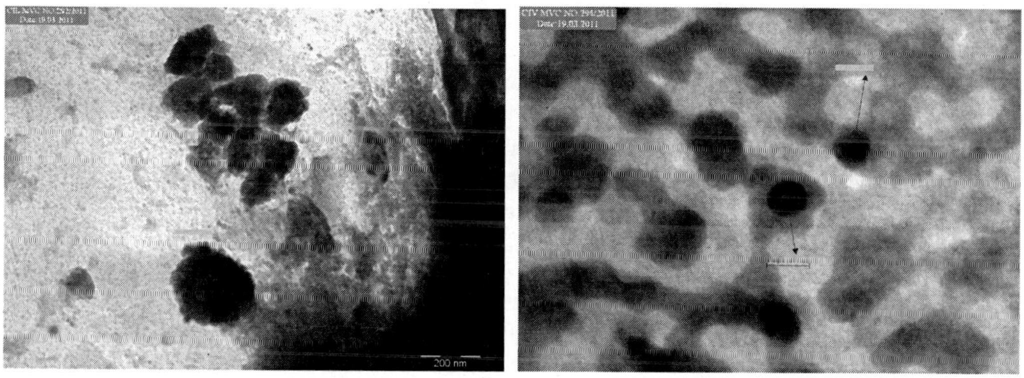

Fig. 4: TEM Images of Prednisolone Magnetic Nanosuspension

Table 2: Encapsulation, Drug Loading Efficiency and *in-vitro* Drug Release

Formulation Code	*Feed Drug (mg)*	*Encapsulation Efficiency (%)*	*Drug Loading Efficiency (%)*	*In-vitro Release after 8 hours (%)*
F1	5	42.48 ± 0.69	7.22 ± 0.17	88.23 ± 0.77
F2	5	72.30 ± 0.49	10.3 ± 0.34	80.12 ± 0.44

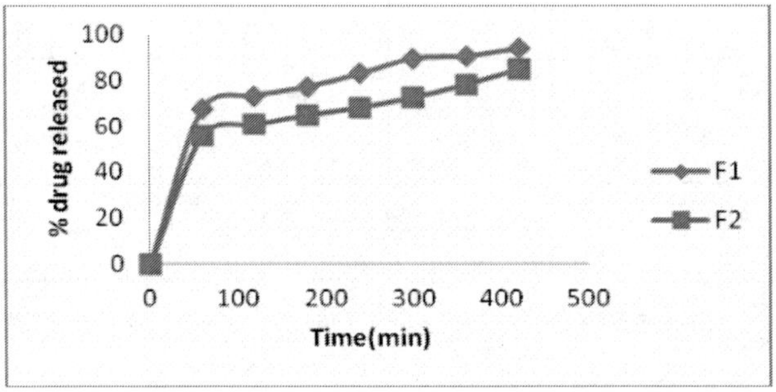

Fig. 5: *In-vitro*drug Release Profile of Prednisolone Loaded Magnetic Nanoparticles

DISCUSSION

The results of various experimental evaluation of the formulated prednisolone magnetic nanosuspension presented above confirmed that the particle size, polydispersity index, zeta potential falls between the acceptance range. The results confirm better absorption of drug due to decreased particle size, increased surface area, dissolution velocity and higher saturation constant. Therefore the bioavailability is expected increase. The particles in the formulated magnetic nanosuspensions were uniform with size range of about 244–288 nm and possessed ideal electro kinetic properties i.e. zeta potential between –5.87 mV to –9.12 mV. TEM image of the formulation confirmed the surface morphology and the particles were present in nanodimensions. Further, the analytical parameters like FT-IR and XRD that there is no drug excipient interaction and the crystallinity of the drug in formulation was confirmed. The information from FT-IR and XRD spectrum revealed that the drug, polymer and magnetite were encapsulated without any chemical interaction and remains physically stable magnetic nanosuspension.The drug release from the formulated prednisolone magnetic nanosuspensions followed the kinetic model of Higuchi's equation and the release mechanism was found to be linear with matrix type obeying Higuchi's equation.

Confirmation of Synthesized Polymers

The above Figure 1 reveals the FT-IR spectrum of the magnetite, prednisolone (pure drug). Prednisolone showed major peaks at 3401.77 (heterocyclic secondary amine), 3023.43 (aromatic C-H stretching), 2748.89 (Methoxy group), 1900.03 (aromatic combination band), 1625.31

(COO stretching), 1281.68 (C-O stretching), 762.60 (N-H wagging).The spectrum of magnetite showed the major peak at 570.63 and 441.61 (stretching and torsional vibration) of Fe-O. The peak position obtained in the formulated suspension confirmed that the drug was capped with the polymer and magnetite via hydrogen bonding (3436.31) and this was confirmed by the broadening of the peak width. There was not any new peak appeared in the formulations.

Compatibility Studies

The FT-IR spectrum of individual polymers, prednisolone, magnetite and mixture of all ingredients were compared and no significant drug excipient interactions were noted. The results confirmed the components used in the formulations were compatible.

Particle Size

The particle sizes of the formulations were in the range of 200–300 nm with low polydispersity which illustrated from particle size histograms (Figure 2). Smaller particles have higher surface area/volume ratio, which makes it easier for the encapsulated drug to be released from the particles via diffusion and surface erosion and also have the added advantage for the magnetic nanosuspensions to penetrate into, and permeate through the physiological drug barriers at the targeted site restricted by applying external magnetic field. The smaller particles will have greater ease of entry in the Arthritic site. Thus, control of particle size provides a means of tuning drug release rates.

Surface Charge

Surface charge reflects the electrical potential of particles and is influenced by the composition of the particle and the medium in which it is dispersed. Nanoparticles with a zeta potential above ±30 mV have been shown to be stable in colloidal dispersion, as the surface charge prevents aggregation of the particles. The prednisolone loaded magnetic nanoparticles had negative charge within normal value –5.87 mV and –9.12 mV (Figure 3).

Particulate Morphology

The particles were analyzed by TEM to obtain further information of their surface morphology were shown in Figure 4. The particles appear cluster like aggregates which were spherical in nature. The TEM images also substantiate the particle size of those aggregates are within nanoscale dimensions.

In vitro Release

Release studies of prednisolone magnetic nanosuspension was performed in USP model Type II dissolution apparatus (paddle) using dialysis membrane (Himedia, 2.4 nm pore size) Phosphate buffer pH - 7.4 (alkaline) as the dissolution medium. A sample quantity of 1 ml was withdrawn at regular intervals up to 8 hrs and absorbance was measured by UV-Visible spectrophotometer at 246 nm. the concentration of prednisolone was determined from the calibration graph.

CONCLUSION

The formulated prednisolone magnetic nanosuspensions confirmed that they were having the particle size, polydispersity index, zeta potential of acceptance criteria which was necessary for better absorption because decrease in particle size would have more surface area, dissolution velocity, higher saturation constant therefore bioavailability also increases. The polymer PEG4000 had efficiently coated the drug particle and also linked the drug with the magnetite. The magnetic susceptibility of the developed formulation that contains magnetite imparts superparamagnetic behaviour on application of external magnetic field. Thus it is well evidenced that, the formulated prednisolone magnetic nanosuspension offers a promising targeted drug delivery module for better anti-arthritic activity at a much lower dose.

REFERENCES

[1] Kipp, J.E., Microprecipitation method for preparing submicron suspension, US patent 6,607, 784, 2003.

[2] Rabbinow, B.E., Nanosuspension as drug delivery, *Natural, Review,* 3, 785–796, 2004.

[3] Alexiou, C. *et al.*, targeting cancercells: magnetic nanoparticles as drug carriers, European. *Biophysics Journal with Bio Physics Letters*, 35 (5), 446–450, 2006.

[4] Yeom, M.J. *et al.*, Anti-arthritic effects of *Ephedra sinica* STAPF herb-acupuncture: inhibition of lipopolysaccharide-induced inflammation and adjuvant-induced polyarthritis, *Journal Pharmacolgy Science*, 100, 41–50, 2006.

TOXICITY ASSESSMENT OF ORMOSIL NANOPARTICLES

DHIRAJ KUMAR, C.K. PRASHANT[1], AMIT KUMAR DINDA[1] AND SUSMITA MITRA[2]

NIBEC, University of Ulster, Northern Ireland, U.K.
[1]All India Institute of Medical Sciences, New Delhi
[2]Amity Institute of Nanotechnology, Amity University, Noida
E-mail: smitra@amity.edu

ABSTRACT

The physicochemical characteristics of nanoparticles can alter cell interaction during in vitro and in vivo application. We have carried out an in vitro study using 'ormosil' nanoparticles to determine any cytotoxicity of the nanoparticles. Spherical and rod-shaped nanoparticles having surface hydroxyl and amine groups were synthesized in Tween 80 micelles using organosilanes. Nanoparticles of similar size and morphology were considered for comparative assessment. Characterization was carried out using zetasizer, TEM and FTIR to establish the above parameters. MTT and PI-DAPI assay were carried out using mouse macrophage (J774) cell line. Super-nano (<100 nm) particles and rod-shaped nanoparticles with surface hydroxyl groups demonstrate lowered cell viability. On the other hand, 'ormosil' nanoparticles of similar size and shape, with surface amine groups and low net negative potential, do not exhibit any in vitro cytotoxicity.

INTRODUCTION

Nanoparticles of organically modified silane 'ormosils', have properties that has led to their application in various fields.[1] Biomedical applications such as the delivery of DNA[2], photodynamic therapy[3], *in vitro* detection of cells, microbes and biomolecules[4] have been carried out using fluorescent dye tagged nanoparticles where required. The nanoparticles are hydrophilic, chemically inert, do not show swelling or porosity changes with alteration in pH, and are less prone to microbial attack. In addition, their surface can be easily modified with

V. Rajendran, P. Prabu and K.E. Geckeler (eds.)
Nano Biomaterials, pp. 83–88 (2012).

different functional groups for attachment of targeting ligands.[5] Presently an area of concern in the application of engineered nanoparticles is the potential adverse environmental and health effects due to the unique chemical and physical properties at the nanoscale.[6] Considerable variation in cell behavior has been noted with change in size, shape and surface properties of synthesized nanomaterials.[7] In the present study we have addressed this problem through a systematic investigation of the cytotoxicity of ormosil nanoparticles having variable size, morphology and surface functionality.

EXPERIMENTAL

Nanoparticles were synthesized in Tween 80 micelles in deionized water as reported earlier.[8] Measured volumes of neat organosilane, octyltriethoxysilane (OTES)/vinyltriethoxysilane (VTES) was added to this micellar system to obtain nanoparticles of variable dimension. The resulting dispersion was stirred vigorously till translucent, followed by hydrolysis and polycondensation under alkaline conditions. For amine functionalization, aminopropyltriethoxysilane (APTES) was incorporated in the system.

From the nanoparticles synthesized six samples were selected and designated as 'A, B, C' (from OTES/VTES) 'D, E, F' (with APTES). The nanoparticle size and zeta (ζ) potential were obtained using zetasizer, and shape of all nanoparticles and the size of the high aspect ratio nanoparticles was determined by transmission electron microscopy (TEM). The functional groups of the nanoparticles were determined by FT-IR spectroscopy.

The nanoparticles of different size, shape, and surface charge were tested using J774 (mouse macrophage) cell line for assessing cell viability. Cells were cultured in RPMI-1640 supplemented with 10% FBS, at 37°C with 5% CO_2 in a humidified incubator. J774 cells were seeded in microplates and when confluent were exposed to calculated concentration of nanoparticles for 3 hours and 6 hours. Effect was determined by MTT and PI-DAPI assay. For MTT assay, cells were over-layered with fresh media and MTT and incubated overnight. The media was removed and 100 μl of DMSO was added to the wells to dissolve the MTT formazan crystals. Following incubation at 37°C, the absorbance was measured at 490 nm. For PI-DAPI staining, the cells grown on cover slips were exposed to nanoparticles and then fixed in 3.7% paraformaldehyde, washed with PBS and incubated with stain for 5–10 minutes. The apoptotic cells were viewed under fluorescence microscope and imaged. Control experiments were run alongside and the data expressed as mean of nine individual readings from three separate experiments.

RESULTS AND DISCUSSION

The samples of ormosil nanoparticles assessed for cell viability were either spherical or rod-shaped, with size ranging from 6 nm to 70 nm (Figure 1). Samples A, B, D, E are spherical having diameter 6 ± 0.5 nm, 70 ± 8 nm, 18 ± 5 nm, 60 ± 8 nm respectively. Samples C and F are rod-shaped, of length 60–70 nm and 70–80 nm. The ζ-potential of the particles was estimated at pH 7.0. Those with surface (-OH) groups (Table 1: A, B, C) have ζ-potential in excess of –4.5

mV. For the spherical particles (A and B) it is –6.20 mV and –4.56 mV respectively, while for the rod-shaped (C) it is –4.82 mV. With an increase in the surface to volume ratio of the nanoparticles, the number of free –OH groups also increase.[8] This can result in the elevated negative potential observed. In addition, the number of nanoparticles in the sample also affects the overall estimated potential. In the case of the nanoparticles with surface $-NH_2$ groups derived from APTES (Figure 1: D, E, F), the spherical nanoparticles (D and E) have lower ζ-potential of –2.90 mV and –2.53 mV in comparison to –3.85 mV of the rod-shaped ones (F) the negative charge is partially off-set due to the presence of protonated amine resulting in lower net negative potential.

Surface Functional Group: (-OH)

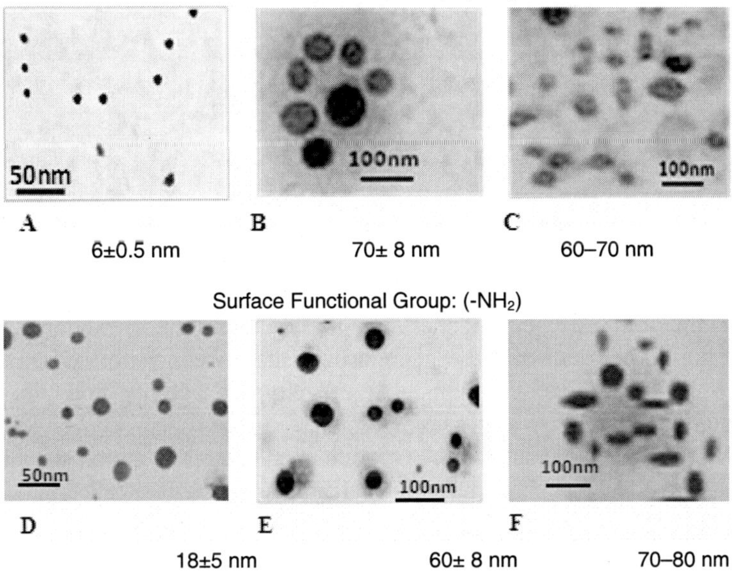

Fig. 1: TEM Images of Ormosil Nanoparticles

Table 1

Sample Name	*A*	*B*	*C*	*D*	*E*	*F*
ζ-potential	–6.20 mV	–4.50 mV	–4.82 mV	–2.90 mV	–2.53 mV	–3.85 mV

The FTIR spectra Figures 2a and 2b represent samples A, B, C and samples respectively. As the nanoparticles were D, E, F synthesized from organosilanes, the representative peaks at 763.47cm^{-1} and 843.54 cm^{-1} due to C-H bending out of plane (C=C-H) from vinyl group are observed. In Figure 2a peak at 949.14 cm^{-1} represents O-H out of plane bending, and the O-H in plane bending gives peak at 1410.63 cm^{-1}. In Figure 2b, peak at1644.94 cm^{-1}(triplet) is given by N-H stretching associated with free NH_2. The presence of (-OH) and ($-NH_2$) groups are confirmed from the representative spectra.

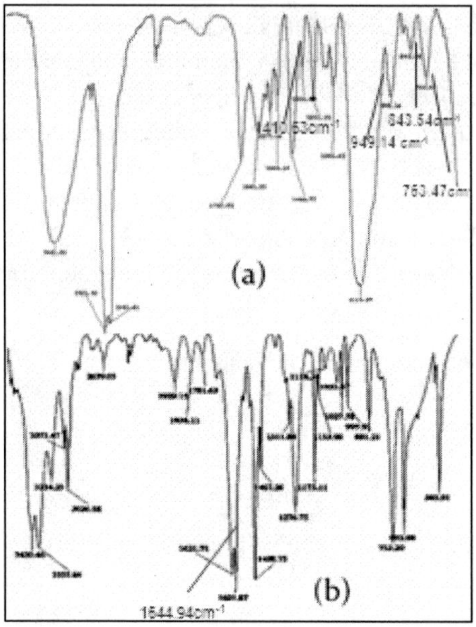

Fig. 2: FTIR

Considering the omnipresence of nano-silica in the environment, extensive *in vitro* and *in vivo* studies on the toxic effect of these nanoparticles have been reported in literature.[9, 10] In the light of current knowledge, it seems that the biologic activity and biokinetics of the nanoparticles are dependent on many parameters: size, shape, chemistry, crystallinity, surface properties (area, porosity, charge, and surface modifications), agglomeration state, biopersistence, and dose.[11]

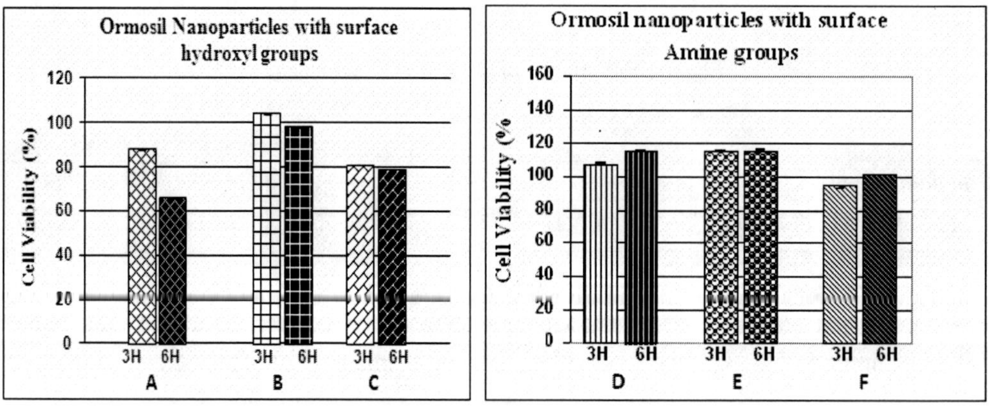

(a) (b)

Fig. 3: MTT Assay

In an investigation using super-nano sized (7 nm) silica particles, Waters *et al.*[12] have demonstrated low cell survival relative to the larger particles (500 nm) studied. On the basis of an *in vitro* study using J774 cell line, Rabolli *et al.*[13] reported enhanced cytotoxicity with reduction in size and increase in the surface area of the nanoparticles. In the present investigation, the results reveal that the super-nano (6 nm) 'ormosil' particles decrease viability of macrophage cells (J774 cell line) in a time dependent manner [Figure 3a (A)]. Since the ormosil nanoparticles have the same surface chemistry as the nanoparticles discussed, it appears that in this case also the important parameter determining cell viability is the total surface area of the super-nano particles.

The rod-shaped nanoparticles having surface hydroxyl groups demonstrate 80% cell viability (Figure 3a (C)). As reported in literature[11] high aspect ratio of nanoparticles can lead to adverse cell interactions resulting in apoptotic cell death. Apoptotic cell death is observed in both (Figures 4b and 4c). After six hours exposure to sample A (6 nm spherical nanoparticles), larger number of apoptotic cells are observed relative to sample C (70 nm rod shaped) ones. Thus it seems that size is more important than shape in determining the fate of the cells. On the other hand, the nanoparticles, both rod-shaped and spherical, with surface amine functionality do not reduce viability of the cells studied (Figure 3b, D, E, F).

This strongly reflects cellular interaction with the nanoparticles that is based on surface charge. And in the case of these ormosil nanoparticles the lower surface potential (Table 1: D, E, F) is the determining factor for the lack of any adverse effect on the cells.

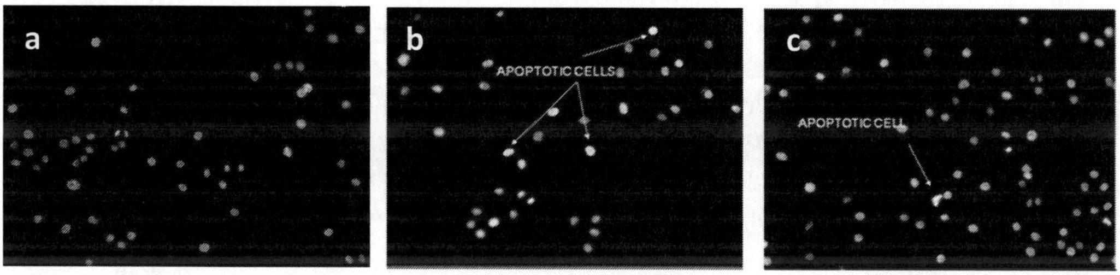

Control Sample-A Sample-C

Fig. 4: Images of PI-DAPI Stained Cells (following 6 hours exposure to the nanoparticles) Showing Apoptotic Cells (arrows)

CONCLUSIONS

Cell Injury and apoptosis appears to be associated with surface functionality, size and shape of 'ormosil' nanoparticles. Surface functionalization with amine group enhances cell-compatibility, thus these can be used for the development of different cellular assays and detection systems *in vivo*.

REFERENCES

[1] Sharma, R.K., Das, S. and Maitra, A., Surface modified ormosil nanoparticles, *Journal of Colloid and Interface Science,* 2004; 277(2): 342–346.

[2] Roy, I., Stachowiak, M.K. and Bergey, E.J., Nonviral gene transfection nanoparticles: function and applications in the brain, *Nanomedicine*, 2008; 4(2): 89–97.

[3] Ohulchanskyy, T.Y., *et al.,* Organically Modified Silica Nanoparticles with Covalently Incorporated Photosensitizer for Photodynamic Therapy of Cancer, *Nano Lett.* 2007; 7 (9), 2835–2842.

[4] Solokov, S. Naik, Novel fluorescent silica nanoparticles: towards ultra bright silica nanoparticles, *Small,* 2008; 4(7): 934–9.

[5] Park, S. *et al.* Cellular toxicity of various inhalable metal nanoparticles on human alveolar epithelial cells, *Inhal Toxicol.* 2007;19 (1): 59–65.

[6] Nishimori, H. *et al.,* Histological analysis of 70 nm silica particles-induced chronic toxicity in mice, *Eur J. Pharm Biopharm.* 2009; epub.

[7] Kumar, D., Dinda, A.K. and Mitra, S., Synthesis of hydrophobic dye-doped variable core-shell nanoparticles using organosilane precursors, *NANO Today 2009.*

[8] Jin, C., Jin, Y., Wang, J. and Zhao, C., Comparative study of the effect on oxidative damage in rats inhaled by nano-sized and micro-sized silicon dioxide, *Wei Sheng Yan Jiu.* 2008, 37(1): 16–8, 36.

[9] So, S.J., Jang, I.S. and Han, C.S., Effect of micro/nano silica particle feeding for mice. *J Nanosci. Nanotechnol*, 2008, 8(10): 5367–71.

[10] Napierska, D. *et al.,* The nanosilica hazard: another variable entity, *Particle and Fibre Toxicology,* 2010, 7: 39.

[11] Waters, K.M., *et al.* Macrophage responses to silica nanoparticles are highly conserved across particle sizes, *Toxicol Sci.*, 2009, 107(2): 553–569.

[12] Rabolli, V., Influence of size, surface area and microporosity on the in vitro cytotoxic activity of amorphous silica nanoparticles in different cell types, *Nanotoxicology* 2010, 4: 307–318.

LUMINESCENCE STUDIES ON Er^{3+}/Yb^{3+} DOPED TELLURO-BORATE GLASSES

S. ARUNKUMAR, K. MAHESHVARAN AND K. MARIMUTHU

Department of Physics, Gandhigram Rural University, Gandhigram
E-mail: mari_ram2000@yahoo.com

ABSTRACT

The present work reports the structural and luminescence studies on $50TeO_2 + 15B_2O_3 + (5\text{-}x\text{-}y) La_2O_3 + 15MgO + 15SrO + xEr_2O_3 + yYb_2O_3$ (x = 1; y = 0, 0.1, 1 and 2 in wt%) telluro-borate glasses. The influence of the Yb^{3+} sensitizer on Er^{3+} ion have been studied by measuring FTIR, absorption and luminescence spectra. The intensity of the $^4I_{15/2}\rightarrow{}^4I_{11/2}$:($Er^{3+}$) and $^2F_{7/2}\rightarrow{}^2F_{5/2}$:($Yb^{3+}$) transitions increases linearly with the increase in Yb^{3+} ion concentration. The bonding parameters $\overline{\beta}$ and δ for the prepared Er^{3+}/Yb^{3+} doped telluro-borate glasses were calculated and reported. To investigate the local structure around the Er^{3+} ions, the Judd-Ofelt intensity parameters Ω_λ (λ = 2, 4 and 6) have been derived following least squares fitting procedure and the JO parameters values are used to estimate the various radiative properties such as effective line width, stimulated emission cross-section, branching ratios and radiative lifetime of the emission transitions.

INTRODUCTION

Recently, rare earth ions (REI) doped telluroborate glasses have become attractive because of their structural and optical properties and due to potential applications in the field of lasers, amplifiers and optical devices.[1] Among the RE ions Er^{3+} ions are more interesting due to the weak $^2H_{9/2}\rightarrow{}^4I_{15/2}$ (blue) and strong $^4S_{3/2}\rightarrow{}^4I_{15/2}$ (green) stoke emissions which is useful to fabrications of photonic devices.[2–3] Yb^{3+} ion is also an important RE ions known as sensitizer and it can be used to improve the absorption and emission cross-sections of the Er^{3+} ions.[4] Many research works have been devoted to structural and optical properties of Er^{3+} ion single

V. Rajendran, P. Prabu and K.E. Geckeler (eds.)
Nano Biomaterials, pp. 89–94 (2012).

doped and codoped with Yb^{3+} ion in the different glass matrices.[5,6] Generally the glass host with low phonon vibrational energy could be ideal for the codoping Er^{3+}/Yb^{3+} ions due to the reduction of non-radiative rate namely multiphonon relaxation deexcitation which is reason for the improved quantum efficiencies of the radiative transitions.[7] Telluroborate glasses have the advantages of good chemical and thermal stability, high transparency and high mechanical strength of borate host and the higher refractive index and low vibrational energy of tellurite host which is useful to enhance the luminescence properties.[2,3] In the present work reports the structural and luminescence behaviors of Er^{3+}/Yb^{3+} doped telluroborate glasses.

EXPERIMENTAL

The Er^{3+}/Yb^{3+} doped telluro-borate glasses were prepared by following melt-quenching technique following the procedure reported in literature.[2,3] The starting materials used in the present work are high purity analytical grade (from Sigma Aldrich 99.99% purity) chemicals such as H_3BO_3, TeO_2, Sr_2CO_3, $MgCO_3$, Er_2O and Yb_2O_3. Fourier transform infrared (FTIR) spectra of the glasses were measured using Perkin Elmer Peragon 500 FTIR spectrophotometer with a spectral resolution of ± 4.0 cm^{-1}. Optical absorption spectra were recorded using Perkin Elmer Lambda 35 UV-Vis-NIR spectrometer in the wavelength range 350–1650 nm with a spectral resolution of ± 0.1 nm. The luminescence measurements were made using Perkin Elmer LS55 spectrometer in the wavelength range 500–600 nm with a resolution of ± 0.1 nm. All these measurements were carried out at room temperature (RT) only.

RESULTS AND DISCUSSION

FTIR Analysis

Figure 1 shows the FTIR spectra of the prepared Er^{3+}/Yb^{3+} doped telluro-borate glasses. The band centered at 430 cm^{-1} is mainly attributed to the bending vibrations of Te–O–Te or

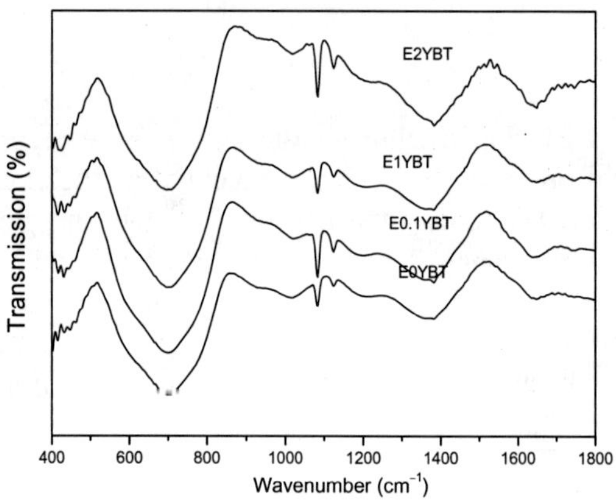

Fig. 1: Infrared Spectra of the Er^{3+}/Yb^{3+} Doped Telluro-Borate Glasses

O–Te–O linkages.[2] The band around 700 cm^{-1} is may be combined vibrations of the O_3B–O–BO_3 bond bending and Te–O bending in TeO_3 and TeO_6 units.[3] The region 740–770 cm^{-1} is mainly due to the vibrations of TeO_2 in the TeO_4 group and the TeO_3 vibrations of the TeO_3 groups. The bands around 920 cm^{-1} is attributed to the B–O stretching vibration of BO_4 units in the various borate network. The strong vibrations at 1380 cm^{-1} are attributed to the presence of B–O stretching vibration of BO_3 units in the borate network. The strong absorption around 1650 cm^{-1} is attributed due to the B–O^- bond in the isolated pyroborate groups.[1]

Absorption Spectra and Bonding Parameter

Figure 2 represents the absorption spectra of Er^{3+}/Yb^{3+} doped telluro-borate glasses and their band assignments are presented in the table 1. Thirteen absorption transitions such as $^4I_{13/2}$, $^4I_{11/2}$(Yb^{3+}: $^2F_{7/2}\rightarrow{}^2F_{5/2}$), $^4I_{11/2}$, $^4I_{9/2}$, $^4F_{9/2}$, $^4S_{3/2}$, $^2H_{11/2}$, $^4F_{7/2}$, $^4F_{5/2}$, $^4F_{3/2}$, $(^2G,^4F,^2H)_{9/2}$, $^4G_{11/2}$ and $^4G_{9/2}$ were observed.[3–6] The intensity of the $^4I_{15/2}\rightarrow{}^4I_{11/2}$ transition increases linearly with the increase in Yb^{3+} ion concentration. It is further observed that, while increasing the concentration of the Yb^{3+} ion in the host matrix, $^2F_{7/2}\rightarrow{}^2F_{5/2}$ transition pertaining to the Yb^{3+} ion could also be observed along with the Er^{3+} ion transition $^4I_{15/2}\rightarrow{}^4I_{11/2}$. The bonding parameters $\bar{\beta}$ and δ for the prepared Er^{3+}/Yb^{3+} doped telluro-borate glasses were calculated to estimate the nature of the metal–ligand bond. It is observed the table 1 that, the Er^{3+}–ligand bond is of ionic in nature and the degree of ionic character changes with the change in chemical composition.[3]

The oscillator strengths of the observed transitions are to determine the intensity parameters Ω_λ ($\lambda = 2, 4$ and 6) using least squares fitting approximation method by the expressions as follows:[9,10]

$$f_{cal} = \left[\frac{8\pi^2 mcv}{2h(2J+1)}\right]\left[\frac{(n^2+2)^2}{9n}\right]\Sigma_{\lambda=2,4,6}\,\Omega_\lambda\left(\psi J\|U^\lambda\|\psi' J'\right)^2$$

where ν is the wave number (in cm^{-1}) of the transition from the ground state (ΨJ) to the excited state (Ψ'J'), n is the refractive index, $(n^2+2)^2/9n$ is the Lorentz local field correction accounts for the dipole–dipole transition, J is the total angular momentum of the ground state, Ω_λ ($\lambda = 2, 4$ and 6) are the JO intensity parameters and $\|U^\lambda\|^2$ are the squared doubly reduced matrix elements of the unit tensor operator which are evaluated from the intermediate coupling approximation for a transition from ΨJ to Ψ'J'.[9,10] The higher oscillator strength and Ω_2 intensity parameter values lead to have lower symmetry around the RE ion site.[7] In the present case, the hypersensitive transitions (HST) $^4I_{15/2}\rightarrow{}^2H_{11/2}$ and $^4I_{15/2}\rightarrow{}^4G_{11/2}$ are observed to be more intense than the other absorption transitions for all the prepared glasses. The JO parameter follow the trend as $\Omega_2>\Omega_4>\Omega_6$ for all the prepared glasses. Among the prepared glasses, E2YBT glass possess higher spectroscopic quality factor (Ω_4/Ω_6) value as 1.284 and the trend follows as E2YBT>E1YBT>E0.1YBT>E0YBT in the present study.

Table 1: Observed band positions (cm^{-1}) and Bonding parameters ($\overline{\beta}$ and δ) of the Er^{3+}/Yb^{3+}:ExYBT Boro-Tellurite Glasses

Transitions $^4I_{15/2}\rightarrow$	*E0YBT*	*E0.1YBT*	*E1YBT*	*E2YBT*	*Aquo-ion [8]*
$^4I_{13/2}$	6527	6523	6523	6523	6600
$^4I_{11/2}$	10268	10241	10241	10241	10250
$^4I_{9/2}$	12527	12506	12488	12489	12400
$^4F_{9/2}$	15338	15338	15338	15338	15250
$^4S_{3/2}$	18340	18387	18366	18356	18350
$^2H_{11/2}$	19152	19150	19151	19150	19150
$^4F_{7/2}$	20486	20475	20475	20475	20450
$^4F_{5/2}$	22141	22128	22162	22161	22100
$^4F_{3/2}$	22507	22564	22562	22564	22500
$(^2G, ^4F, ^2H)_{9/2}$	24556	24550	24551	24550	24550
$^4G_{11/2}$	26503	26435	26435	26387	26400
$^4G_{9/2}$	27368	27315	27314	27315	27400
$\overline{\beta}$	1.0011	1.0006	1.0005	1.0003	–
δ	–0.1086	–0.0612	–0.0517	–0.0331	–

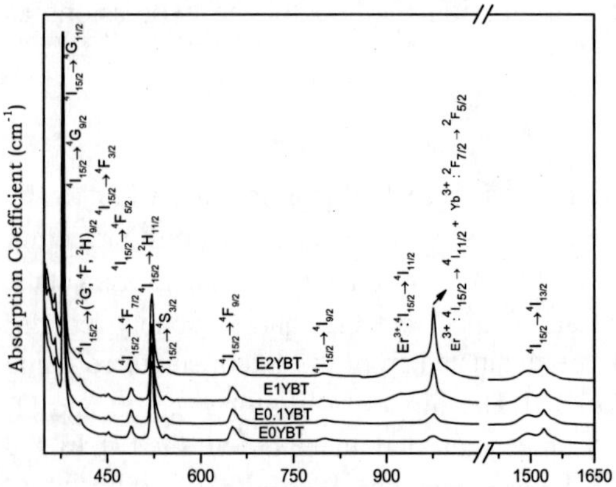

Fig. 2: Absorption Spectra of Er^{3+}/Yb^{3+} Doped Telluro-Borate Glasses

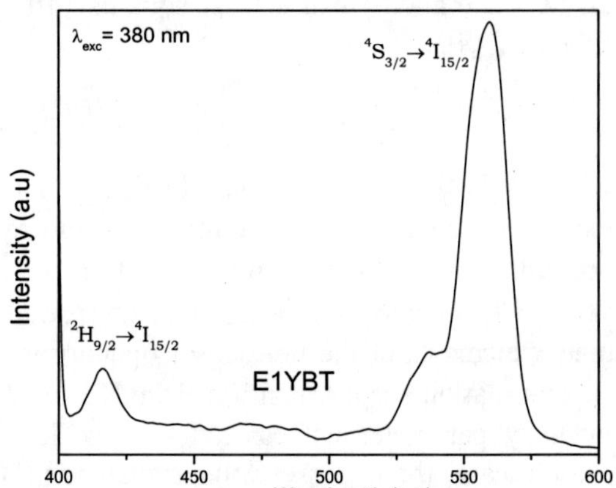

Fig. 3: Emssion Spectrum of Er^{3+}/Yb^{3+} Doped E1YBT Telluro-Borate Glass

Emission Spectra and Radiative Properties

The emission spectra of the title glasses have been recorded in the region 400–600 nm monitoring an excitation at 380 nm and is g shown in Figure 3. These spectra contain two emission transitions such as $^2H_{9/2}\rightarrow^4I_{15/2}$ and $^4S_{3/2}\rightarrow^4I_{15/2}$ corresponds to the band position at 411 nm and 554 nm respectively and among these transitions, $^4S_{3/2}\rightarrow^4I_{15/2}$ transition possess more intensity.[2–6] The luminescence intensity changes with the change in glass composition, and it is observed from the emission spectra that, E2YBT glass exhibits strongest emission of all the studied glasses due to the effective energy transfer between Yb^{3+} and Er^{3+} ions. No change in the shape or peak position of the broad emission could be observed. Using the J–O intensity parameters various radiative parameters such as transition probability (A_R), radiative lifetime (τ_R), branching ratio (β_R), effective bandwidth ($\Delta\lambda_{eff}$), and stimulated emission cross-section $\sigma(\lambda_p)$ have been calculated for the ExYBT telluro-borate glasses and the results are presented in table 2. The total radiative transition probability (A_T) is the sum of radiative transition probability for over all terminal levels and is expressed as:[2]

$$A_T(\Psi J) = \Sigma A(\Psi J, \Psi' J')$$

The branching ratios (β_R) have been calculated form the following expression:[3]

$$\beta_R(\Psi J, \Psi' J) = \frac{A(\Psi J, \Psi' J')}{A_T \Psi J}$$

The experimental branching ratio is the relative area under the emission curve. The radiative lifetime is the reciprocal of the total radiative transition probability, which is defined as:

$$\tau_R(\Psi J) = [A_T(\Psi J)]^{-1}$$

The laser characteristic parameter, the stimulated emission cross-section have been determined from the following expression:

$$\sigma_P^E = \frac{\lambda_p^4 A}{8\pi c n^2 \Delta\lambda_{eff}}$$

Where n is the refractive index and λ_p is the emission peak wavelength and $\Delta\lambda_{eff}$ is the effective bandwidth. The $\Delta\lambda_{eff}$ is given as:

$$\Delta\lambda_{eff} = \frac{1}{I_{max}} \int I(\lambda) d\lambda$$

Where I is the fluorescence intensity and Imax is the maximum intensity of the band.

The stimulated emission cross-section value is found to be 42.684, 48.738, 41.362 and 49.710 for the $^4S_{3/2}\rightarrow{}^4I_{15/2}$ transition corresponding to the E0YBT, E0.1YBT, E1YBT and E2YBT boro-tellurite glasses respectively. The stimulated emission cross-section value for E2YBT glass possess higher value as 65.353. The branching ratio values are found to follow the order as $^2H_{9/2}$<$^4S_{3/2}$→$^4I_{15/2}$ corresponding to the E0YBT, E0.1YBT, E1YBT and E2YBT boro-tellurite glasses respectively. Of all the studied glasses, E2YBT glass possess higher stimulated emission cross-section and branching ratio value which is suitable for green laser applications.

Table 2: Radiative parameters effective bandwidth ($\Delta\lambda_{eff}$, nm), transition probability (A, s^{-1}), peak stimulated emission cross-section ($\sigma_P^E \times 10^{-22}$ cm^2), experimental branching ratios (β_R), calculated (τ) lifetime for the Er^{3+}/Yb^{3+}:ExYBT boro-tellurite glasses

Parameters		*E0YBT*	*E0.1YBT*	*E1YBT*	*E2YBT*
JO parameters	Ω_2	5.557	5.266	5.487	5.624
	Ω_4	0.718	0.807	0.690	0.932
	Ω_6	0.692	0.688	0.551	0.726
	Ω_4/Ω_6	1.038	1.173	1.252	1.284
$^2H_{9/2}\rightarrow{}^4I_{15/2}$	$\Delta\lambda_{eff}$	9.79	10.66	10.84	13.12
	A	947.98	981.29	813.81	1125.25
	σ_p	11.452	10.902	8.702	9.783
	β_R(Exp)	0.1512	0.0283	0.1369	0.1041
	τ_{cal} (ms)	307	300	322	260
$^4S_{3/2}\rightarrow{}^4I_{15/2}$	$\Delta\lambda_{eff}$	8.51	7.68	8.61	8.13
	A	922.68	953.48	933.76	1078.94
	σ_p	42.684	48.738	41.362	49.710
	β_R(Exp)	0.8088	0.9717	0.6841	0.8959
	τ_{cal} (ms)	729	702	855	621

REFERENCES

[1] Karunakaran, R.T., Marimuthu, K., Surendra , S. Babu and Arumugam, S., *Physica B*, 404, 3995, 2009.

[2] Selvaraju, K. and Marimuthu, K., *J. Lumin.* 132, 1171–1178, 2012.

[3] Selvaraju, K. and Marimuthu, K., *Physica B*, 407, 1086–1093, 2012.

[4] Yung, S.W., Hsu, S.M., Chang, C.C., Hsu, K.L., Chin, T.S., Hsiang, H.I. and Lai, Y.S., *J. Non-Cryst. Solids*, 357, 1328–1334, 2011.

[5] Zhanga, Yan, Lu, Chunhua, Feng, Yijun, Sun, Liyan, Ni, Yaru and Xu, Zhongzi, *Mater. Chem. Phys.* 126, 786–790, 2011.

[6] Yung, S.W., Lin, H.J., Lin, Y.Y., Brow, R.K., Lai, Y.S., Horng, J.S. and Zhang, T., *Mater. Chem. Phys.* 117, 29–34, 2009.

[7] Jayasankar, C.K. and Babu, P., *J. Alloys Compd.*, 307, 82–95, 2000.

[8] Carnall, W.T., Fields, P.R. and Rajnak, K., *J. Chem. Phys.*, 49, 4424–4442, 1968.

[9] Judd, B.R., *Phys. Rev.* 127, 750–761, 1962.

[10] Ofelt, G.S., *J. Chem. Phys.* 37, 511–521, 1962.

STUDIES ON BIOSYNTHESIS OF IRON NANOPARTICLES AT ROOM TEMPERATURE USING AQUEOUS EXTRACT OF AQUATIC PLANT (WATER HYACINTH) AND ITS APPLICATION FOR CHROMIUM (VI) REMOVAL

R. VASANTHA GOKILAM, M. ANJALI NAIR, A. KANAKALAKSHMI, C. SUKUMAR AND K. SHANTHI

Department of Environmental Science, PSG College of Arts and Science, Coimbatore
E-mail: shanthivis@gmail.com

ABSTRACT

The biosynthesis of nanoparticles by plant extracts has been an exciting discovery in the field of bionanotechnology. These nanoparticles find extensive applications in a wide spectrum of scientific, engineering and industrial areas. Iron nanoparticles were synthesized using a rapid, single step, and completely green biosynthetic method employing aqueous Water hyacinth extracts as both the reducing and capping agent. The biosynthesis of nanoparticles has been proposed as a cost effective environmental friendly alternative to chemical and physical methods. Recently, silver and iron nanoparticles of various sizes and morphologies have been synthesized using coffee and green tea extract. Iron nanoparticles of various sizes and morphologies were also formed instantaneously using Water Hyacinth extracts. The study has been carried out for biosynthesis of iron nanoparticles using aqueous aquatic plant Water Hyacinth extracts. Fe^0 nanoparticles were synthesized by the ferric chloride and were characterized by UV–Vis (UV-Visible) and FTIR (Fourier Transform Infrared spectroscopy), VSM (Vibrating sample magnetometer), XRD (X-Ray Diffraction), Scanning Electron Microscopy (SEM) and EDS. To study the efficiency of the biosynthesized Iron Nanoparticles in the removal of cr(VI) from aqueous solution. Various kinetic studies have been carried out. This environmentally friendly method of biological iron nanoparticles production provides rates of synthesis faster or

V. Rajendran, P. Prabu and K.E. Geckeler (eds.)
Nano Biomaterials, pp. 95–100 (2012).

comparable to those of chemical methods and can potentially be used in human contacting areas such as cosmetics, foods and medical application.

INTRODUCTION

Nanotechnology is the engineering and art of manipulating matter at the nanoscale (1–100 nm). Biosynthesis of nanoparticles by plant extracts is under exhaustive research however, an exact mechanism for the synthesis of nanoparticles using plants has not yet been elucidated although several hypothetical mechanisms have been proposed (Rai *et al.* 2004). The use of plant extracts for synthesis of nanoparticles is potentially advantageous over microorganisms due to the ease of scale up, the biohazards, and elaborates process of maintaining cell cultures (Sathishkumar *et al.*, 2009). Chromium is a commonly identified contaminant in soils and groundwater. Much research has focused on the remediation of Cr (VI) and many treatment processes have been developed. The study has been carried out to biosynthesize iron nanoparticles using plant at room temperature. This study also includes the properties of iron nanoparticles and how the adsorption processes of chromium occurs by biosynthesized iron nanoparticles.

EXPERIMENTAL

Iron nanoparticles were synthesized by of ferric chloride ($FeCl_3$), with the water hyacinth plant extract. (Njagi *et al.*, 2010). Iron nanoparticles prepared by the above method was characterized with X-Ray Diffraction (XRD), UV-Vis Spectroscopy, Scanning Electron Microscopy with energy Dispersion X-Ray Method (SEM & EDX),and Fourier Transform Infrared Spectroscopy (FTIR). Hexavalent chromium was determined spectrophotometrically by diphenyl carbazide method. Biosorption studies were carried with different chromium (VI) concentrations, pH, and Dosage. Various Kinetics and Isotherms studies were also carried out.

RESULTS AND DISCUSSION

The biosynthesized iron nanoparticles from water hyacinth were characterized for its properties and the results were shown in Figures 1, 2, 3 and 4. The Fe^0 nanoparticles showed its absorption maxima at 260 nm. This observation is similar to that obtained by Morgada *et al.* 2009. In the FTIR band at 3,421 cm^{-1} was ascribed to OH stretching vibration and the one at 1,641 cm^{-1} to the OH bending vibration of surface-adsorbed water which, suggests the formation of ferricoxy hydroxide (FeOOH) layer on Fe^0 nanoparticles A single peak was observed at around 3350.95 cm^{-1} and 3350.95 cm^{-1}, which is due to absorptions peak of – OH or – NH_2 in the iron nanoparticles reported by (Harajyoti Mazumdar *et al.,* 2011). The basal spacing of mount in XRD Figure 2 was 1.99 nm, 3.14nm indicates that the iron nanoparticles consisted mostly of iron (reflection at 2θ of 45°, d = 1.99 nm, 2θ of 28°, d – 3.14 nm) and small amounts of crystalline iron (hydr) oxides (reflections at the 2θ of 35° and 31°, d = 2.81 nm). The SEM images and EDS spectrum of iron nanoparticles loaded with chromium are shown in Figures 3 and 4. These images suggest that the particles agglomerate to form irregular clusters and showed peak of the iron and chromium. The spectrum also reveals the presence of chloride ions, Zirconium, Potassium, and manganese impurities with oxygen that indicates the surface of oxidation of the iron nanoparticles (Njagi, *et al.,* 2010).

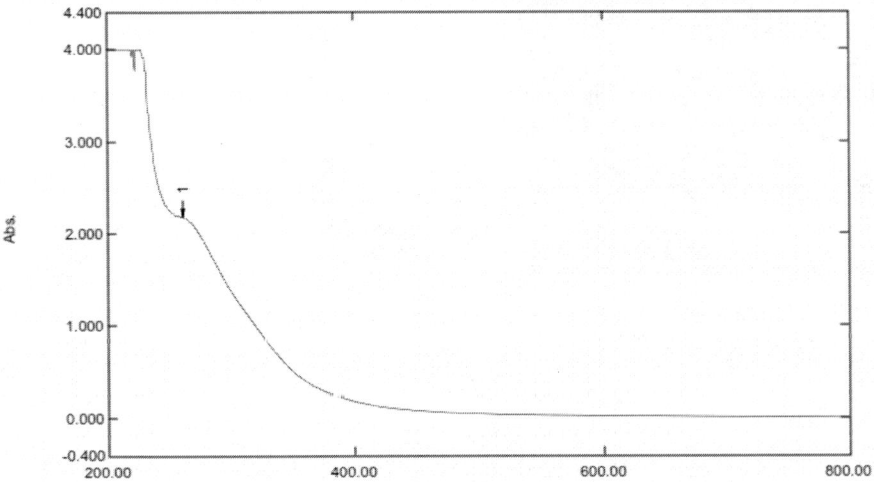

Fig. 1: UV Spectrum of Biosynthesized Iron Nanoparticle

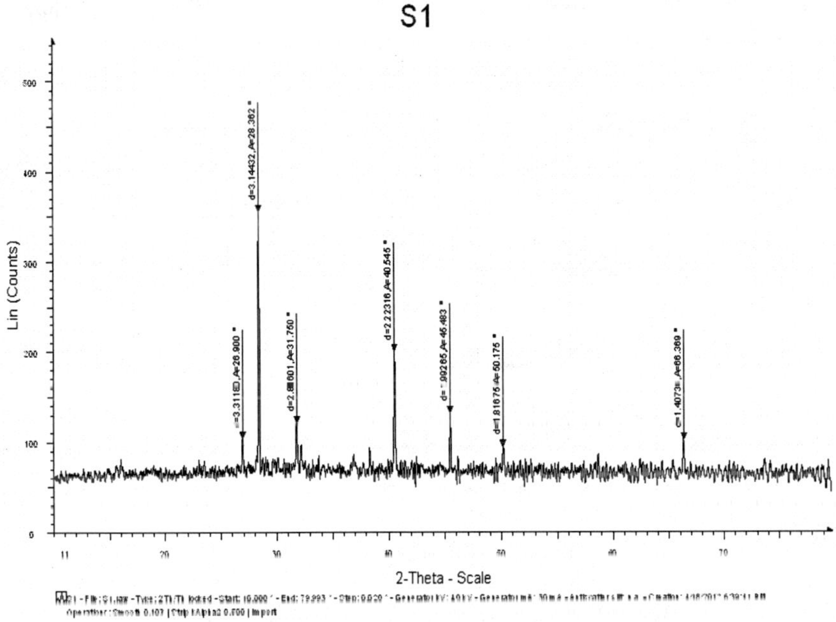

Fig. 2: XRD of Biosynthesized Iron Nanoparticle

Fig. 3: SEM of Chromium Loaded Iron Nanoparticles

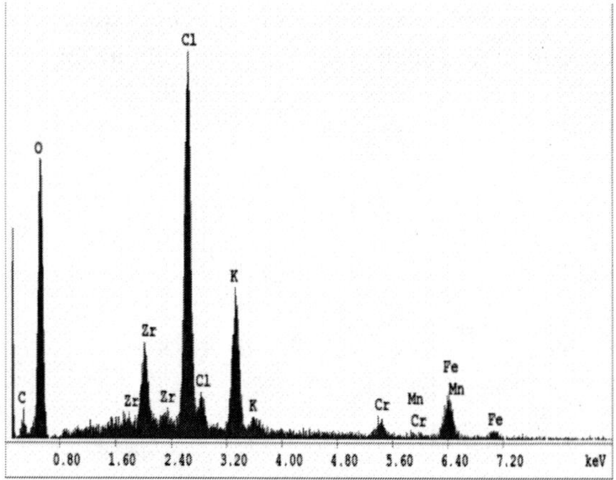

Fig. 4: EDS of Chromium Loaded iron Nanoparticles

Sorption Studies

Studies on the effecicieny of Iron nanoparticles in the removal of chromium using different dosage, different pH, different chromium concentration and contact time were carried out and the results were presented in (Table 1). The optimum condition for the maximum removal of chromium by iron nanoparticle synthesized from plant extract was 100mg of particle dosage, at pH 2.0 with contact time of 150 minutes from the 100 mg/L of chromium solution.

Table 1: Effect of pH, Nanoparticle Dosage and Chromium Concentration on Chromium Removal by Biosynthesized Iron Nanoparticles

pH	*% Removal*	*Iron Nanoparticle Dosage, gm/100 ml*	*% Removal*	*Chromium Concentration, ppm*	*% Removal*
2	99	0.05	78	25	99.5
3	81	0.10	99	50	99.25
4	56	0.15	99.8	100	99
5	44	0.20	100	150	98
6	71			200	97
7	76				
8	31				
9	19				

Isotherm Studies

In this study, Langmuir, Freundlich and Temkin models were used to determine the adsorption equilibrium between the adsorbent and metal ions. Taking into account the correlation coefficients for Langmuir isotherm (R^2) of Iron nanoparticles was 0.995 for chromium removal which confirms the strong monolayer sorption onto specific sites. The Freundlich biosorption constants "n" values was 1.9. Same results reported by (Iram, *et al.*, 2010). Now the R^2 value for all isotherm showed the trend of Langmuir (0.995) > Freundlich (0.979) > Harkins Jura (0.72) >Temkin (0.9773)> D-R (0.888) (Table 2). From this study it was observed that the pseudo second order kinetic model fitted very well than the pseudo first order kinetic model for chromium biosorption by Iron nanoparticles studied. The pseudo second-order equation fitted the data verywell in a large quantity of literature for biosorption (Ho, 2006) (Table 3).

Table 2: Isotherm Studies

Freundlich			*Langmuir*			*Tempkin*		
K_f	*n*	R^2	Q_{max}, *mg/gm*	*b, L/mg*	R^2	A_t	b_t, *l/gm*	R^2
81.7	1.9	0.98	226	0.8	0.995	10.9	43	0.98
Harkins Jura			*DR-Isotherm*					
A	*B*	R^2	Q_{DR}, *mg/gm*	*β*	*E, kJ/mole*	R^2		
1193	0.52	0.72	141	5.86	0.29	0.89		

Table 3: Kinetic Studies

Kinetics	*Pseudo Second Order*			*Pseudo Second Order*			
Cr (VI), mg/L	*K, L/min*	Q_e *(cal), mg/gm*	R^2	K^2, *gm/mg/ min*	Q_e *(cal), mg/gm*	*h*	R^2
25	10.4	24.5	0.885	0.03	0.76	0.025	0.991
50	13	49	0.839	0.005	2.00	0.011	0.941
100	13	98	0.904	0.002	4.16	0.010	0.941
150	11	145	0.851	0.003	5.00	0.019	0.974
200	11	191	0.849	0.003	6.25	0.024	0.981

CONCLUSIONS

A "green" one-step approach has been developed for the synthesis of nanoscale iron particles at room temperature.The Biosynthesized Iron nanoparticles samples were assessed with Chromium removal from aqueous solution. This simple method offers agglomerate and irregular clusters of iron nanoparticles with unique size and shape when compared with control samples. The study reveals that Fe^0 nanoparticles play a key role in Cr (VI) removal through reduction. Fe^0 nanoparticles and factors like concentration, pH and time of treatment facilitate the process, which may lead to an innovative remediation technology that is likely more cost effective and less environmentally disruptive. This biological method is potentially attractive for large scale synthesis of metallic and metal oxide nanomaterials.

REFERENCES

[1] Harajyoti Mazumdar and Nabanita Haloi A study on Biosynthesis of Iron nanoparticles by *Pleurotus* sp. *J. Microbiol. Biotech. Res.*, 2011, 1(3): 39–49.

[2] Ho, Y.S. (2006). Review of second-order models for adsorption systems. *J Hazard Mater.*, 136:103–111.

[3] Iram M, Guo C., Guan Y.P. Ishfaq, A. Liu, H.Z., Adsorption and magnetic removal of neutral red dye from aqueous solution using Fe_3O_4 hollow nanospheres *J. Hazard. Mater.*, 181 (2010), 1039–1050.

[4] Morgada, M.E., Levy, I.K., Salomone, V., Farias, S.S., Lopez, G. and Litter, M.I. (2009). Arsenic(V) removal with nanoparticulate zerovalent iron: effect of UV light and humic acids. Catal Today 143: 261–268.

[5] Njagi, Eric C. Hui Huang, Lisa Stafford, Homer Genuino, Hugo M. Galindo, Biosynthesis of Iron and Silver Nanoparticles at Room Temperature Using Aqueous Sorghum Bran Langmuir 2011, 27(1), 264–271.

[6] Rai, D.D.A. Moore, Hess, N.J., Rao, L. and Clark, S.B., Chromium (III) hydroxide solubility in the aqueous system: a thermodynamic model, *J. Solution Chem.*, 33 (10), (2004), 1213–1242.

[7] Sathishkumar, M., Sneha, K., Kwak, I.S., Mao, J., Tripathy, S.J. and Yun, Y.-S., Phyto-crystallization of palladium through reduction process using *Cinnamom zeylanicum* bark extract *J. Hazard. Mater.*l, 2009, 171, 400–404.

STUDIES ON THE REMOVAL OF CHROMIUM (VI) BY IRON NANOPARTICLE SYNTHESISED FROM FUNGI

A. KANAKALAKSHMI, M. ANJALI NAIR, R. VASANTHA GOKILAM, C. SUKUMAR AND K. SHANTHI

Department of Environmental Science, PSG College of Arts and Science, Coimbatore
E-mail: shanthivis@gmail.com

ABSTRACT

The illegal disposal of industrial wastewater into the environment results in the accumulation of heavy metals. These heavy metals are released by many industries such as textiles, electronic, paper, plastic, leather and electroplating industries. Unlike other organic contaminants this heavy metals are non-degradable and can accumulate in living organisms and cause various health disorders. The most common heavy metals that are released from the industries are chromium, Lead, copper, etc. However these ubiquitous elements which occur naturally in water, soil, rocks and air. Usually these compounds released mainly due to the anthropogenic sources from various industries. The concentration of these heavy metals in the environment has been increased significantly. Among this the concentration of Cr (VI) in liquid effluents released such source varies from 0.5–270,000 mg/l. Iron nanoparticles are an attractive component for heavy metal remediation. The use of iron nanoparticle is increasingly becoming a mature technology to solve the environmental problems such as coagulation of sewage, adsorption of organic dyes, and of heavy metal contamination. The synthesis of iron nanoparticle was usually carried out by physical or chemical method. It employs use of toxic chemicals. Recently, biological methods by using fungi for the synthesis of iron nanoparticles are gaining importance since they occur in aqueous medium of under ambient experimental conditions temperature and pressure. In this study, the iron nanoparticle was synthesised by using fungi and it was characterized by using UV-Spec, FTIR. The magnetic property was analyzed and confirmed by VSM and XRD. These nanoparticles were checked for the removal of heavy metals like Chromium and the percentage of removal was calculated.

V. Rajendran, P. Prabu and K.E. Geckeler (eds.)
Nano Biomaterials, pp. 101–106 (2012).

INTRODUCTION

Nanotechnology has been extended to the waste treatments in the recent years (Yang-Hsin Shih *et al*., 2010). Nanoparticle exhibit good adsorption efficiency especially due to higher surface area and greater active sites for interaction with metallic species. Nanoscale iron particles are increasingly attracting interest as efficient sorbents for various types of pollutants from aqueous solutions and effluents.

These nanoscale iron particles are low cost materials having a high surface area for adsorption (Aswin Sidhaarth *et al.,* 2012). Green approach for the synthesis of nanomaterials utilizes of biological components, primarily prokaryotes and eukaryotes. The second most abundant organisms in our biosphere are fungi. Slight climate changes can potentially be disastrous to the life processes of micro organisms; this can result in the prolific advantage for the production of nanoparticles (Anuradha *et al.,* 2010). Although heavy metal is an essential micronutrient for animals and takes part in synthesis of vitamin B12, its higher concentration cause cancer of the lungs, nose, and bones and it may also cause nausea, rapid respiration, headache, cyanosis, and dry cough. It is thus necessary to treat industrial effluent rich in heavy metal contaminant before their discharge. The study has been carried out to biosynthesize iron nanoparticles using fungus at room temperature under totally aerobic conditions. This study also includes the properties of iron nanoparticles and how the adsorption processes of chromium occurs by biosynthesized iron nanoparticles.

EXPERIMENTAL

Synthesis and Characterization of Iron Nanoparticle

The *penicillium* sp., was used for the synthesis of iron nanoparticle by using the procedure followed by Harajyoti *et al*., 2011 and the synthesized iron nanoparticle was futher characterized with UV-Vis spectroscopy, Fourier Transform Infrared spectroscopy (FTIR), Scanning Electron Microscopy (SEM)-EDAX

Batch Experiment

Chromium reduction was analyzed using the synthesized iron nanoparticle with diphenyl carbozide and it shows reduction. Effect of dosage, concentration, and pH was observed. Various kinetics and isotherm was studied.

RESULTS AND DISCUSSION

Iron nanoparticle was synthesized from *Penicillium* sp using $FeSO_4$ and was characterized using UV-Vis spectroscopy, Fourier Transform Infrared spectroscopy (FTIR), Scanning Electron Microscopy (SEM)-EDAX. Formation of iron nanoparticles were detected by spectral analysis under UV-Vis Spectrophotometer. The absortion spectra were observed at 273nm. Iron nanoparticles mainly show peaks at 216 nm and 268 nm in distilled water

(Guo *et al.,* 2001). The shifts in peaks of nanoparticles may be due to media compositions or size of particles (Figure 1).

FTIR analysis was mainly done to find out the possible biomolecules interactions with Iron nanoparticles in fungal cells. The FTIR spectra of fungal iron nanoparticle result in absorption peak which located in the regions 2316 and 2376.30 cm^{-1}. Similar absorption peaks in the range 2352.66 cm^{-1} and 2378.09 cm^{-1} were observed by Harajyoti *et al.,* 2011. The SEM images of the iron nanoparticle showed that the morphology of the particles were uniformly spherical in structure and showed in (Figure 2). The EDAX shows the chemical and elemental analysis of iron nanoparticle. Results of the EDAX analysis showed that iron nanoparticle consist of iron, chlorine, sulfur, sodium, lead, rubidium and Oxygen (Figure 3).

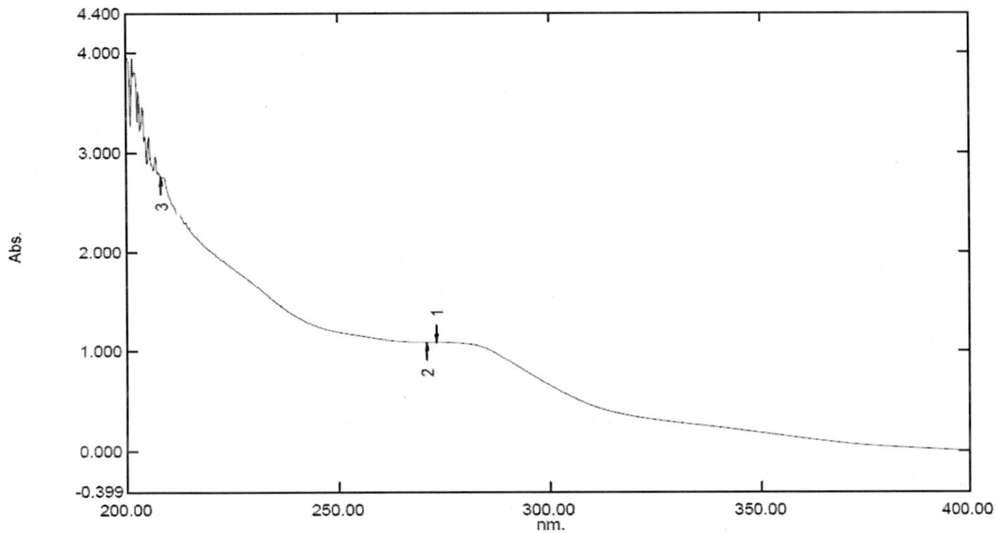

Fig. 1: UV-Vis Spectrum of Iron

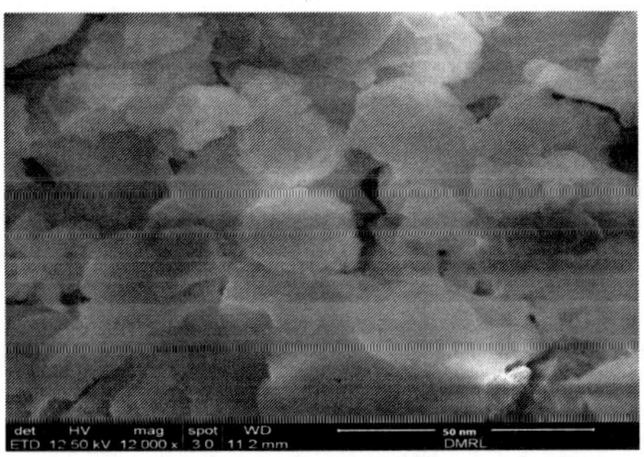

Fig. 2: SEM of Iron Nanoparticle

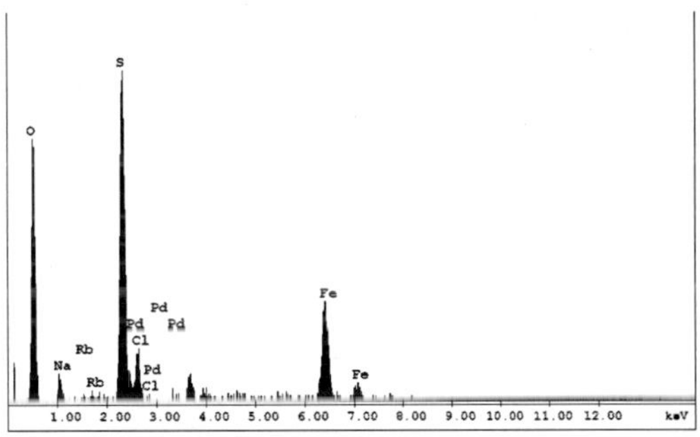

Fig. 3: EDAX Image for Iron Nanoparticle

Batch Experiment

Results of the batch experiments conducted on Cr (VI) using different parameters like effect of chromium concentration, effect of nanoparticles dosage, and effect of pH using iron nanoparticle and the results were shown in (Table 1). All the experiments were carried for the period 4hours. The Cr (VI) removal was calculated based on the final and initial concentration of Cr(VI) in the solution. The difference between the initial concentration of Cr (VI) and final concentration found after the treatment is attributed to the reduction by iron nanoparticles.

The results showed that addition of 0.10 g L-1 of iron nanoparticles resulted in reduction of Cr (VI) concentration. It was further noticed in batch experiments that the reduction of Cr (VI) in all cases increased with increase in iron nano particle concentration. Similar trend have been observed by others researchers (Franco *et al.,* 2009 and Singh *et al.,* 2011).

Table 1: Effect of pH, Nanoparticle Dosage and Chromium Concentration on Chromium Removal by Biosynthesized Iron Nanoparticles

pH	*% Removal*	*Iron Nanoparticle Dosage, gm/100 ml*	*% Removal*	*Chromium Concentration, ppm*	*% Removal*
2	99	0.02	66.25	25	99.5
3	75	0.04	76.25	50	99.25
4	48.75	0.06	78.75	100	99
5	41.25	0.08	81.25	150	97.9
6	58.75	0.1	90	200	96.8
7	76.25	0.12	92.5	250	95
8	37.5	0.14	95		
9	31.25	0.16	95		

KINETIC AND ISOTHERM STUDIES

The kinetics and isotherms study shows that pseudo first order was poorly fitted to the the study when compared to the pseudo second order. The R^2 values for the langumir, frendlich, tempkin isotherms show 0.989, 0.975, and 0.977 respectively Tables 2 and 3.

Table 2: Isotherm Studies

Freundlich			*Langmuir*			*Tempkin*		
K_f	n	R^2	Q_{max} mg/gm	b, L/mg	R^2	A_t	b_t, l/gm	R^2
79.4	2	0.98	263.1	0.5	0.99	2.65	46.73	0.98

Table 3: Kinetic Studies

Kinetics	*Pseudo First Order*			*Pseudo Second Order*			
Cr (VI), mg/L	*K, L/min*	Q_e *(cal), mg/gm*	R^2	K^2*, gm/mg/min*	Q_e *(cal), mg/gm*	*h*	R^2
25	10	26	0.760	0.0033	1.316	0.0043	0.831
50	0.01	8.5	0.827	0.0024	2.32	0.0056	0.877
100	11	98	0.870	4.24	18.41	0.0007	0.113
150	6	145	0.895	0.0048	4.474	0.0216	0.99
200	6	191	0.871	0.0039	5.872	0.0231	0.993
250	5.5	229.08	0.736	0.0035	7.007	0.0247	0.995

CONCLUSION

The heavy metals like chromium in the environment is a serious threat to human and animals. Thus the present study reveals that iron nanoparticle synthesized by the biological method using fungi can be used for the chromium removal up to 98% was observed. Factors like dosage, concentration, pH will facilitate the removal process. The study further suggests that it can also be used for column studies or insitu studies. This work using iron nanoparticle will provide a promising practical solution for the electroplating industries at a very low cost and environmentally friendly method.

REFERENCES

[1] Anuradha Prakash, Seema Sharma, Naheed Ahmad, Ashok Ghosh and Preety Sinha (2010). Bacteria mediated extracellular synthesis of metallic nanoparticles *International Research Journal of Biotechnology,* Vol. 1(5), pp. 071–079.

[2] Aswin Sidhaarth, K.R., Jeyanthi and Suryanarayanan, J. (2012). Comparitive Studies of Removal of Lead and Zinc from Industrial Wastewater and Aqueous Solution by Iron Oxide Nanoparticle: Performance and Mechanisms, *European Journal of Scientific Research* Vol. 70, No. 2, 169–184.

[3] Franco, D.V., Da silva L.M. and Jardium, W.F. (2009). Reduction of hexavalentchromium in soil and ground waterusing zero valent iron under batch and semi batch conditions, *Water Air Soil Pollut,* 197; 49–60.

[4] Guo, L., Huang, Q., Li, X. and Yang, S. (2001). *Phys. Chem. Chem. Phys*. 2001. 3, 1661–1665.

[5] Harajyoti Mazumdar and Nabantita Haloi (2011). A study on Biosynthesis of Iron nanoparticles by *Pleurotus* sp. J. Microbiol. Biotech, Vol. 1(3): 39–49.

[6] Singh, R., Misra, V. and Singh, R.P. (2011). Synthesis, characterization and role of zero-valent iron nanoparticle in removal of hexavalent chromium from chromium-spiked soil. *J Nanopart Res.,* 13: 4063–4073.

[7] Yang-Hsin Shi, Chih-Ping Tso and Li-Yuan Tung (2010). Rapid degradation of Methyl orange with nanoscale Zerovalent Iron particles, *J. Environ. Eng. Manage.,* 20(3), 137–143.

COMPARISON OF DIFFERENT POLYMERIZATION PROTOCOLS ON THE MOLECULAR RECOGNITION PERFORMANCE OF MOLECULARLY IMPRINTED POLYMERS

SUSHMA PARDESHI, RITA DHODAPKAR[1] AND ANUPAMA KUMAR

Department of Chemistry, Visvesvaraya National Institute of Technology (VNIT), Nagpur
[1]National Environmental Engineering Research Institute (NEERI), CSIR, Nagpur
E-mail: drkumaranupama@rediffmail.com

ABSTRACT

In this comparative study, Molecularly imprinted polymers (MIPs) with a similar composition were synthesized by bulk, precipitation and reverse micro-emulsion polymerization for the template Gallic acid (GA). The resulting imprinted materials were characterized physically and their capacity to recognize the template GA under identical analytical conditions was assessed from the real sample. The results showed that the MIPs are obtained in good yield in a single preparative step by following precipitation and reverse emulsion protocol. Moreover, the product consisted of spherical and smooth particles. When the extraction capability form real matrix was measured, the trend was observed as reverse emulsion (68.6%) > precipitation (55.5%) > bulk (44.5%).

INTRODUCTION

Molecularly imprinted polymers (MIPs) are synthetic polymeric materials widely used as sorbents for solid-phase extraction.[1] With the latest developments in the technique of molecular imprinting, MIPs developed can be used in the detection of drugs, toxins, pesticides, food components, and other molecules that would be difficult to isolate otherwise. Considering the versatility, high level of specificity and recognition properties, the future of these materials is very promising.[2] Molecular recognition is the most important property of

V. Rajendran, P. Prabu and K.E. Geckeler (eds.)
Nano Biomaterials, pp. 107–112 (2012).

MIPs.[3] The different potential applications demand diverse properties in the MIPs. This has led to development of different polymerization protocols for MIP preparation. In order to establish a reasonable comparison between the different polymerization protocols, performance evaluation of the MIPs prepared by using these methods should be done under similar synthetic conditions. The present work compares molecular recognition ability of the MIPs synthesized by bulk (MIP_B), precipitation (MIP_P), and reverse micro-emulsion polymerization method (MIP_R) for the template Gallic acid (GA) to assess the advantages and drawbacks that these approaches bring to imprinted polymers. Comparison is made on the basis of yield, morphology and extraction capacity to study the effect of the polymerization method on the performance of the imprinted material. GA is chosen as template due to its pharmaceutical and industrial importance.[4] Acrylic acid is selected from the virtual library of eighteen functional monomers using computational modeling methods.[5] The morphological characterization of the imprinted material is done by BET and SEM and their capacity to rebind GA is determined in the aqueous extracts of real sample.

EXPERIMENTAL METHODS

GA, acrylic acid, Span 80 and sodium dodecyl benzene sulphonate (DBS) were purchased from Sigma-Aldrich (Buchs, Switzerland). Ethylene glycol dimethacrylate (EGDMA) was purchased from Merck (Darmstadt, Germany), and 2, 2-azoisobutyronitrile (AIBN) from National Chemicals (India). All the solvents were reagent grade or HPLC-grade and purchased from Merck (India) and used without further purification.

Synthesis of MIP_B by Bulk Polymerization Protocol

1 mmol of GA and 4 mmol of the acrylic acid were added into 10 ml acetonitrile, followed by 40 mmol of cross-linker EGDMA and initiator AIBN (0.05g). The prepolymerization solution was sonicated for 15 min, cooled in ice and purged with N_2 for 5 min. The flask was sealed and polymerization was performed in a water bath maintained at 60°C for 24 hours. The resultant rigid bulk polymers were washed with water and dried at 55°C. The product was ground into powder and passed through sieves (ASTM No. 200 and 240) to obtain particle size fractions between 63 and 75 μ. Non-imprinted polymer was prepared and treated following the same procedure as for the imprinted polymer but, without addition of the print molecule (GA).

Synthesis of MIP_P by Precipitation Polymerization Protocol

Similar composition of precursors was dissolved in 100 ml mixture of acetonitrile: toluene (75/25 v/v). The solution was sparged with N_2 for 5 min, sealed and left to polymerize on rotamantle (Remi, India) at 70° for 24 hours under the mild starring. NIP_P was also synthesized using the same protocols without the addition of GA. The particles were collected by centrifugation at 4500 rpm for 10 minutes and dried at 70°C.

Synthesis of MIP_R by Reverse Emulsion Polymerization Protocol

0.5 mL Span 80 as emulsifier was dissolved in 20 mL acetonitrile before adding 7.5 mL of EGDMA. 10 mL aqueous solution containing 1mmol GA and 4 mmol of acrylic acid was added to the acetonitrile solution. The mixture was sonicated for 5 min and 7 mmol DBS as well as 150 mL magnesium sulfate solution (0.015 $molL^{-1}$) was added and sonicated for further 5 min. After the addition of 0.5 g AIBN as the initiator, polymerization was carried out at 70°C for 6 h under nitrogen protection. The resultant polymer was collected by centrifugation at 4500 rpm for 10 minutes and dried at 70°C. NIP_R was also prepared by using the same synthetic protocol.

All the MIPs were washed using methanol: acetic acid (80:20 v/v) to extract GA from the polymeric matrix followed by washing with solution of 0.1 $mmolL^{-1}$ Na_2CO_3 and water to remove the residual acetic acid. Finally, polymers were dried at 60°C and stored at ambient temperature for further experiments and characterization.

Application Potential of MIPs

The fresh fruits of *E. officinalis* (1 kg) were dried and powered (6 gm). For batch rebinding experiments, the solutions were freshly prepared by adding 10 mg of powder in 10 mL of distilled water. These solutions were shaken with MIPs and NIPs with optimum dose and time and analyzed for GA by using UV-visible spectrophotometer.

RESULTS AND DISCUSSIONS

Comparison on the Basis of Yields and Morphology

The yield of MIP_B was 96% which reduced to only 47.5% after post processing stages. Thus, useful particles are lost during the post processing stages of grinding, sieving and sedimentation. The yields of MIP particles obtained by precipitation and reverse emulsion polymerization were 80% and 91% respectively which means that good yield can be obtained in a single preparative step in the precipitation and reverse emulsion protocols.

The morphology and surface area of MIPs prepared by different polymerization technique is presented in the Table 1.

For bulk polymerization, the morphology of the crushed polymer is shown. In all the other procedures, polymers are in the form of beaded particles. The MIP_B and MIP_P have higher surface area and porosity as compared to the MIP_R. MIP_R has less surface area and pore volume which may be due to the fact that the porogen was present in small quantity while, system was diluted with large quantity of water during polymerization process. However, it does not appear that binding and selectivity in MIPs is dependent on a particular porosity.[9]

Comparison on the Basis of Recognition Capabilities

Application potential of these MIPs in the real sample matrices were explored for the extraction of GA from the fruits of Emblica officinalis (amla). The real sample extract was equilibrated with the MIPs and NIPs and the concentration of solution was determined by UV-visible spectrophotometer. The decrease in the peak intensity of the initial solution of real sample extract due to binding of GA is shown in the Figure 1 for the solutions equilibrated with the MIPs. The binding capacity and the percent extraction capabilities of the MIPs are presented in the Table 2.

The results showed that MIP_R prepared following reverse polymerization protocol has highest rebinding capacity followed by MIP_P prepared by precipitation and MIP_B prepared by bulk polymerization protocol.

Table 1: Comparison of Bulk, Precipitation and Reverse Emulsion Polymerization Protocol and their Influence on Morphology of MIPs

Polymerization Method	*Influence on Morphology*	*Merits/Demerits*
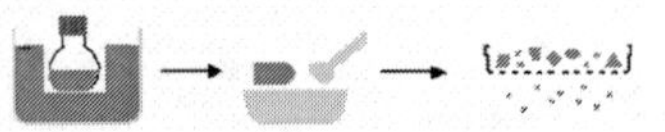 Bulk polymerization method All the precursors (template, monomer, solvent and initiator) are mixed and polymerized. Resultant polymeric block needs to be crushed and ground to obtain MIP particles.	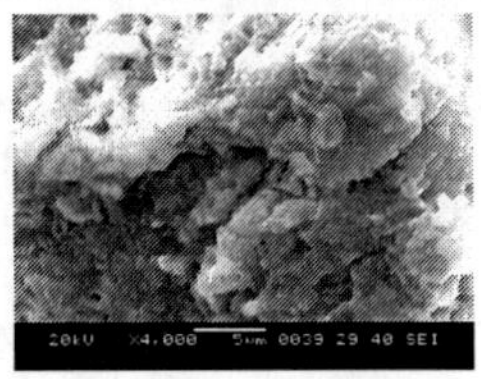 MIP_B Surface area – 413.27 m^2/g Total pore volume – 0.6935 (cm^3/g) Ave. pore diameter– 67.1324 (Å)	Simple and versatile, Time consuming laborious, Irregular particle size.[6]
 Precipitation polymerization method All the precursors (template, monomer, and initiator) are mixed in large quantity of solvent and polymerized.	 MIP_P Surface area – 337.19 m^2/g Total pore volume – 0.9911 (cm^3/g) Ave. pore diameter– 61.4219 (Å)	Simple and convenient, good control of particle size, grinding and sieving steps are avoided, Larger amount of porogen is required.[7]
 Reverse emulsion polymerization method An aqueous solution of monomer is emulsified in non-polar organic solvent and the polymerization is initiated with initiator which is soluble in organic solvent	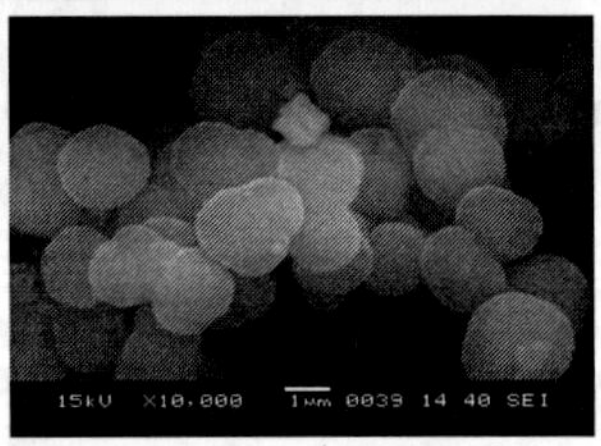 MIP_R Surface area – 205.64 m^2/g Total pore volume – 0.228 (cm^3/g) Ave. pore diameter– 134.35 (Å)	Spherical morphology of MIP particles Synthesis in presence of large quantity of water, expected to perform well in the aqueous environment,[8] Emulsifiers and stabilizers are required in synthesis.

The probable reason for the highest binding capacity of MIP_R is the presence of the large quantity of water during the synthesis procedure. As the interactions between the template and the monomer are formed in presence of water, it can recognize the template in the aqueous conditions.

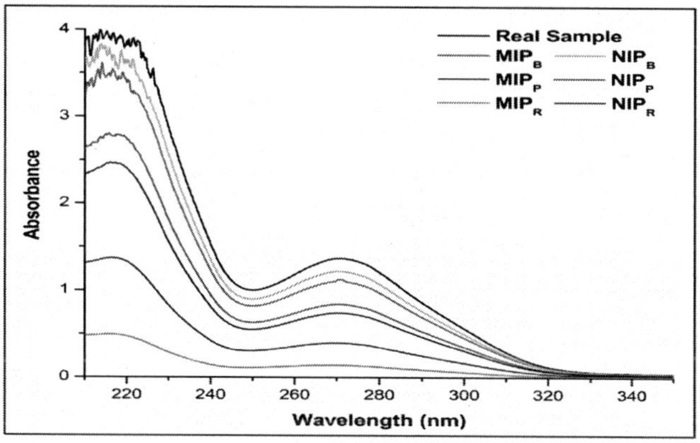

Fig. 1: The Overlapped UV-Visible Spectra of Real Sample Extract in Water Equilibrated with MIPs and NIPs Prepared Using Different Protocols Showing Variation in Peak Intensity

Table 2: The Extraction Capability of MIPs Prepared Using Different Polymerization Techniques

	Extraction Capacity (mg/g)	*Specific Rebinding MIP/NIP*	*% Extraction Capabilities*
MIP_B	9.45	2.03	44.5
NIP_B	4.61		
MIP_P	11.79	2.06	55.5
NIP_P	5.72		
MIP_R	14.56	2.11	68.6
NIP_R	6.90		

CONCLUSION

The specific and effective rebinding of template is a major consideration for MIP, but the yield and morphology of the material produced and the ease and scale-up potential of the method may also be key factors in selecting a particular method. All the different methods discussed here have advantages and disadvantages in these respects. The polymerization method selected for preparation of particular MIP may depend upon its application.

REFERENCES

[1] Pichon, V. and Chapuis-Hugon, F., *Anal Chim Acta,* 622 (2008), 48–61.

[2] Liu, Y., Hoshina, K. and Haginaka, J., *Talanta*, 80 (2010) 1713–1718.

[3] Pardeshi, S., Dhodapkar, R. and Kumar, A., *Adsorpt Sci Technol*, 30 (2012) 23–34.

[4] Pardeshi, S., Kumar, A. and Dhodapkar, R., *Mater Sci Forum*, 665–667, (2011), 512–520.

[5] Pardeshi S., Dhodapkar R. and Kumar A., *J Mol Model,* 2012, DOI: 10.1007/s00894-012-1481–5. Article online 15–6–2012.

[6] Liu, Y., Hoshina, K. and Haginaka, J., *Talanta* 80 (2010) 1713–1718.

[7] Beltran, A., Marcé, R.M., Cormack, P.A.G. and Borrull, F., *J Chromatogr* A, 1216 (2009), 2248–2253.

[8] Pardeshi, S., Dhodapkar, R. and Kumar, A., *J. Mol. Recognit* (Under review), 2012.

[9] Spivak, D.A., Optimization, evaluation, and characterization of molecularly imprinted polymers, *Adv. Drug Delivery Rev.*, 57 (2005), 1779–1794.

HYDROGEN BONDED SUPRAMOLECULAR ARCHITECTURES IN ORGANIC DICARBOXYLATE SALTS

D. SATHYA, R. PADMAVATHY[1], R. JAGAN AND K. SIVAKUMAR

Department of Physics, Anna University, Chennai
[1]Department of Physics, Quaid-E-Millath Govt. College for Women, Chennai
E-mail: ksivakumar@annauniv.edu

ABSTRACT

Three new organic dicarboxylate salts, namely Benzimidazolium hydrogen maleate (1) Hexadecylaminium hydrogen maleate (2) and 4-chloroanilinium hydrogen malonate (3) have been synthesized and characterized by FT-IR, TG/DTA and single crystal X-ray diffraction. The crystal structure of all the compounds display O-H...O and N-H...O intermolecular bonds. The supramolecular structures consist of different motifs generating the different architectures.

INTRODUCTION

In the field of supramolecular chemistry, the directional properties of non-covalent interactions like hydrogen bond, halogen bond and other weak interactions are useful in the construction of molecular self-assemblies.[1–4] Hydroxyl, imides and carboxylic acid groups are the commonly used functional groups to construct the molecular assemblies.[5–7] Carboxylic acids form salts or co-crystals with various organic compounds like amines, amides and amino acids which are useful in pharmaceutical industry for preparation of drugs.[8–10] Different types of hydrogen bonded motifs are observed due to the interaction of aminium cations and carboxylate anions. Herein, the supramolecular frameworks of three organic acid-base salts of dicarboxylic acid with various amines namely, Benzimidazolium maleates (I), Hexadecylaminium maleate (II), 4-chloroanilinium malonate (III) are reported.

V. Rajendran, P. Prabu and K.E. Geckeler (eds.)
Nano Biomaterials, pp. 113–118 (2012).

EXPERIMENTAL SECTION

Benzimidazolium Hydrogen Maleate (1)

A 1:1 mixture of Maleic acid (0.5 mol) and Benzimidazole (0.5 mol) were dissolved in hot methanol. This solution was then warmed in a water bath for 5 minutes and kept for crystallization at room temperature by slow evaporation method. Colorless crystals were collected after five days.

Hexadecylaminium Hydrogen maleate (2)

Stiochiometric amounts of Hexadecylamine (0.5 mol) and Maleic acid (0.5 mol) were dissolved in ethanol. The combined solution was warmed in a water bath to evaporate the solvent which yields colorless crystals.

4-Chloroanilinium Hydrogen malonate (3)

A methonalic solution of Malonic acid (0.5 mol in 5ml MeOH) was added to a methanolic solution of 4-chloroaniline (0.5 mol in 5 ml MeOH) and the solution was allowed to warm over a water bath. The resulting mixture was kept for slow evaporation at room temperature. After three days light green crystals were obtained.

Physical Measurements

Characterization of the samples was performed using FT-IR Spectra, Thermo gravimetric analysis (TGA) and single crystal X-ray diffraction methods. The FT-IR Spectra of the samples were recorded with a Perkin Elmer Spectrum one FT-IR Spectrophotometer in the spectral region 4000–450 cm^{-1} using KBr pellets. Thermo gravimetric experiments were carried out on a Sii Nano instruments (TGDTA 6300) from 20 to 600°C under nitrogen atmosphere at heating rate of 20°C/min.

X-ray Crystallographic Studies

Suitable single crystals of compounds 1–3 were mounted on glass fibers for X-ray measurement. X–ray diffraction data for compounds 1–3 were collected at room temperature with MoKα radiation (λ = 0.7107Å) using Broker AXS Kappa APEX II single crystal CCD diffractometer equipped with a graphite monochromator. The APEX2 software was used for data collection, indexing the reflections and determining the unit cell parameters. Data reduction and cell refinement were performed using XPREP, SAINT software. All structures were solved by direct methods using the SHELXS program and refined with SHELXL by full-matrix least-squares. The final refinement was performed by full matrix least-squares method with anisotropic thermal parameters for non-H atoms on F^2. Diagrams were generated using Mercury 2.4 software. Crystallographic data and structural refinement parameters for compounds (1–3) are given in Table 1.

Table 1: Crystallographic Data and Refinement Parameter of Compounds 1–3

Parameters	*1*	*2*	*3*
Empirical formula	C11 H10 N2 O4	C20 H39 N O4	C9 H10 Cl N O4
Formula weight	234.21	357.52	231.63
Crystal system, space group	Monoclinic $P2_1/n$	Triclinic P-1	Monoclinic $P2_1/c$
Unit cell dimensions			
a (Å)	12.7945(6)	5.5834(2)	12.9579(4)
b(Å)	5.4746(2)	10.0556(3)	9.2197(3)
c(Å)	15.8379(7)	19.5139(6)	8.5060(3)
α(°)	90°	91.421(2)	90°
β(°)	92.734(3)	91.144(2)	94.338(2)°
γ(°)	90°	97.676(2)	90°
Volume $Å^3$	1108.10(8)	1085.15(6)	1014.27(6)
Molecules/Unit cell, Z	4	2	4
θ range for data collection (°)	2.10 to 27.38	2.04 to 30.04	2.71 to 32.08
Reflections collected/unique	11414/2500	26839/6296	13690/3529
Final R indices [I > 2sigma(I)]	R1 = 0.0382, wR2 = 0.0987	R1 = 0.0459, wR2 = 0.1253	R1 = 0.0408, wR2 = 0.1099
R indices (all data)	R1 = 0.0572, wR2 = 0.1106	R1 = 0.0849, wR2 = 0.1459	R1 = 0.0527, wR2 = 0.1185
Goodness of fit F^2	1.002	1.033	1.006

RESULTS AND DISCUSSION

The supramolecular crystal structures of three dicarboxylate salts were analysed. In compound 1–3, the hydrogen bonds are formed between the cationic acceptor (NH_3^+) and the anionic donor (COO^-). The schematic representations of compound (1-3) hydrogen bonded motifs are summarized in Scheme 1. In the structure of 1, the maleate anion forms a S (7) motif through intramolecular O-H…O hydrogen bond. The successive anions and cations are interconnected to two N-H…O hydrogen bonds resulting in one dimensional infinitive zig-zag chain along [0 0 1] direction. The maleate anion links with benzimidazolium cation via N-H…O and C-H…O hydrogen bonds with the graph set notation $R^2_2(8)$. The benzimidazolium cation are centrosymmetrically paired through the couple of C-H…O hydrogen bonds involving the maleate oxygen resulting in the formation of the ring motif $R^2_2(10)$ in the alternative chains. Due to the centrosymmetry between the chains, the secondary C-H…O interactions also arise from the C_3, C_5 carbons of benzimidazolium cation and O_4, O_2 oxygen atoms of the carboxylate anion (C_3-H_3...O_2 and C_5-H_5...O_4). This repeats in the structure forming two dimensional corrugated sheet parallel to (1 1 -1) plane. These sheets are further linked by weak intermolecular C-H...O hydrogen bonds, making a three dimensional network. In compound 2, the anions are bridged with hexadecylaminium cation through two hydrogen bonds (N1-H1B...O_3 and N1-H1C...O1). The centrosymmentric adjacent chains are interconnected via two N-H...O interactions (N1-H1C...O1 and N1-H1A...O_2) to form fused R^6_6 (16) and R^4_4(12) motifs resulting in the formation infinite tape extending along [0 1 0] direction. In addition to these strong hydrogen bonds, the weak C_3-H_3...O_4 interaction is responsible for the interconnection between adjacent tapes thus contributes to the formation of two dimensional molecular channels. In structure 3, the anions repeats itself along the chain through O_3-H_3...O1 hydrogen bond forming C(5) motif resulting one dimensional infinite zig-zag chain. The resulting motif undergoes syn-syn contacts with the neighbouring anionic carboxyl groups. Further the one dimensional anionic chain incorporated with 4-chloroanilinium cations via N1-H1B...O4 and N1-H1A...O2 hydrogen bonds generate R^4_4 (16) motif runs parallel to (0 0 1) plane.

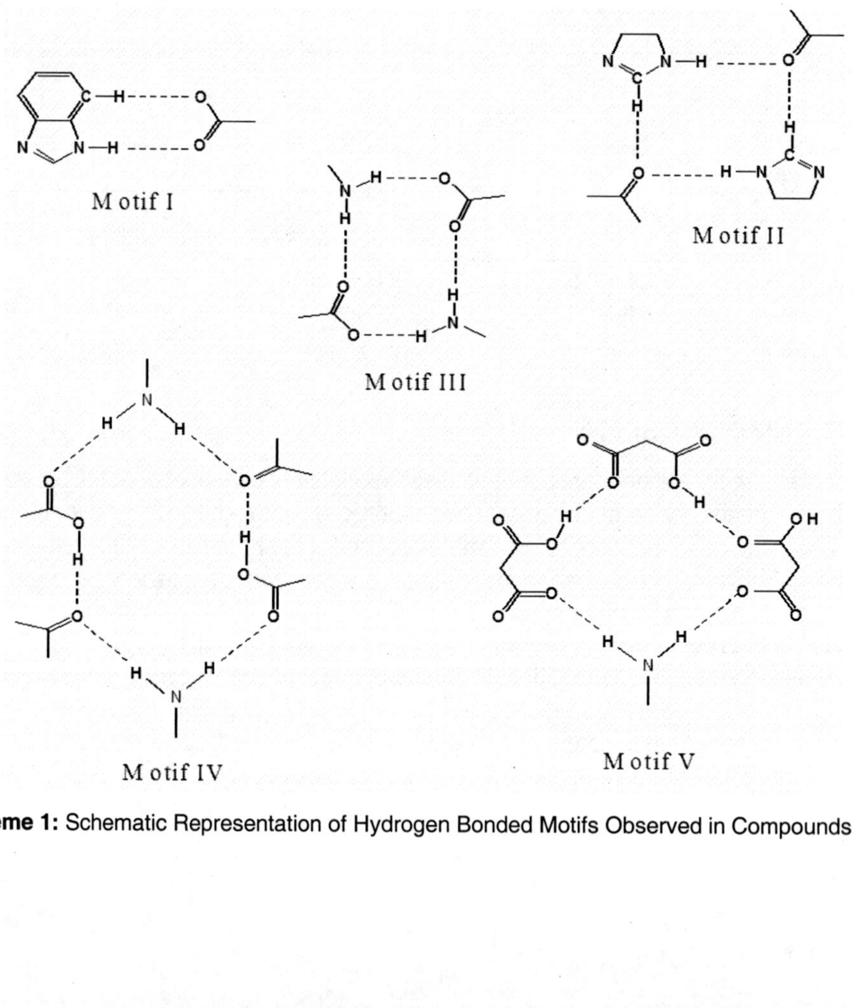

Scheme 1: Schematic Representation of Hydrogen Bonded Motifs Observed in Compounds 1–3

Infrared Spectra

The IR spectra of the organic salts (1–3) confirm the formation of carboxylate ion and protonated amine group. The asymmetrical stretching bands in the region 1558–1619 cm^{-1} show the presence of COO^{-} group. The broad bands in the region 3130–2579 cm^{-1} reveals the presence of both N-H and O-H group.

In salt 1 and 2, the sharp peaks observed at 3114, 2918 cm^{-1} and 869,870 cm^{-1} due to the strong hydrogen bonded O-H stretching and bending.

Thermal Analysis

Comparison of the thermal stability of all the salts show the melting takes place above 100°C. The salts decompose with single step weight loss exhibit the simultaneous decomposition of the anion and cation. The variations of the melting points of the salts are due to the number and strength of intermolecular hydrogen bonds.

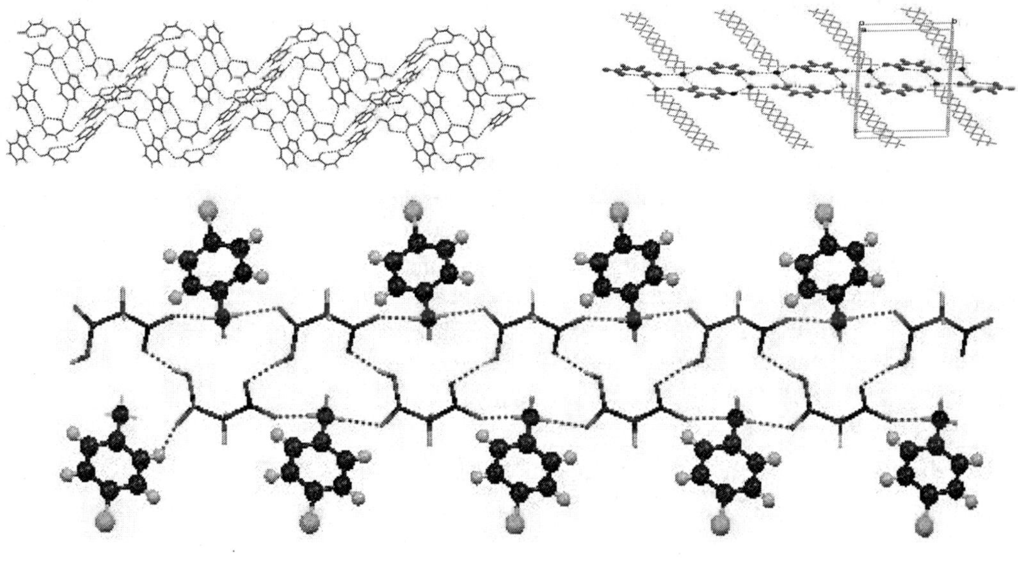

Fig. 1: 2D Structure of (a) Corrugated Sheet Structure of 1 (b) Tape Structure of 2 (c) Chain Structures of 3

CONCLUSION

The crystal structures of three different organic salts were determined and the patterns of each structure were analysed. In all the three salts the cationic nitrogen involves direct hydrogen bonding between the anions. The salt 1 dictates two dimensional corrugated sheet through motifs (I and II) via weak C-H…O interactions, while the motifs (III, IV& V) of salts 2 and 3 consist of classical N-H…O and O-H…O interactions. However due to twisting of phenyl ring in salt 3 lacks C-H...O and π...π interaction and lose connectivity between the layers. Therefore the compounds yields to different diverse architectures ranging from 1D in 3, 2D in 2 and 3D in 1. This result indicates that the relative orientation of the anions and cations decides the supramolecular assembly.

REFERENCES

[1] Jing, L.V. *et al.*, Self-assembly and properties of low dimensional nanomaterials based on π-conjugated organic molecules, *Pure Appl. Chem.*, 80, 639–658, 2008.

[2] Hofmeier, H. and Schubert, U.S., Combination of orthogonal supramolecular interactions in polymeric architectures, *Chem. Commun.*, 2423–2432, 2005.

[3] Feng, Xing *et al.*, Design and Synthesis of Self-assembly Supramolecular Entities Based on Noncovalent Interaction of Cucurbit [5]uril, Metal Ions, and Hydroxybenzene or Its Derivatives, *Cryst. Growth Des.*, 10, 2901–2907, 2010.

[4] Tang, L. *et al.*, Two Novel 3D Hydrogen-bonded Architectures Constructed from Maleic Acid and N-donor Ligands: Structures, Magnetic Properties and Theoretical Studies, *Chin. J. Chem.*, 27, 273–280, 2009.

[5] Tsaggeos, K. *et al.*, Crystal Structure, Thermal Behavior, and Photochemical Reactivity of a Series of Co-Crystals of trans-1,2-Bis(4-pyridyl) Ethylene with Dicarboxylic Acids, *Cryst.Growth Des.*, 12, 2187–2194, 2012.

[6] Baoming, Ji *et al.*, Network Formation of Bis(2-benzimidazylmethyl)amine in Salts with Carboxylic Acids: The Structures Based on the Interplay of Strong and Weak Intermolecular Interactions, *Cryst. Growth Des.*, 11, 4090–4100, 2011.

[7] Jian-Rong, L. *et al.*, Hydrogen-Bonded Supramolecular Architectures of Organic Salts Based on Aromatic Tetracarboxylic Acids and Amines, *Cryst. Growth Des*, 6, 2493–2500, 2006.

[8] Harjyoti Thakuria *et al.*, A one-pot synthesis and self-assembled superstructure of organic salts of a 1,5-benzodiazepine derivative, *Tetrahedron Letters,* 47, 3135–3138,2006.

[9] Haynes, D.A. *et al.*, Cocrystallisation of succinic and fumaric acids with lutidines: a systematic study, *CrystEngComm.*, 8, 830–840, 2006.

[10] Veerendra, K. *et al.*, New Trends in the Co-crystallization of Active Pharmaceutical Ingredients, *Journal of Applied Pharmaceutical Science,* 08, 01–05, 2011.

SYNTHESIS OF GOLD NANOPARTICLES USING *PUNICA GRANATUM* EXTRACT AND EVALUATION OF THEIR ANTIMICROBIAL ACTIVITIES

S. LOKINA AND V. NARAYANAN[1]

Rajalakshmi Engineering College, Thandalam, Chennai
[1]Department of Inorganic Chemistry, University of Madras, Chennai
E-mail: loksiv182@gmail.com; vnnara@yahoo.co.in

ABSTRACT

In the present study, bio-inspired gold nanoparticles are rapidly synthesized at room temperature using highly energetic Punica Granatum fruit extract which is commonly known as pomegranate. A green and low cost synthesis is effective in the formation of stable crystalline gold nanoparticles in the solution. Polyphenols which is present in the pomegranate fruit extract function as reducing as well as stabilizing agent to produce shape controlled gold nanoparticles. Surface plasmon resonance confirmed the formation of gold nanoparticles in UV-Visible (UV-Vis) spectra at 500 nm. The Fourier transform infrared spectroscopy (FTIR) analysis was carried out to identify and study the functional groups responsible for the bio-reduction of gold ion. High resolution transmission electron microscopy (HRTEM) images showed the spherical crystalline gold nanoparticles with a diameter of 0.9–1.2 nm. Thermo gravimetric analysis (TGA) measures the weight loss of gold nanoparticles as a function of temperature under a controlled atmosphere. The synthesized gold nanoparticles showed excellent antimicrobial activity towards most of the tested fungal and bacterial culture.

INTRODUCTION

The researchers show much interest in synthesizing nanoparticles and study their size with various applications because of their unique physicochemical characteristics. Different techniques

V. Rajendran, P. Prabu and K.E. Geckeler (eds.)
Nano Biomaterials, pp. 119–124 (2012).

can also be employed to develop a new substance with safety and low toxicity.[1] The green chemistry particularly utilizes nontoxic chemicals and environmentally friendly solvent water for synthesizing nanoparticles [2] which clearly shows a different path in eliminating wastes before it is generated. A variety of preparation routes have been reported for synthesizing metallic nanoparticles. Recently, biosynthetic route using plant and fruit extracts [3–5] has been proved superior to other physical and chemical methods.[6–9] The current work presents the overview of the synthesis of gold nanoparticles from abundantly available pomegranate fruit extract. Water soluble organic materials present in the fruit extract were mainly responsible for the reduction of gold ions to gold nanoparticles. The most abundant polyphenol in pomegranate juice are the ellagitannins which formed when ellagic acid binds with carbohydrate.

Gold nanoparticles have high potential towards antimicrobial and antifungal activity.[10] The antibacterial efficacy of gold nanoparticles increases because of their larger total surface area per unit volume.[11] The biocompatible gold nanoparticle synthesized from pomegranate extract showed good antibacterial activity against the tested organisms.

EXPERIMENTAL

Preparation of Pomegranate Extract

The arils from the pomegranate fruit were removed, washed thoroughly with distilled water and made into aqueous extract. It was centrifuged and filtered through Whatmann 41 Filter paper to get the clear extract. The extract was used immediately after preparation. The filtrate was used as reducing and stabilizing agent for 0.01M of Chloro auric acid ($HAuCl_4$).

Synthesis of Gold Nanoparticles

For the synthesis of gold nanoparticles, 50 ml of 0.01 M of $HAuCl_4$ was added into 25 ml of aqueous extract of pomegranate. The reaction mixture was then shaken to ensure thorough mixing and allowed to settle at room temperature. The color change indicates the formation of gold nanoparticles. The physical, chemical and biological properties of gold nanoparticles are strongly depending on the method of preparation. The aqueous solutions of gold nanoparticles are not stable and rapidly undergo agglomeration due to the high reactivity of gold nanoparticles. Hence the lyophilization process was used to preserve this perishable material.

RESULTS AND DISCUSSION

UV-Visible Spectral Analysis of Gold Nanoparticles

As the pomegranate extract was mixed with the aqueous solution of the chloro auric acid, slowly it started to change the colour due to the reduction of gold ion to gold which indicates the formation of stable gold nanoparticles The formation of gold nanoparticles was monitored periodically using UV-Vis spectroscopy which has proved to be a useful technique. The broad plasma resonance peak around 500 nm can be assigned to gold nanoparticles (Figure 1).

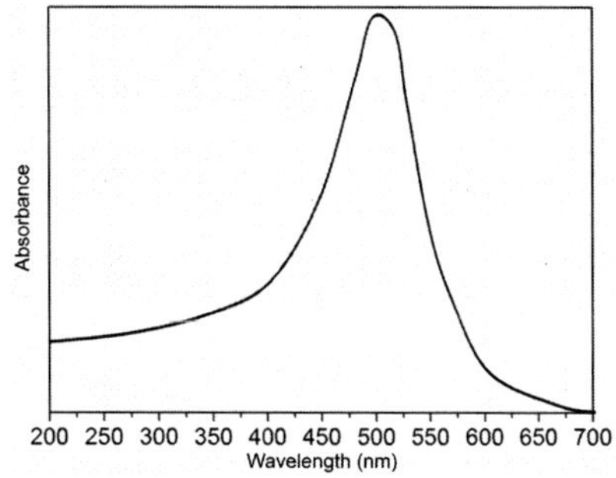

Fig. 1: UV- Vis Absorption Spectrum of Au Nanoparticles in Aqueous Solution of Pomegranate Fruit Extract

FTIR Spectral Analysis

FTIR spectrum of biosynthesized gold nanoparticles shows absorption bands at 3358, 2268, 1644, 1435, 1262, 1161, 1052, 991 and 701 cm^{-1} (Figure 2). The peak at 3358 cm^{-1} assigned to O-H stretch due to the presence of surface adsorbed alcohols (carbohydrates, etc.) and phenols (Polyphenols) of the fruit extract. The absorption bands at 2268 and 1644 cm^{-1} correspond to C-N and C=O (amide) stretching vibrations respectively. The band at 1262 and 1052 cm^{-1} was assigned for aliphatic amines.[12] The presence of esters was confirmed by the presence of absorption band at 1161 cm^{-1}. The band at 701 cm^{-1} assigned for C-H stretching and the band at 1435 cm^{-1} corresponds to C-C stretching vibrations for aromatic ring. FTIR spectroscopic study confirmed the presence of proteins [13] and showed that the protein along with the carbohydrates and polyphenols has the ability to form a layer covering the gold nanoparticles to prevent the agglomeration and stabilizing the gold nanoparticles.

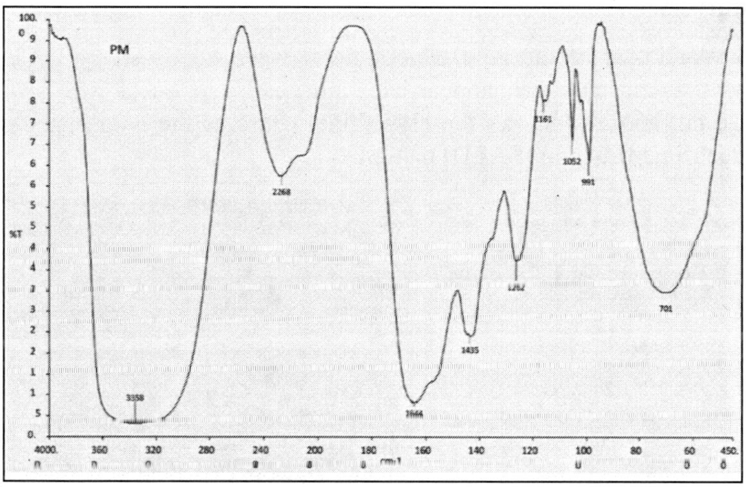

Fig. 2: FTIR Spectrum Recorded by Making a KBr Pellet with Synthesized AuNp

Thermogravimetric (TGA) Analysis

The thermal stability of the synthesized gold nanoparticles was monitored by TG-DTA analysis and the thermogram is given in Figure 3. The biosynthesized gold nanoparticles were subjected to heating from 50°C to 800°C. According to TGA curves, the thermal decomposition process is represented by multiple falls in weight over the temperature ranges from 40 to 180, 180 to 340, 340 to 660 and 660°C. The weight loss between 40 and 180°C was attributed to residual water evaporation, whereas the rest are due to the decomposition of organic contents. The residual mass (680°C), in sample was 50%, indicating the high amount of organic contents in synthesized sample.

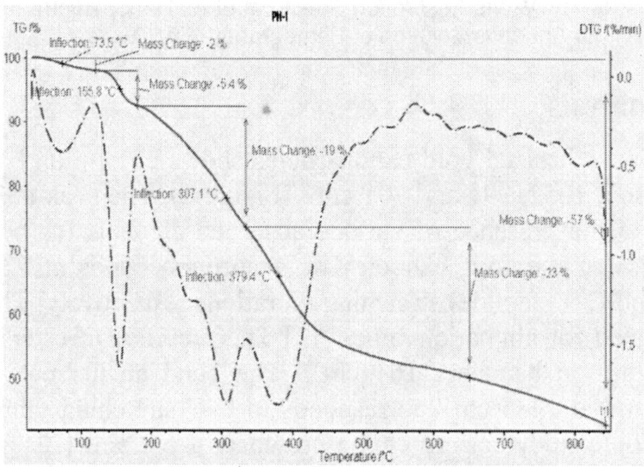

Fig. 3: TGA Thermogram of Gold Nanoparticles

HRTEM Analysis

The morphology and size of the gold nanoparticles were examined by HRTEM analysis. Figure 4 shows that most of the nanoparticles are nearly spherical with size as small as 0.94 nm to 1.20 nm and also shows the crystalline nature of the gold nanoparticles by the selected area electron diffraction (SAED) pattern.

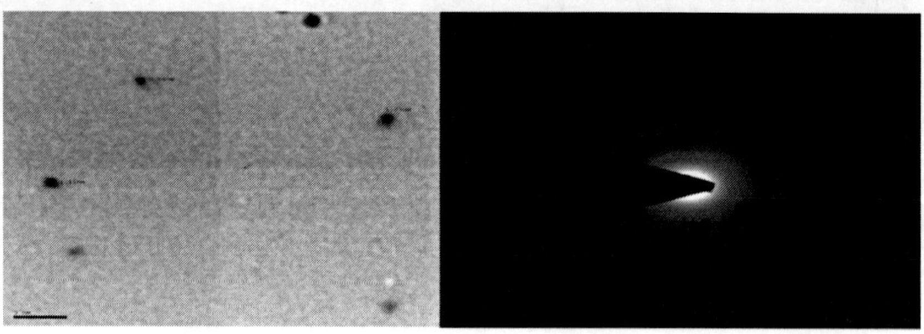

Fig. 4: HRTEM Micrograph and SAED Pattern of Gold Nanoparticles

Antimicrobial Activities of Gold Nanoparticles

Minimum inhibitory concentration of gold nanoparticles was conducted to investigate the antimicrobial activity of gold nano particles synthesized using pomegranate extract against bacterial strains such as *Staphylococcus aureus*, *Citrobacter koseri*, *Bacillus cereus*, *Pseudomonas aeruginosa*, *Escherichia coli* and against the fungus *Candida albicans*. The MIC of gold nanoparticles against various microorganisms was clearly compared by Figure 5. It showed the higher activity against *Pseudomonas aeruginosa* (MIC: 0.07825 mg/ml), followed by *Bacillus cereus* (MIC: 0.15625 mg/ml), *Staphylococcus aureus* (MIC: 0.3125 mg/ml), *Escherichia coli* (MIC: 0.3125 mg/ml) and *Citrobacter koseri* (MIC: 1.25) mg/ml).[14] The nanoparticle did not show considerable antifungal activity against *Candida albicans*.

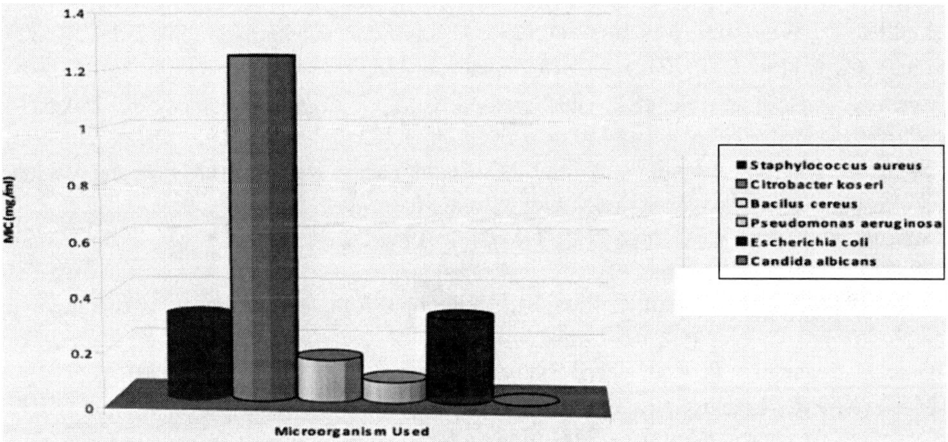

Fig. 5: Determination of MIC for Nano Au Synthesized Using *Punica Granatum*

CONCLUSION

In this paper, we reported the extraordinary and fastest bio-reducing method for synthesizing gold nanoparticles using pomegranate fruit extract. These biologically synthesized gold nanoparticles play a crucial role in protecting our environment as green. UV-Vis spectroscopy reveals the surface plasmon property, while HRTEM images reveal the nano nature of the prepared samples with a diameter of 0.9–1.2 nm. The stability of gold nanoparticles confirmed by UV-Vis and HRTEM analyses after one month. It is proven that the gold nanoparticles synthesized from pomegranate extract seem to be promising and effective antibacterial agent against the multidrug resistant strains of various bacterias.

ACKNOWLEDGEMENT

We gratefully acknowledge, Centre for Nanoscience and Nanotechnology, University of Madras for HRTEM analysis.

REFERENCES

[1] Nie, S. and Emory S.R., *Science,* 275, 1102–1106, 1997.

[2] Mukherjee, P., Ahmad, A., Mandal, D., Senapati, S., Sainkar, S.R., Khan, M.I., Parishcha, R., Ajaykumar, P.V., Alam, M., Kumar, R., and Sastry, M., *Nano Lett.,* 1,515–519, 2001

[3] Begum, N.A., Mandal, S., Basu, S., Laskar, R.A., and Mandal, D., *Colloids Surf. B Biointer.,* 71,113–118, 2009.

[4] Bar, II., Bhui, D.K., Sahoo, G.P., Sarkar, P., De, S.P., and Misra, A., Colloids Surf. A. Physicochem. Eng. Aspects, 339,134–139, 2009.

[5] Parasher, V., Parasher, R., Sharma, B., and Pandey, A.C., *Digest J. Nanomater. Biostruct,* 4, 45–50, 2009.

[6] Yu, D.G., *Colloids Surf. B*, 59, 171–178, 2007.

[7] Mallick, K., Witcombb, M.J. and Scurrella, M.S., *Mater. Chem. Phys.,* 90, 221–224, 2005.

[8] Liu, Y.C., and Lin, L.H., *Electrochem. Commun.*, 6, 1163–1168,2004.

[9] Smetuna, A.B., Klabunde, K.J., and Sorensea, C.M., *J. Colloid Interface Sci.*, 284 521–526, 2005.

[10] Mohamed Fayaz, A., Balaji, K., Girilal, M., Ruchi Yadav, Kalaichelvan, P.T. and Venkatesan, R., *Nanomedicine: Nanotechnology, Biology and Medicine*, 6, 103–109, 2010.

[11] Sun, R .W., Chen, R., Chung, N.P., Ho, C.M., Lin, C.L. and Che, C.M., *Chem. Commun.,* 40, 5059–5061, 2005.

[12] Sathyavathi, R., Balamurali Krishna, M., Venugopal Rao, S., Saritha, R., and Narayana, D., Rao, *Adv. Sci. Lett*., 3, 1–6, 2010.

[13] Gole, A., Chaudhari, P., Kaur, J. and Sastry, M., *Langmuir*, 17, 5646–5656, 2001.

[14] Morones, J.R., Elechiguerra, J.L., Camacho, A., Holt, K., Kouri, J.B., Ramirez, J.T. and Yacaman, M.J., *Nanotech*, 16, 2346–2353, 2005.

MODIFICATION OF CARBON NANOTUBES FOR BIO-APPLICATIONS AND TOXICITY EVALUATION

P. KHALID, M.A. HUSSAIN[1], P.D. REKHA[2], V.B. SUMAN[3] AND A.B. ARUN[2]

P. A. College of Engineering, Mangalore, VTU, India
[1]King Abdulaziz University, Jeddah, KSA
[2]Yenepoya Research Centre, Yenepoya University, Mangalore
[3]Kasturba Medical College, Mangalore, Manipal University, India
E-mail: bhagwatharun@hotmail.com; dydirectoryrc@yenepoya.edu.in

ABSTRACT

Multi Walled Carbon Nanotubes (MWNTs) were produced by Chemical Vapour Deposition (CVD) method. Covalent functionalization was accomplished with three types of oxidants: (1) 70% Nitric acid under reflux (2) Nitric acid and sulfuric acid (v/v = 3/1) under sonication (3) 20% Nitric acid and sulfuric acid by heating. The non-covalent functionalization was accomplished via di-imide activated amidation using N-(3-Dimethylaminopropyl)-N'-ethylcarbodiimide hydrochloride (EDAC) and Bovine Serum Albumin (BSA). The covalent and Non-Covalent functionalized Carbon Nanotubes were readily dispersed in water, DMF and cell culture media to investigate their toxicity. Characterization of functionalized and dispersed Nanotubes were carried out by using FTIR, FESEM, UV-vis spectroscopy and light microscopy. The results revealed that different functional groups could be introduced when the Nanotubes are treated with different oxidants. Refluxing in dilute nitric acid can be considered as a mild oxidation for MWNTs, introducing the carboxylic acid groups only at those initial defects that already exist. The abundance of the carboxylic acid groups generated with this oxidant remained constant along with time. In contrast, sonication of SWNTs in H_2SO_4/HNO_3 increased the incidence of carboxylic acid groups not only at initial defect sites but also at newly created defect sites along the walls of MWNTs. The carboxylic acid groups generated with such type of oxidants decreased along with time. Similar type of result was obtained when MWNTs were heated at 100°C with diluted H_2SO_4/HNO_3.

V. Rajendran, P. Prabu and K.E. Geckeler (eds.)
Nano Biomaterials, pp. 125–130 (2012).

INTRODUCTION

Due to their exceptional electronic, optical, mechanical and chemical properties,[2] Carbon Nanotubes has become the first choice for various applications, e.g. nanocomposites, nanoelectronics, field emission devices, probe tips, optical filters and biomedical applications like bio-sensors, bio-delivery, bio-imaging, tissue-engineering.[1] Functionalization and dispersion of these nanotubes is essential for their processing in various applications. They tend to form bundles and agglomerates in solution because of the van der Waals attraction forces between the nanotubes. Various covalent [3] and non-covalent [4] modifications has been applied for functionalization and dispersion of the nanotubes, both in organic and highly polar solvents. The covalent approaches include controlled oxidation with strong acid,[5] organic peroxides,[6] cyclo addition of nitrines,[7] diazonium reaction,[8] and ball-milling.[9] The covalent functionalization alters the electrical as well as chemical properties of nanotubes. The non-covalent approach includes dispersion of nanotubes in sodium dodecyl sulphate (SDS), sodium dodecyl benzene sulphate (SDBS),[10] Tween surfactant,[11] dimethylformamode (DMF) [12] and Bovine Serum Albumin.[13] The non-covalent functionalization does not alter the internal electronic structure of the nanotubes. Based on the above background information, this study was focused on covalent and non-covalent functionalization and characterization of MWCNT using FTIR, FESEM, UV-vis spectroscopy and light microscopy.

EXPERIMENTAL

The Multiwalled Carbon Nanotubes (> 95 wt%) were obtained from Nanoshel USA. In the first set of experiment, the pristine Multiwalled Carbon Nanotubes (P-MWCNT) 2.5 mg/ml were refluxed under 70% HNO_3 at 120°C in oil bath for 6h (R6-MWCNT), the refluxed MWCNT 10 mg/40 ml were then sonicated under acid mixture (3:1 ratio of H_2SO_4 and HNO_3) in bath sonicator (350 W, 40 KHz) for 3 h (R6S3-MWCNT). In a separate reaction, the P-MWCNT was sonicated under acid mixture (3:1 ratio of H_2SO_4 and HNO_3) in bath Sonicator (350 W, 40 KHz) for 3h. In the second set of experiment, the Multiwalled Carbon Naotubes 100mg/60ml were heated at 100°C in 3:1 ratio of 20% H_2SO_4 and 20% HNO_3. The sample is drawn at 10, 20, 40, 60, 80 and 100 min. and named as H10-MWCNT, H20-MWCNT, H40-MWCNT, H60-MWCNT, H80-MWCNT and H100-MWCNT. In another set of experiment, the MWCNT were functionalized using BSA and EDAC [13] and named as PB-MWCNT. The H-MWCNT are actually H60-MWCNT which is heated for 40-60 min with stirring. All these treated MWCNTs were washed and dried.[9] The appended functional group on MWCNTs was investigated by Fourier Transform Infrared (FTIR) spectroscopy using KBr pellet. The morphology of the MWCNTs were studied by using Field Emission Scanning (FESEM) Electron Microscopy on carbon tape with a thin layer of gold coating.

RESULTS AND DISCUSSION

FTIR spectra of reflux treated multi-walled carbon nanotubes for 6h R6-MWCNT shows that the nanotubes are not appended well with the functional groups. It is almost similar to pristine multi-walled carbon nanotubes P-MWCNT Figure 2. At the same time SEM image also confirms that the R6-MWCNT and P-MWCNT are physically similar except that some breakage in the R6-MWCNT Figure 1. When the 6h reflux treated MWCNT were sonicated for 3h R6S3-MWCNT, the FTIR shows the attachments of functional group Figure 2, but SEM

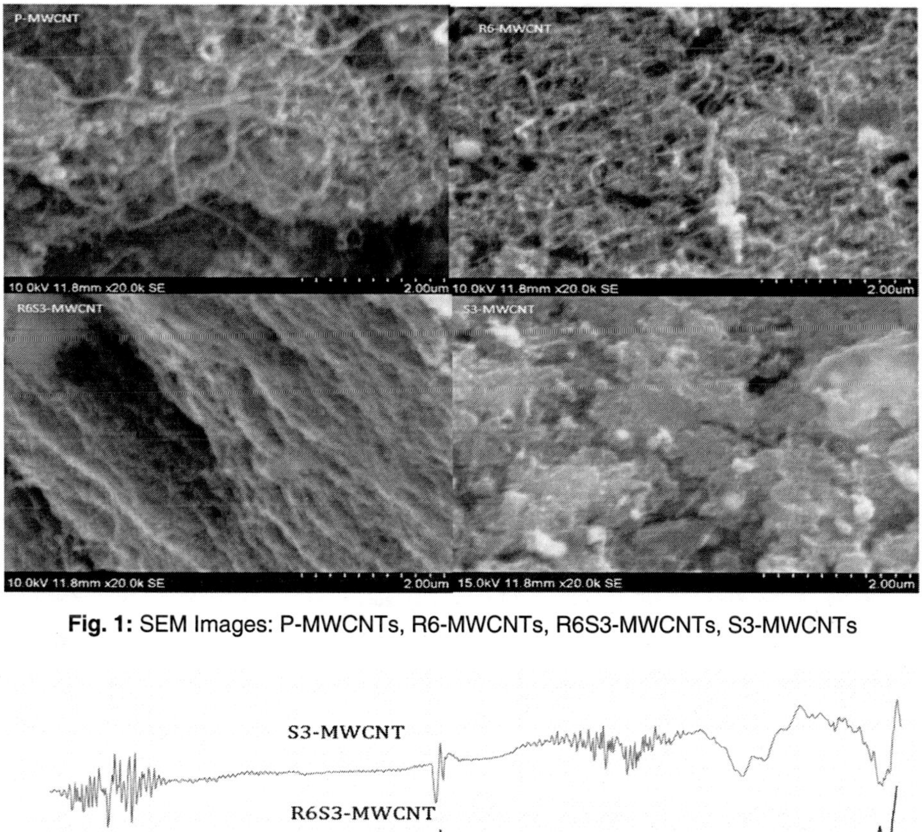

Fig. 1: SEM Images: P-MWCNTs, R6-MWCNTs, R6S3-MWCNTs, S3-MWCNTs

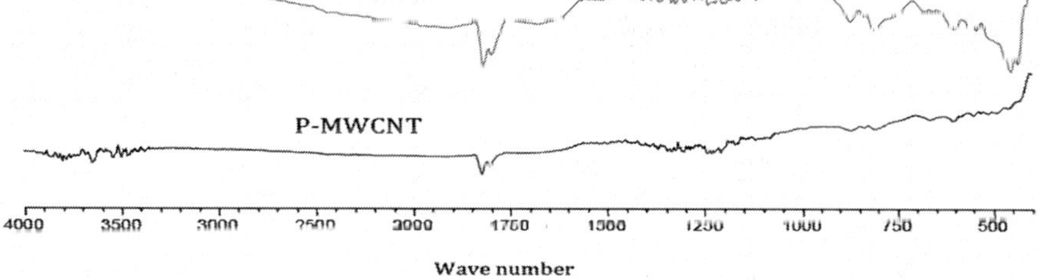

Fig. 2: FTIR Spectra: P-MWCNTs, R6-MWCNTs, R6S3-MWCNTs, S3-MWCNTs

image shows the change in the morphology of the nanotubes Figure 1. The nanotubes started aligning together. When the muti-walled carbon nanotubes are sonicated for 3h S3-MWCNT, the FTIR shows that the functional groups are appended Figure 2, but SEM images revealed that there is a complete damage of the tubes Figure 1. Figure 3 shows 60 min of heating H60-MWCNT result in well modification of the tubes, more number of functional groups appended to the nanotubes. Whereas heating for 80 min and 100 min caused the reduction of functional group attachments. That may be because of the reduction of surface area of the nanotubes due to breakage. The separation and dispersion of nanotubes primarily depend on the carbon nanotubes functionalization and dispersion methods adopted. Thin-long and curved structure and van der Waals forces are the factors result in the formation of nanotubes agglomerate. Figure 4 shows the di-imidation can well modify the MWCNTs which is readily soluble in water.

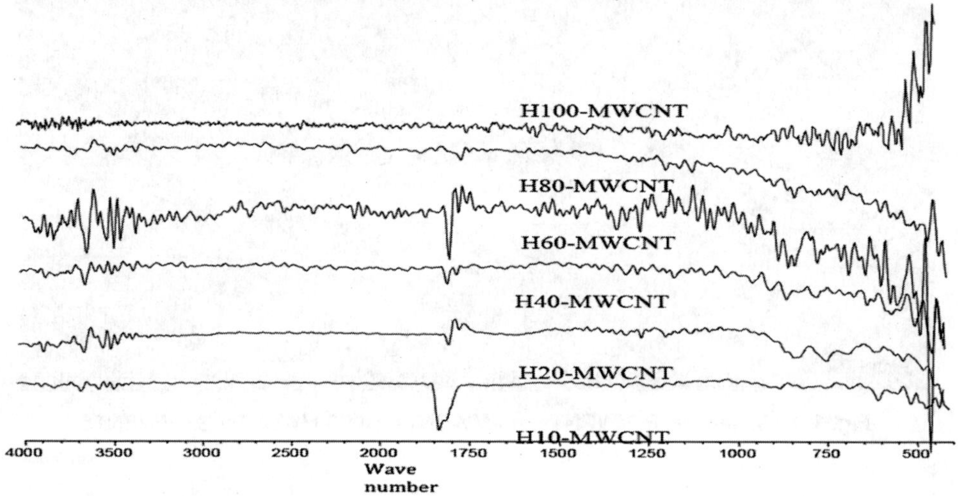

Fig. 3: FTIR Spectra: H10-MWCNTs, H20-MWCNTs, H40-MWCNTs, H60-MWCNTs, H80-MWCNTs, H100-MWCNTs

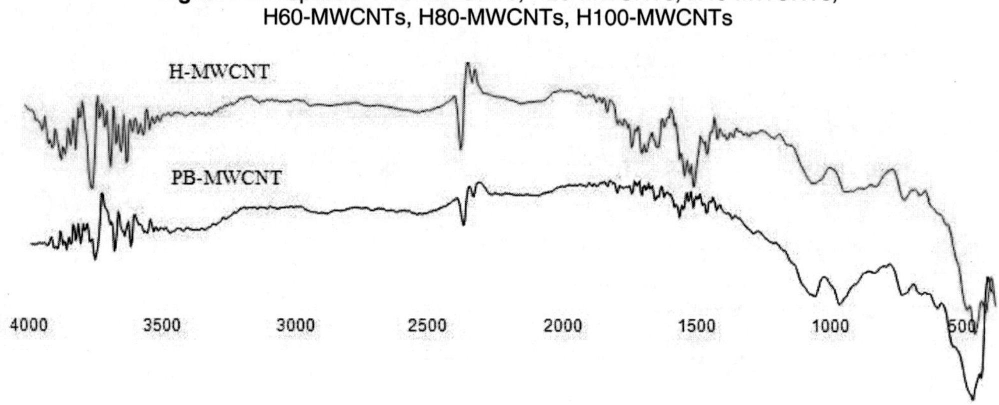

Fig. 4: FTIR Spectra of H-MWCNT and PB-MWCNT

CONCLUSIONS

The pristine carbon nanotubes entangled into aggregates of up to a few tens of mm. Entangled CNTs are difficult to disperse without damage but damage can be minimized by mild treatment. Reflux treatment of MWCNTs in 70% HNO_3 at 120°C can be considered as the mild oxidative treatment with less damage. On the other hand, 60 min. heating of the MWCNTs at 100°C in 3:1 ratio of 20% H_2SO_4 and 20% HNO_3 can also be the alternative option for the mild treatment of carbon nanotubes. Even one more alternative method we have tested for dispersion of MWCNTs called as di-imide activated amidation. We found that this method is better than the above mentioned treatment methods. Di-imidation of MWCNTs leads to better modifications and well dispersion in water. Since these modified carbon nanotubes are readily dispersed in water, and the culture media, after suitable modifications can be used for their toxicity testing and other bio-applications.

REFERENCES

[1] Niraj, S., John and Yeow, T.W., Carbon Nanotubes for Biomedical Applications, *IEEE Trans. Nanobiosci*, 4, 180–195, 2005.

[2] Dresselhaus, M.S., Dresselhaus, G. and Avouris, P., Carbon Nanotubes: Synthesis, Properties and Applications, *Springer-Verlag*, Berlin, 2001.

[3] Christopher, A.D. and James M.T., Functionalization of Single Walled Carbon Nanotubes for Material Applications, *J. Phys. Chem.* 108, 51, 2004.

[4] Robert, J.C., Sarunya, B., Katerina, A.D., Nadine, W.S.K., Moonsub, S. *et al.,* Noncovalent Functionalization of Carbon Nanotubes for Highly Specific Electronic Biosensors, *PNAS,* 100, 4984–4989, 2003.

[5] Jin, Z., Hogling, Z., Quan, Q., Yanlian, Y., Qingwen, L. *et al.,* Effect of Chemical Oxidation on the Structure of Single Walled Carbon Nanotubes, *J. Phys. Chem.* 107, 3712–3718, 2003.

[6] Datsyuk, V., Kalyva, M., Papagelis, K., Parthenios, Tasis, D. *et al.,* Chemical Oxidation of Multiwalled Carbon Nanotubes, *Carbon*, 46, 833–840, 2008.

[7] Christopher, A.D. and James M.T., Solvent Free Functionalization of Carbon Nanotubes, *J. Am. Chem. Soc.,* 125, 1156–1157, 2003.

[8] Christopher, A.D., Michael, P.S., Francisco, M. and James, M.T., Diazonium Based Functionalization of Carbon Nanotubes. XPS and GC-MS Analysis and Mechanistic Implications, *Temp. Synlett. Synth*, 1–6, 2003.

[9] Yao, W., Jun, W. and Fei, W., A Treatment Method to give Separated Multi Walled Carbon Nanotubes with High Purity, High Crystallization and Large Aspect Ratio, *Carbon*, 41, 2939–2948, 2003.

[10] Bystrzejewski, M., Huczko, A., Lange, H., Gemming, T. and Buchner, B., Dispersion and Diameter Separation of Multi walled Carbon Nanotubes in Aqueous Solution, *J. Coll. Interface Sci.*, 345, 138–142, 2010.

[11] Valerie, C.M., Michael, S.S., Erik, H.H., Robert, H.H. and Richard, E.S., Individually Suspended Single Walled Carbon Nanotubes in Various Surfactants, *Nano Lett.,* 3, 1379–1382, 2003.

[12] Nguyen, T.T., Nguyen, S.U., Phuong, D.T., Nguyen, D.C. and Mai, A.T., Dispersion of Denatured Carbon Nanotubes by using Dimethylformamide Solution, *Adv. Nat. Sci.: Nanosci. Nanotechnol.*, 2, 035015, 2011.

[13] Weijie, H., Shelby, T., Kefu, F., Yi, L., Donghui, Z. *et al.,* Attaching Proteins to Carbon Nanotubes via Diimide Activated Amidation, *Nano Lett.,* 2, 311–314, 2002.

GROWTH AND STRUCTURAL STUDY OF CADMIUM ARACHIDATE MULTILAYERS ON SOLID SUBSTRATE LIKE SILICON AND GLASS SUBSTRATE BY LANGMUIR BLODGETT TECHNIQUE

DHRUBOJYOTI ROY, NAYAN MANI DAS AND P.S. GUPTA

Department of Applied Physics, Indian School of Mines, Dhanbad
E-mail: dhrubojyoti27@gmail.com

ABSTRACT

Different Langmuir–Blodgett (LB) multilayer films of Cadmium Arachidate (CdAA) deposited at normal subphase pH (6.5) on solid substrate like hydrophilic Glass (SiO_2) and Silicon(100)(Si) substrate have been studied by X-ray diffraction and X-ray reflectivity techniques which reveals ordered layer by layer deposition with uniform packing of CdAA molecules, and with a small tilt angle of long alkyl chain of CdAA molecule equal to 6.8°. Electron density profiles (EDP) reveals that the coverage of films remains almost constant with increase in bilayer thickness which indicate very little presence of pinhole defects. Fourier transform infrared (FTIR) studies are also consistent with XRD and XRR studies of ordered deposition of CdAA molecule. FTIR studies confirms orthorhombic packing in CdAA LB films and the formation of unidentate bridging metal-carboxylate coordination type headgroups consistent with one cadmium metal ion between two carboxylic (COO) groups in each headgroup.

INTRODUCTION

Langmuir- Blodgett (LB) technique, has been used extensively to prepare highly ordered thin films of various organic materials having amphiphilic property, because of its great efficiency in controlling film thickness and uniformity of the ordered multilayers.[1,2] Hence LB films have a good possibility to be artificial molecular assemblies with planned structure and

V. Rajendran, P. Prabu and K.E. Geckeler (eds.)
Nano Biomaterials, pp. 131–136 (2012).

properties. In multilayers, the alkane chain and carboxylic acid moieties of fatty acids are arranged in regular two-dimensional layers and the number of layer can be controlled at any desired value. Because of this layered and molecular ordered, LB multilayers are suitable precursor for the growth of nanoparticles, nanowires and microparticles on solid substrate having better control over the size, shape and distribution of them.[3]

In recent years, metal-organic multilayers such as Langmuir-blodgett films has gain much importance among the researchers because of their potential applicability in molecular electronics, non-linear optics, biosensors and as a templates for growing functional nanomaterials like quantum dot.[4–5] This is because, incorporation of different divalent cations of transition metals having considerable electronegativity, such as Cd2+, Ca2+,Mn2+, Zn2+, Ba2+, etc.) into their headgroup structure of long-chain fatty acids plays an important role in forming stable multilayer LB films of fatty acid and hence increases its ease of deposition on solid substrate, which led to better ordered organization of LB multilayers on solid substrates.

In the present work, different LB multilayer films of Cadmium Arachidate(CdAA) has been synthesized by inducing divalent cation Cd2+ of transition metals into the arachidic acid at pH~6.5 of water subphase, as the relative fractions of free acid and their divalent metal salt are sensitive functions of pH.[6] The growth and interfacial structure of this LB multilayer films have been characterized by X-ray diffraction and X-ray reflection technique. FTIR spectroscopy is used to study the defects in the LB films and hence helps in determining the conformational orders in the LB monolayers.

EXPERIMENTAL

3, 5 and 9 monolayer (ML) LB films of cadmium arachidate (CdAA) were prepared by the LB deposition technique using APEX INSTRUMENTS CO. Arachidic acid ($CH_3(CH_2)_{18}$ COOH) monolayer was spread from a 1 mg/ml chloroform solution onto the Langmuir trough containing 0.5 mM cadmium chloride salt (Merck, 98%) dissolve in Milli-Q water (resistivity 18.2 MΩ cm). Water subphase pH was maintained at 6.5 using sodium bicarbonate at a temp of about 25°C. Subsequently the monolayer was compressed slowly at a barrier speed of 3mm/min to the required transfer pressure of ~30 mN/m. The monolayer was then transferred onto the hydrophilic silicon and glass substrate by subsequent up-down strokes of the substrate through the air-water interface at a 3mm/min dipping speed. The substrates were made hydrophilic before deposition. The glass substrates were made hydrophilic by placing them inside the degreasing unit filled with iso-propanol. Similarly silicon substrates were made hydrophilic by treating with a solution containing ammonium hydroxide (Rankem, Analytical Reagent), hydrogen peroxide (Qualigens), and Milli-Q water ($H_20:NH_4OH:H_2O_2$ in ratio 2:1:1 by volume) for about 10–15 minutes at 80°C. Drying time for the each upstroke, i.e. above subphase was kept 10 min and similarly for downstroke, i.e. below subphase was 5 min. The arachidic acid LB film showed good reproducibility.

For required structural study of the films X-ray diffraction studies were performed using Bruker D8 Advance X-ray diffractometer which uses Cu-K_α radiation ($\lambda = 1.543$Å). FTIR Study was performed in absorption mode using Perkin Elmer D850 Instrument on silicon (100) substrate. XRR study was performed using Bruker D8 Discover Diffractometer using Cu-K_α wavelength for measuring film thickness and Electron Density Profile.

RESULTS AND DISCUSSION

X-ray Diffraction Study

The XRD measurements, shown in Figure 1, shows that as the number of monolayers increases, the intensity as well as the prominent (00*l*) Bragg's peaks number goes on increasing which indicates the formation of regular ordered layered structure. The other lower number stacks of multilayers i.e for 3ML and 5ML showed reduced intensity of peaks as well as lesser count of peak numbers. The average bilayer period of CdAA molecule calculated from the XRD graph using Bragg's equation modified for refraction [7] and is found to be 5.5 nm which corresponds to alkyl chains packed nearly perpendicular to the glass substrate. The typical length of CdAA molecule is 5.54 nm, therefore the alkyl chain tilt of the CdAA molecule in the multilayered stack on the glass substrate can be calculated from the formulae, $\cos\theta = d/d_o$, where d is the measured bilayer length and d_o is the typical length of CdAA molecule. And the alkyl chain tilt found to be ~6.8°.

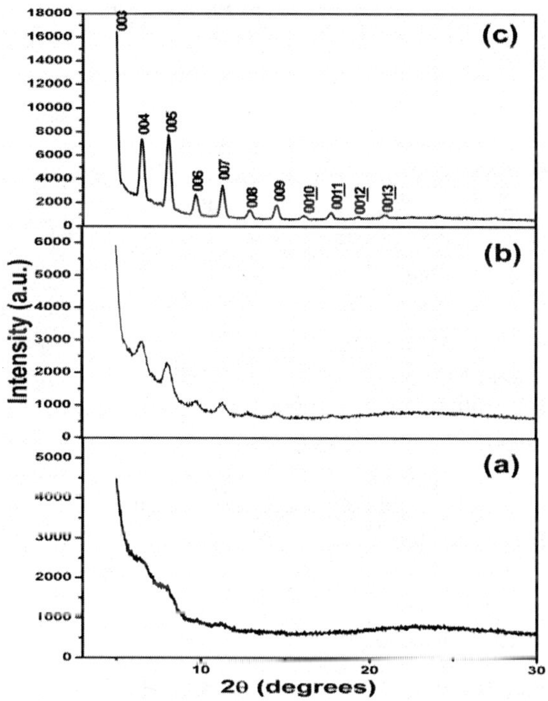

Fig. 1: X-ray Diffraction Pattern from Cadmium Arachidate (CdAA) LB Film with Varying Multilayer Deposition (a) 3ML, (b) 5 ML and (c) 9 ML

FTIR Study

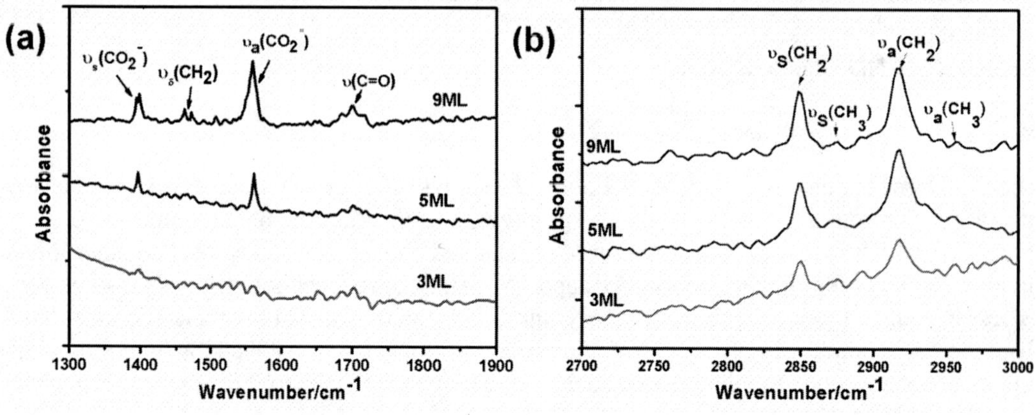

Fig. 2: FTIR Absorbance Spectrum of 3ML, 5ML and 9ML LB film of Cadmium Arachidate on Silicon Substrate from Region (a) 1300 cm^{-1} to 1800 cm^{-1}, and (b) 2700 cm^{-1} to 3000 cm^{-1}

The FTIR absorption spectrum obtained from 3ML, 5ML and 9 ML LB film of CdAA deposited on the silicon(100) substrate is shown in Figure 2 (a and b). The peaks at 1560 cm^{-1} and 1397 cm^{-1} corresponds to carboxylate asymmetric $\nu_a(CO_2^-)$ and symmetric $\nu_s(CO_2^-)$ stretch modes,[8] which indicates the large conversion of fatty acid to its metal salt. The coordination of metal ion with carboxylate group is measured from the difference in $\nu_s(CO_2^-)$ and $\nu_a(CO_2^-)$ stretching wavenumber ($\Delta\nu$). Therefore $\Delta\nu = (\nu_s - \nu_a) = (1560–1397) = 163$ cm^{-1} is ascribed to unidentate coordination, having one metal ion per carboxylate group.[9] The presence of peaks at 1462 cm^{-1} and 1471 cm^{-1} corresponding to $\delta(CH_2)$ scissoring band progression confirms orthorhombic packing in CdAA LB films.[10]

Figure 2(b) shows region corresponding to the CH_2 and CH_3 stretching. The strong bands at 2919 cm^{-1} and 2850 cm^{-1} is assigned to asymmetric $\nu_a(CH_2)$ and symmetric $\nu_s(CH_2)$ stretching vibrations of CH_2 (methylene) group respectively. The other two weak bands at about 2960 cm^{-1} and 2875 cm^{-1} assigned to the symmetric (degenerate) $\nu_a(CH_3)$ and asymmetric $\nu_s(CH_3)$ stretching vibration of the methyl group respectively. From Figure 2(a), it is also noticed that as the film thickness of CdAA increases from 3 ML to 9 ML on silicon substrate the absorption band intensity corresponding to $\nu_s(CH_2)$ and $\nu_a(CH_2)$ methelyne group also increases. Therefore, it can be concluded that the molecular orientation of alkyl chain of CdAA molecule remains unchanged at the silicon substrate with increase in layer numbers, which tells about the perfect layer by layer deposition of CdAA film by LB Technique.

X-ray Reflection Study

The X-Ray Reflectivity of 9 ML LB films of CdAA in the Figure 3 shows prominent and intense Bragg's which indicates uniform layer by layer stacking of long chain of cadmium arachidate molecules upon the glass substrate which lead to the formation of well-defined layered structures and good out of plane crystalline growth along the (00*l*) direction, which is the growth direction of the CdAA multilayer film as interpreted from the XRD results. The reason for good multilayer stacking and growth of CdAA molecule is because of good coherence between air/film and film/substrate interfaces, with minimal presence of pinhole defects.

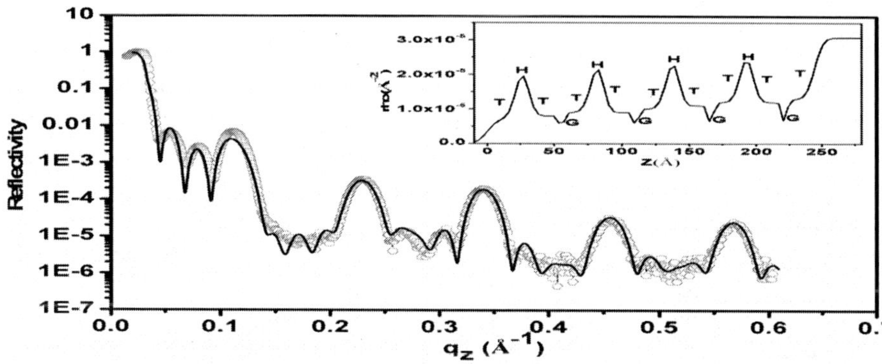

Fig. 3: X-ray Reflectivity Data (open circles) with Corresponding Fits (solid lines), of the Deposited 9 ML LB Films of CdAA on Glass Substrate

The XRR data is analysed with help of Parratt software. The solid lines are fitted curve. In the inset of Figure 4, the Electron Density Profiles (EDPs) of respective films of CdAA are shown. The EDPs (ρ-electron density Vs z-film thickness graph) of the respective films have almost a constant ρ with increase in bilayer thickness which shows no such decrease in coverage of film with increase of bilayer thickness. This is because of presence of few pinhole defects. From the data of fitted curve, the total thickness of CdAA calculated to be 247.3Å. The average thickness of the CdAA molecule comes out to be 27.48 Å. Using the relation of $d/d_0 = \cos(\theta)$ the tilt angle of the molecule with respect to the normal on the substrate is calculated to be 7.2°, which is almost similar to the tilt angle calculated from XRD Graph.

CONCLUSION

XRR and XRD results clearly indicates well ordered packing of CdAA molecules on silicon (100) and glass substrate which lead to the formation of uniform multilayered structured of CdAA, with excellent out-of-plane crystalline growth along (00*l*). Also shows alkyl chain of CdAA molecule having small tilt of 6.8° with normal to the substrate. FTIR study justify the uniform multilayer growth. It also suggest that the CdAA multilayer is having unidentate bridging metal-carboxylate coordination type headgroups consistent with one cadmium metal ion between two carboxylic (COO) groups in each headgroup.

ACKNOWLEDGEMENTS

We gratefully thank Dr. Mukul Gupta for his support in taking XRD measurements and and Dr. V.R. Reddy for providing XRR facility.We acknowledge UGC-DAE CSR, Indore for giving opportunity to carry out characterization of the samples under the CRS project. Two of the authors thank UGC for financial support for carrying out the research work.

REFERENCES

[1] Petty, M.C., Langmuir–Blodgett Films: An Introduction, Cambridge University Press, Cambridge, 1996.

[2] Blodgett, K.B., Films Built by Depositing Successive Monomolecular Layers on a Solid Surface. *J. Am. Chem. Soc*, 57, 1007, 1935.

[3] Parhizkar, M., Kumar, N., Nayak, P.K., Singh, S., Talwar, S.S., Major, S.S. and Srinivasa R.S., Nanocrystalline ZnO films prepared by pyrolysis of Zn-arachidate LB multilayers, *Colloids and Surface A: Physicochem. Eng. Aspects*, 257–258, 445–449, 2005.

[4] Nabok, A., Ray, Asim K., Iwantono, Hassan, A., Simmonds, M., Optical Study on II–VI Semiconductor Nanoparticles in Langmuir–Blodgett Films, *IEEE Transactions on Nanotechnology*, Vol. 2, No. 1, 2003.

[5] Azam, A., Muhamed, S.M., Ahmed, A.S., Singla, M.L. and Naqvi, A.H., The kinetics of Langmuir–Blodgett ZnO nano-film deposition for sensing electrodes, *Materials Letters*, 65, 2721–2723, 2011.

[6] Kundu, S., Datta, A. and Hazra, S., Manipulating headgroups in Langmuir-Blodgett films through subphase pH variation, *Chemical Physics Letters,* 405, 282–287, 2005.

[7] Agarwal, B.K., X-ray Spectroscopy, Springer: Berlin, 2005.

[8] Kimura, F., Umemura, J. and Takenaka, T., FTIR-ATR studies on Langmuir-Blodgett films of stearic acid with 1–9 monolayers, *Langmuir*, 2, 96, 1986.

[9] Nakamoto, K., Infrared and Raman Spectra of Inorganic and Coordination Compounds, Wiley and Sons, New York, 1970.

[10] Rabolt, J.F., Burns, F.C., Schlotter, N.E. and Swalen, J.D., Anisotropic orientation in molecular monolayers by infrared spectroscopy, *J. Chem. Phys,* 78, 946, 1983.

INTERGROWN AND MICROPOROUS ZEOLITE-A CRYSTALS TO FORM MEMBRANE LAYER ON POLYSULFONE HOLLOW FIBER SUPPORT

SANJAY G. CHAUDHRI, ANKIT M. KANSARA AND PUYAM S. SINGH

CSIR-Central Salt and Marine Chemicals Research Institute, RO Membrane Discipline, G.B. Marg, Bhavnagar, Gujarat
E-mail: puyam@csmcri.org

ABSTRACT

We report preparation of zeolite A membrane coated on hollow fiber polymeric support by in situ crystallization of Zeolite-A on the fiber surface using a clear liquid synthesis mixture composition $1SiO_2$:$0.15Al_2O_3$:$9Na_2O$:$200H_2O$. The hydrothermal crystallization was carried out at 60°C for 15 hrs. A layer thickness of about 3–4 μm comprised of zeolite A crystal with globular shape was obtained. It was observed that the coated zeolitic-like material as layer still consist of some amorphous evident from the globular shapes rather than cubic faces of zeolite A crystals. The layer comprised of zeolite particles on a polysulfone substrate did not make a compact, interconnected zeolite film because of the intercrystalline porosity (open space between the particles) which must be filled in order to form a continuous zeolite A layer. On the other hand, prolong hydrothermal reaction time increased the formation of other zeolitic phases. When it was subjected to repeated hydrothermal crystallization with the same condition as above, an intergrown and ordered zeolite-A layer over the fiber was obtained.

INTRODUCTION

Zeolite membrane unlike polymeric membrane can withstand extreme conditions such as high temperature, strong acidic, alkaline and chemical solvents. Further, if the zeolite membrane can be made using cheaper polymeric support, it will extend its merits. Normally zeolite membrane is coated on expensive inorganic support like ceramic, α-Al_2O_3, silica, etc. The

V. Rajendran, P. Prabu and K.E. Geckeler (eds.)
Nano Biomaterials, pp. 137–142 (2012).

problem with the polymeric support for zeolite membrane is the temperature instability as normally the zeolite is calcined over 400–500°C to remove organic which are filled in zeolite pores. Fortunately some zeolites, such as, zeolite-A can be prepared without any structure directing organic template. Therefore for such zeolite membrane, calcination at high temperature is not required. Zeolite-A membrane coated on inorganic support like ceramic, α-Al_2O_3, silica and stainless steel using different techniques have been widely studied for various separation applications.[1–11]

We prepare organic free zeolite A membrane coated on hollow fiber polymeric support. The zeolite A membrane was prepared by in situ crystallization of Zeolite-A on polysulfone hollow fiber using a clear liquid synthesis mixture having composition $1SiO_2$:$0.15Al_2O_3$: $9Na_2O$:$200H_2O$.[12] The hydrothermal crystallization was carried out at 60°C for 15 hrs. The layer comprised of zeolite particles on a polysulfone substrate did not make a compact, interconnected zeolite film because of the intercrystalline porosity (open space between the particles) which must be filled in order to form a continuous zeolite A layer. On the other hand, prolong hydrothermal reaction time increased the formation of other zeolitic phases. To prepare continuous, intergrown and highly ordered film it was subjected to secondary growth with same condition mention above using freshly prepared synthesis mixture in which presence of seeds provides for direct growth through local epitaxy or re-nucleation on the precursor films, but also allows for incorporation of crystals formed in solution leading to high coverage.[2] The zeolite film coated on hollow fiber, further subjected to crystallization for the third time and fourth time with same condition using freshly prepared synthesis mixture for the comparison studies.

EXPERIMENTAL

Polysulfone (Udel P-3500, Solvay Advanced Polymers, USA); polyvinylpyrrolidone (MW40000, Sigma-Aldrich), dimethylformamide and hexane (Qualigens Fine chemicals, India), sodium lauryl sulfate (SD Fine Chem Limited, India) and deionized water were used for the preparation of polysulfone hollow fiber ultrafiltration substrate. A transparent polymer dope solution of polysulfone and polyvinylpyrrolidone in dimethylformamide was prepared for spinning hollow fibers. The spinning dope was extruded under nitrogen environment through a double orifice spinneret using a semi-automated spinning unit. Water was flowed through the inner nozzle of the spinneret to form the lumen by causing the phase inversion in the inner part of the polysulfone hollow fiber. The fibers were washed thoroughly with water at room temperature and then preserved in deionized water containing 10% glycerol and 1% formalin. The details of the preparation procedure can be found in our patented literature.[13]

Formation of zeolite-A membrane over the hollow fiber support was performed by in-situ hydrothermal crystallization of zeolite-A over the support. A clear synthesis solution for the zeolite-A film formation in this study had molar composition $1SiO_2$: $0.15Al_2O_3$: $9Na_2O$: $200H_2O$. To prepare the clear synthesis solution appropriate amount of aluminum (S.D. Fine, India) and silica gel (S.D. Fine, India) were dissolved separately in aqueous sodium hydroxide solutions. Both solution were filtered and then mixed together before start hydrothermal synthesis. Growth of zeolite-A polycrystalline layer on the fiber support was carried out at 60°C for 15 h. The hydrothermal crystallization procedure was repeated using fresh synthesis mixture in order to develop closely packed, intergrown and highly ordered zeolite-A crystalline structure on the fiber. Surface and cross-sectional membrane microstructure morphology were performed using scanning electron microscopy (SEM) LEO1430VP scanning electron microscope operated at 5–20 kV. Samples were coated gold coating after break it in Liquid nitrogen. Energy dispersive X-ray analysis was used for obtaining chemical composition of zeolite layer. X-ray diffraction analysis of the samples were carried out using a Philips X'pert system using Cu Kα radiation.

RESULTS AND DISCUSSION

Physical structure of hollow fiber support was examined by electron microscopy and nitrogen sorption experiments (Figure 1).

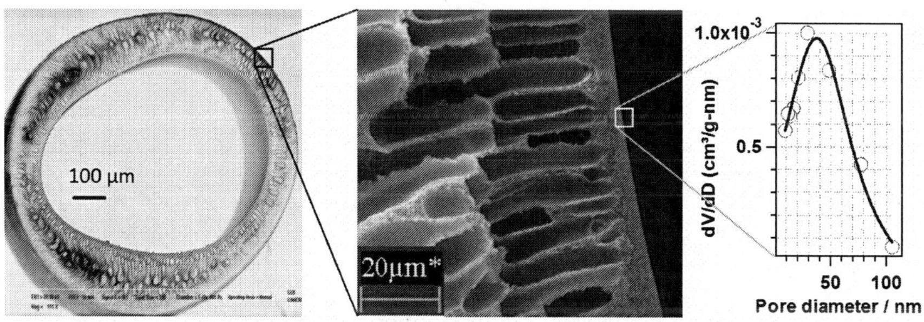

Fig. 1: SEM Images of Hollow Fiber Support and Pore Size Distribution of Surface Pores

The dimension of the hollow fiber was about 1.2 mm as outer diameter and about 0.8 mm as inner diameter. The membrane structure is asymmetric in nature as comprised of narrow pores in the range of 10–100 nm near the surface and large pores of about 1–20 μm in the interior. The large pores in the interior were directly observed by the cross-sectional SEM images whereas the pore size distributions for surface pores were estimated from the N_2 adsorption isotherm curves of the samples. The median surface pore size was determined as about 50 nm.

The zeolite A membrane was prepared by in situ crystallization of Zeolite A on polysulfone hollow fiber using a clear liquid synthesis mixture having composition $1SiO_2$:$0.15Al_2O_3$: $9Na_2O$:$200H_2O$.[1]

The hydrothermal crystallization was carried out at 60°C for 15 hrs. As shown in Figure 2, a layer thickness of about 3–5 μm comprised of zeolite A crystals with somewhat globular shape were obtained implying that the layer still consist of some amorphous particles. The layer consists of amorphous particles did not produce the interconnected zeolite film because of the open space between the particles or crystals even though the excess powder from the synthesis exhibited purely zeolite-A crystals as observed by XRD pattern (Figure 3).

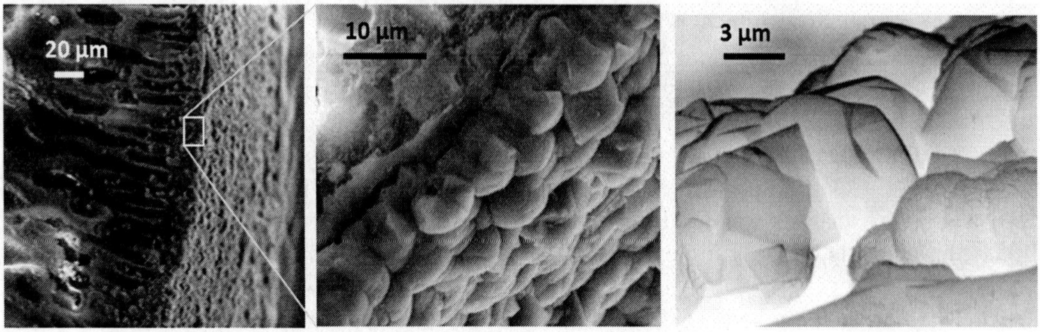

Fig. 2: SEM Images at Different Maginifications Showing the Deposition of Zeolite-A Crystals Over Fiber Support after First Hydrothermal Crystallization Process

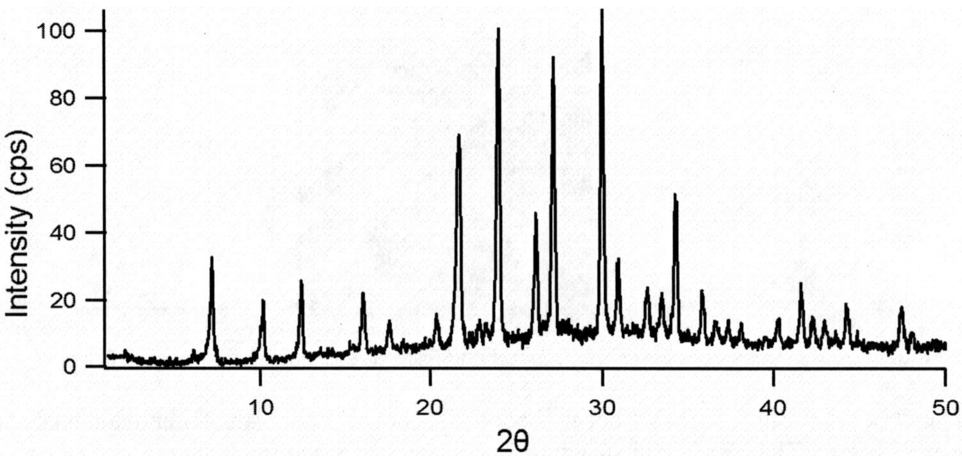

Fig. 3: XRD Pattern of Zeolite-A Structure from Excess Powder Obtained from the Synthesis

The hollow fiber supported zeolite-A membrane thus obtained was again used as support for the next hydrothermal crystallization using a freshly prepared synthesis solution in order to undergo secondary growth of the crystals and at the same interconnecting with the new crystals. The hydrothermal synthesis mixture composition and condition were the same as before. Improved morphology of zeolite-A membrane with predominant cubic like crystals was then observed. More compacted and thicker membrane layer of intergrown crystal were observed by further subjecting hydrothermal crystallization as above. The membrane thickness was increased up to about 15 μm after 5th step of hydrothermal crystallization but developed cracks indicating the limiting thickness of the membrane (Figure 4). Chemical composition of the zeolite layer was examined using EDX and found that all the membranes have similar Si/Al ratio of about 1 similar to that of zeolite-A crystals. These membranes showed potential in the separation of water and ethanol by pervaporation.

Fig. 4: Increase of Membrane Layer Thickness with Hydrothermal Crystallization Steps

CONCLUSIONS

Zeolite-A membrane supported on hollow fiber polymer support was successfully prepared by repeated hydrothermal crystallization of zeolite-A from a clear solution system on the fiber support. Initially the crystals are deposited which on subsequently undergoes secondary growth to form close packed intergrown zeolite membrane suitable for separation application. Increase of zeolite membrane thickness with successive hydrothermal crystallization step was observed. However, the membrane were prone to cracking when the thickness reaches about 15 µm indication thickness limitation.

ACKNOWLEDGEMENTS

The authors are thankful to CSIR for the grant under CSIR EMPOWER project scheme OLP-0048 for carrying out the work. We also gratefully acknowledge Mr. J. Chaudhuri, Analytical Science Division, CSIR—CSMCRI, Bhavnagar, India for his help in SEM measurements of the samples

REFERENCES

[1] Gascon, J., Kapteijn, F, Zornoza, D., Sebastian, V., Casado, C. and Coronas, J., A practical approach to zeolitic membranes and coatings: state of the art, opportunities, barriers and future perspectives. *Chem. Mater.* (in press). 2012.

[2] Jafar, J.J. and Budd, P.M., Separation of alcohol/water mixtures by pervaporation through zeolite A membranes, *Microporous Materials,* 12, 305–311. 1997.

[3] Boudreau, L.C., Kuck, J.A. and Tsapatsis, M., Deposition of oriented zeolite A films: in situ and secondary growth, *J. Membr. Sci.*, 152, 41–59, 1999.

[4] Kumakiri, I., Yamaguchi, T. and Nakao, S., Preparation of Zeolite A and Faujasite Membranes from a Clear Solution, *Ind. Eng. Chem. Res.*, 38, 4682–4688, 1999.

[5] Bronic, J., Subotic, B. and Skreblin, M., Investigation of the influence of seeding on the crystallization of zeolite A in the membrane-type reactor, *Microporous and Mesoporous Materials,* 28, 73–82, 1999.

[6] Aoki, K., Kusakabe, K. and Morooka, S., Separation of Gases with an A-Type Zeolite Membrane, *Ind. Eng. Chem. Res*., 39, 2245–2251, 2000.

[7] Ma, Y.H., Zhou, Y., Poladi, R. and Engwall, E., The synthesis and characterization of zeolite A membranes, *Sep. Purif. Tech*., 25, 235–240, 2001.

[8] Zhang, Y., Li, Y. and Zhang, Y., Supersolubility and induction of aluminosilicate nucleation from clear solution, *Journal of Crystal Growth,* 254, 156–163, 2003.

[9] Zah, J., Krieg, H.M. and Breytenbach, J.C., Layer development and growth history of polycrystalline zeolite A membranes synthesised from a clear solution, *Microporous and Mesoporous Materials,* 93, 141–150, 2006.

[10] Kuzniatsova, T.A., Mottern, M.L., Chiu, W.V., Kim, Y., Dutta, P.K. and Verweij, H., Synthesis of Thin, Oriented Zeolite A Membranes on a Macroporous Support, *Adv. Funct. Mater*., 18, 952–958, 2008.

[11] Ling, W.S., Thian, T.C. and Bhatia, S., Synthesis, characterization and pervaporation properties of microwave synthesized zeolite A membrane, *Desalination*, 277, 383–389, 2011.

[12] Singh, P.S., Dowling, T.L., Watson, J.N. and White, J.W. Light and X-ray scattering from the early growth stages of zeolite A Part I. *Phys.Chem. Chem. Phys.,* 1, 4125–4130, 1999.

[13] Reddy, A.V.R., Ray, P., Singh, P.S., Parshuram, K. and Maurya, S., High flux hollow fiber ultrafiltration membranes and process for the preparation thereof PCT patent WO 2012/035402A1, 2012.

SYNTHESIS AND CHARACTERIZATION OF PECTIN BASED SILVER NANOCOMPOSITE HYDROGELS FOR CONTROLLED RELEASE 5-FLUOROURACIL

P. Ramasubba Reddy, N. Sivagangi Reddy, S. Siva Prasad, S. Eswaramma, K.S.V. Krishna Rao and Yong Ill Lee[1]

Department of Chemistry, Yogi Vemana University, Kadapa, India
[1]Department of Chemistry, Changwon National University, Changwon, South Korea
E-mail: drksvkrishna@yahoo.com

ABSTRACT

In this present contribution we developed a novel silver nanocomposite hydrogels. These are composed from poly(acrylamide-co-N-vinylcaprolactam-co-acrylamidoglycolicacid)/ Pectin (PAVA-PN) and were cross-linked with N,N-methylene-bis-acrylamide. These PAVA-PN hydrogels are used as nano reactors to prepare silver nanocomposite hydrogels (SNH). 5-fluorouracil an anticancer drug has been loaded in situ into PAVA-PN hydrogels. Plain as well as drug loaded PAVA-PN hydrogels have been characterized by Fourier transform infrared spectroscopy and scanning electron microscope studies. The above studies has revealed that a molecular level dispersion of the drug in PAVA-PN hydrogels. In vitro release of 5-fluorouracil from the PAVA-PN hydrogels and SNH has been carried out in GIT fluids.

INTRODUCTION

Hydrogels are three-dimensional polymeric networks with the ability to respond to their environments.[1,2] Bulk (macro) and nano-sized hydrogels, which are known to be smart materials with tunable properties, morphology, size, and hydrophilic/hydrophobic balance, are indispensable materials for numerous applications in biomedical fields such as drug delivery, catalysis, and the development of antimicrobial materials.[3–6] Due to their hydrophilic surface

V. Rajendran, P. Prabu and K.E. Geckeler (eds.)
Nano Biomaterials, pp. 143–148 (2012).

has a low interfacial free energy in contact with body fluids, which results in a low tendency for protein and cells to adhere to these surfaces. Moreover, the soft and rubbery nature of hydrogels minimizes irritation to surrounding tissues. As a result, in comparison to other synthetic materials, hydrogels resemble nature living tissues closely in their physical Properties due to their high water contents and softness, which also contributes to biodegradability and biocompatibility. Therefore, it can find applications in different technological arias such as materials for contact lenses and protein separation, matrices for cell encapsulation and devices for controlled release of drugs and proteins.[7]

MATERIALS AND METHODS

Analytical reagent grade samples of acrylamide (AAm), N-vinylcaprolactam (NVC), acrylamidoglycolic acid (AGA), N, N′-methylene-bis-acrylamide (MBA), potassium persulfate (KPS) N,N,N′,N′-tetramethylethylenediamine (TEMED), and 5-fluorouracil (5-FU) were purchased from Aldrich. All chemicals were used as received and experiments were carried with double distilled water.

PREPARATION OF PAVA-PN HYDROGELS

Dual responsive (Am, NVC, AGA and Pectin) IPN hydrogels were prepared by radical copolymerization. Monomers were dissolved in 3.0 mL of distilled water, to this solution 1.0 mL of cross linking agent (MBA, 1 wt% aqueous solution), and 1.0 mL of APS solution (5wt% aqueous solution). Polymerization was carried out in Petri dishes maintained at 55°C until polymerization was complete. IPN hydrogels obtained washed with double distilled water to remove any unreacted reagents and vacuum dried until attainment of constant weight. The complete conversion of monomers was confirmed gravimetrically. The monomer and polymer compositions are summarised in Table 1. Silver nanocomposite PAVA-PN hydrogels were prepared by reducing the silver ion doped PAVA-PN hydrogels.

Table 1: Chemical Composition of PAVA-PN Hydrogels

Code	*Am (gm)*	*AGA/NVC (gm)*	*Pectin (ml)*	*MBA (mol)% × 10^{-4}*	*APS (mol)% × 10^{-4}*
A	0.5	0.125/0.25	2	6.48	4.40
B	0.5	0.25/0.25	2	6.48	4.40
C	0.5	0.5/0.25	2	6.48	4.40
D	0.5	0.25/0.25	2	12.97	4.40
E	0.5	0.25/0.25	2	0.32	4.40
F	0.5	0.25/0	2	6.48	4.40
G	0.5	0/0.25	2	6.48	4.40
H	0.5	0.25/0.25	2	6.48	4.40

Swelling experiments, 5-FU loading, and 5-FU release studies are fallowed as per the procedure published earlier.[9]

Characterization

Fourier transform infrared spectroscopy (Perkin Elmer, Spectrum Two) analysis was performed to find the chemical structure of the hydrogel. Further more, to investigate the morphology of the PAVA-PN hydrogel were recorded using a MIRA \\TESCAN, scanning electron microscope (SEM),

RESULTS AND DISCUSSION

FTIR Analysis

FTIR spectra of of pure PAVA-PN hydrogel and drug loaded PAVA-PN hydrogel is presented in Figure 1. Two peaks around 1663 and 1608 cm^{-1} are due to amide-I and amide-II of acrylamide units. C-N and C-H stretching bands appear at 1440 and 2922 cm^{-1}, respectively, further confirming the presence of amide groups. A broad band observed between 3600 and 3000 cm^{-1} is due to the overlapping of the O-H stretching bands of carboxylic acid and O-H group of alcohol of AGA and N-H of the amide groups. Moreover, a band is observed at 1694 cm^{-1} as well as a band at 1641cm^{-1} due to the presence of COO^{-} of AGA in the spectrum of PAVA-PN hydrogel. For drug loaded PAVA-PN hydrogel there is high intensive peak observed at 1694 cm^{-1}, indicates that the drug is present in the PAVA-PN hydrogel matrix.

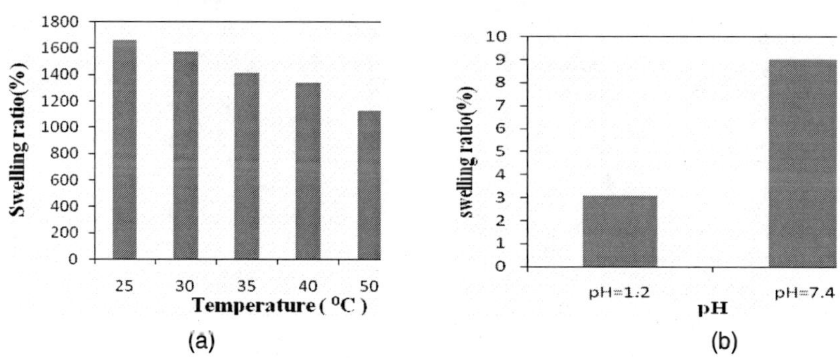

Fig. 1: Temperature Effect of (a) PAVA-PN Hydrogels, and (b) pH Effect of PAVA-PN Hydrogels

One of the most important properties that must be considered is microstructure morphologies. The surface morphology of the samples was investigated by scanning electron microscopy. Figure 3c shows an SEM micrograph of the polymeric hydrogels obtained from the fracture surface. The Figure 3d shows the formation of silver nanoparticles.

Swelling behavior of PAVA-PN hydrogels studied at different temperatures (25, 30, 35, 40 and 50°C). As temperature increase the swelling ratio of different composition of hydrogels decreases the swelling ratio. Swelling behavior of thermo sensitive polymers are influenced by the temperature of swelling media and these polymers display sharp. The equilibrium swelling of hydrogels at 25°C is 1664, 30°C is 1577, 35°C is 1413, 40°C is 1340 and 50°C are 1132.

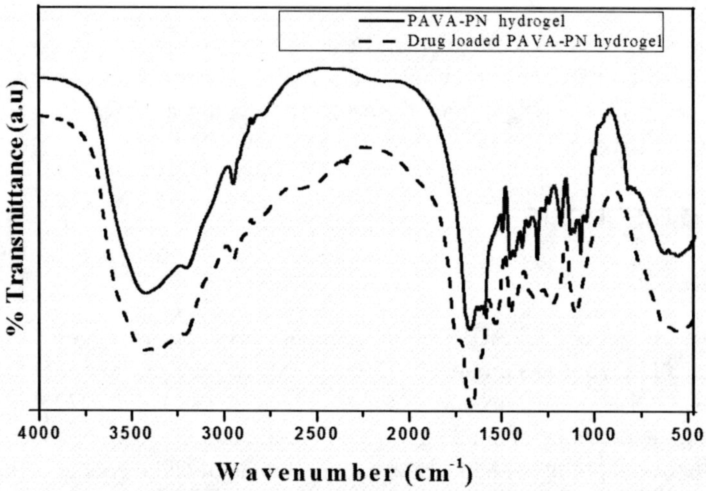

Fig. 2: FTIR Spectra of Pure PAVA-PN Hydrogel and Drug Loaded PAVA-PN Hydrogel

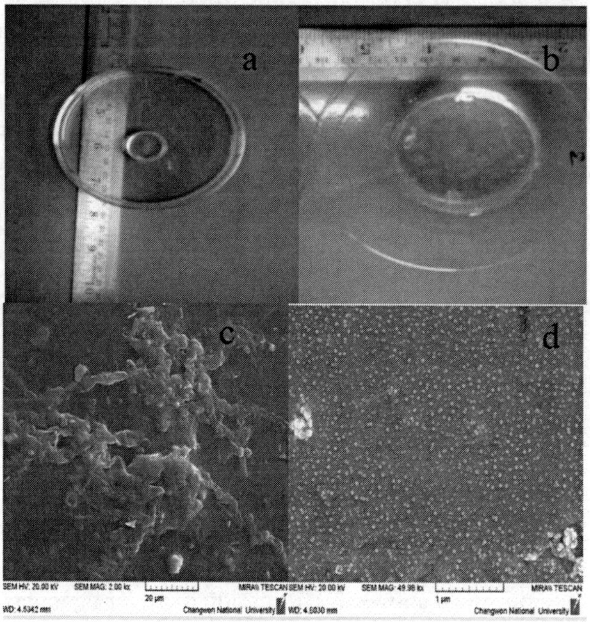

Fig. 3: Photograph of (a) Dry PAVA-PN Hydrogel, (b) Swollen PAVA-PN Hydrogel, and SEM Image of (c) Pristine PAVA-PN Hydrogel, (d) Silver Loaded PAVA-PN Hydrogel

In this study, the effect of pH on the swelling behaviour of the PAVA-PN hydrogel was investigated by varying the pH of the swelling medium in the range of 1.2 and 7.4. The swelling percent increases as a function of pH for C there was a considerable change in the dynamic swelling curves between pH 1.2 and 7.4. However, increase in the equilibrium swelling value of C when the pH of the swelling medium was increased from 1.2 to 7.4.

Drug Delivery

To investigate the release behaviour of a model drug from PAVA-PN hydrogels, 5-FU was incorporated into the hydrogels by soaking the dried hydrogels in a 5-FU saturated solution. As the dried hydrogels absorbed water, the 5-FU was transported with water due to the concentration gradient of 5-FU between the outside and the inside of the hydrogel. The in vitro drug-release studies were performed in pH 1.2 and pH 7.4 buffer media up to 68 h and, hence, the present hydrogel developed is appropriate for the controlled release of 5-FU over an extended period of time up to 58 h. The % cumulative release for the hydrogels cross-linked with 3.2431×10^{-5}% of MBA is lower, while intermediate values were observed for 6.48×10^{-4}% containing MBA hydrogels due to a decrease in matrix swelling as a result of increased MBA content of hydrogels. However, drug release from hydrogels is attributed to three types of mechanisms: (i) release from the surface of hydrogel particles, (ii) diffusion through the swollen rubbery matrix and (iii) release due to hydrogel erosion in the external environment. Initially, higher release rates were observed due to dissolution of the surface-adhered 5-FU. At longer times, the drug release is controlled by diffusion rather than erosion of the matrix, so it becomes much slower when compared to initial release rates. Of all the formulations developed, formulation C and H exhibited 100% release of 5-FU within 12 h compared to other formulations. The PAVA-PN hydrogels showed thermo/pH-sensitive release behaviour. The temperature increases swelling ratio decreases means shrinking takes place. At low pH (1.2) small amounts of 5-FU were released from the hydrogels while at high pH (7.4) relatively large amounts of 5-FU were released from the hydrogels. This pH-sensitive behaviour of hydrogels indicates that the hydrogels can be used as a biological on-off switch for an intelligent drug delivery system triggered by the external pH change in a body. In addition, the PAVA-PN hydrogels released most of the 5-FU within about 30 min, which means the hydrogels can release solute such as drugs and biologically active materials inside the hydrogels for a short period and fastly release the solute from the hydrogels in response to an increase in the external pH above the pKa of the hydrogel (due to the page limitation figures (graphs of 5-Fu release and others) are not provided).

CONCLUSION

Dual PAVA-PN hydrogels having various compositions of VC, AGA, Pectin and MBA, were prepared by free-radical polymerization. The produced PAVA-PN hydrogels were characterized in terms of equilibrium swelling characteristics. It is seen that the swelling of the hydrogels increased with increasing VC and AGA concentrations, but decreased with the cross linker concentration. The diffusion type of hydrogel systems was of a non-Fickian diffusion character. As the hydrogels contained ionisable functional groups, their swelling behaviour was essentially dependent on pH, leading to a typical pH-sensitive swelling behaviour of anionic hydrogels, i.e., low swelling ratios at low pH and high swelling ratios at high pH. The PAVA-PN hydrogels showed pH-responsive release behaviour. At low pH (pH=1.2) small amounts of 5-FU were released while at high pH (pH=7.4) relatively large amounts of 5-FU were released from the hydrogels. These results indicate that the PAVA-PN hydrogels can be used as a biological on-off switch with which the release of materials from the hydrogel can be controlled by the external pH change and pKa of the hydrogels. In addition, the pKa determining the on-off switch can be tailored by changing the type or composition of monomers and cross linkers used to prepare the hydrogels.

ACKNOWLEDGEMENTS

Authors Dr. K.S.V. Krishna Rao and Mr. P. Rama Subba Reddy thank the Department of Science and technology (DST), New Delhi, India, (DST No.SR/FT/CS-047/2009) for a financial support.

REFERENCES

[1] Sahiner, N. and Ilgin, P., *Polymer,* 51 (2010), 3156–3163.
[2] Sahiner, N. and Ilgin, P., *J. Polym. Sci. Part A: Polym. Chem.*, 48 (2010) 5239–5246.
[3] Ozay, O., Akcali, A., Otkun, M.T., Silan, C., Aktas, N. and Sahiner, N., *Colloids Surf. B.*, 79(2010), 460–466.
[4] Sahiner, N., *Eur. Polym. J.*, 43 (2007), 1709–1717.
[5] Sahiner, N., Alb, A.M., Graves, R., Mandal, T., McPherson, G.L., Reed, W.F. and John, V.T., *Polymer,* 48 (2007), 704–711.
[6] Sahiner, N., Turk. *J. Chem.*, 33 (2009), 23–31.
[7] Peppas, N.A., CRC press Boca Raton, Florida, 1986.
[8] Garcia, O., Trigo, R.M., Blanco, M.D. and Teijon, J.M., *Biomaterials*, (1994), 15: 689.
[9] Krishna Rao, K.S.V. and Sik-Ha, Chang (2009). *Polymer Bulletin,* 62(2): 167–181.

SYNTHESIS, CHARACTERIZATION AND INVESTIGATIONS OF ANTIMICROBIAL PROPERTIES OF SILVER NANOPARTICLES VIA A GREEN METHOD USING MULBERRY (MORUS ALBA L) TREE LEAVES

R. RUBIA, YASHIKA, REMY[1], NAMRATA[2], RAJESHWARI, INDU TUCKER SIDHWANI[1], KAVITA VASDEV, MANGALA JOSHI[3], SHUBHA M GOKHALE[4] AND VANDNA LUTHRA

Department of Physics, Gargi College, Siri Fort Road, New Delhi
[1]Department of Chemistry, Gargi College, Siri Fort Road, New Delhi
[2]Department of Microbiology, Gargi College, Siri Fort Road, New Delhi
[3]Department of Textiles, Indian Institute of Technology, New Delhi
[4]School of Sciences, Indira Gandhi National Open University, Maidan Garhi, New Delhi
E-mail: vandna_arora@yahoo.com

ABSTRACT

Synthesis of nanoparticles using green methods [1] especially for biological applications is emerging as new area of research. These methods on one hand reduce the toxic effects of harmful chemicals and on the other hand evolve new chemistry for the synthesis of the nanoparticles. The Synthesis of nanoparticles using plant extracts is being tried for various applications.[2] In this regard, synthesis of nanoparticles such as silver is all the more beneficial due to its intrinsic antimicrobial properties. In the present report, the synthesis of silver nanoparticles has been carried out using mulberry tree leaves (Morus alba L). The various conditions for the synthesis have been optimized for a stable liquid sol. The stabilised silver nanoparticles have been further characterized using various analytical techniques such as UV-Vis spectroscopy and Transmission Electron Microscopy (TEM). The synthesized material was used for studying antimicrobial activity on growth of E.coli. E-coli was grown under shake

V. Rajendran, P. Prabu and K.E. Geckeler (eds.)
Nano Biomaterials, pp. 149–154 (2012).

culture condition and it's growth was monitored by measuring absorbance using UV-Vis spectrophotometer. The antimicrobial activity of nanoparticles was analyzed by incubating E.coli culture with various concentrations of sols/nanoparticles and measuring their growth against E.coli culture without nanoparticles. A decrease in optical density is obtained as the particle size of the nanoparticles decrease. This decrease in optical density is correlated with the antimicrobial activity of silver nanoparticles. These result could be exploited for various other biological applications due to benignity of the developed material in contrast to other harmful materials used for the synthesis of nanoparticles coupled with the inherent potent anti-oxidants or free radical scavengers such as flavonoids and moracins [3] present in mulberry tree leaves which have shown therapeutic effects of lowering blood sugar in animals.[4]

INTRODUCTION

The development of green processes [5] for the synthesis of nanoparticles is evolving as an important branch of nanotechnology. This could be greatly attributed to the safe and ecologically sound methods of synthesis as alternative to other physical and chemical methods. The green methods generally use non-toxic routes of synthesis and solvents such as water, plant extracts as well as microwave assisted synthesis.

There has been a lot of active research on synthesis using plant extract and many materials are being added to the endless list due to their inherent properties. Neem leaf broth (Azadirachta indica), Pelargonium graveolens-Geranium leaves (P. graveolens), Medicago sativa (Alfalfa), Aloe vera, Emblica officinalis (amla, Indian Gooseberry), green tea are some of the common plant extract being used specifically due to easily availability and abundance in India.

In the present investigation, we have synthesized silver nanoparticles using mulberry tree leaves (Sericha Healthcare Pvt. Ltd.) which have medicinal usage for lowering the blood sugar. Silver has got intrinsic antimicrobial and mulberry tree leaves have excellent anti-oxidant properties and are well suited for green synthesis.

Gold and silver plasmonic nanoparticles have optical properties that are sensitive to size, shape, concentration, agglomeration state, and refractive index near the nanoparticle surface, which makes UV/Vis spectroscopy a valuable tool for identifying, characterizing, and studying nanomaterials.[6] The developed materials have been studied using UV-Vis spectroscopy, Transmission Electron Microscopy (TEM) and tested for anti-microbial activity against E.coli.

EXPERIMENTAL

The silver nitrate (.05 M) solution was prepared using distilled water (A). The mulberry leaves (1 g) were soaked in 100 ml deionized water and stirred for half an hour. The extract was collected after filteration with suitable pore size filter paper. Thus obtained extract (B) was mixed drop-wise in various proportions with diluted silver nitrate solutions (.0005 M) as shown in Figure 1. The ratios 1:4, 1:3, 1:2, 1:1 2:1, 1:2, and 3:1 were prepared (silver nitrate to mulberry tree leaves extract). There was a variation in color from pale yellow to brownish yellow.

The sols were stabilized for few hours to 1–2 days before further characterization and for investigations and for antimicrobial activities.

The samples were characterized using UV-Vis spectrophotometer (Systronic T-119) and Transmission Electron Microscopy studies. The samples were also used for antimicrobial activities in which the E-Coli growth was monitored.

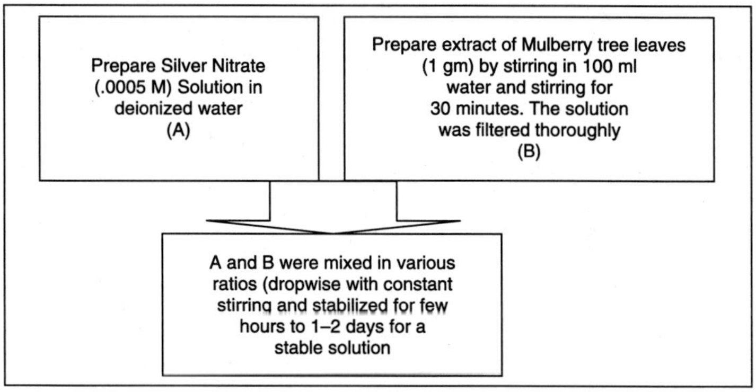

Fig. 1: Flow Chart for the Synthesis of Silver Nanoparticles Using Mulberry Leaves

The growth in absence of these nanoparticles was used as reference. The growth of the E.coli was monitored by measuring optical density of the resultant material in a cuvette using spectrophotometer. This study was done simultaneously with stability of these nanoparticles with time. The coupled results were taken into consideration.

RESULTS AND DISCUSSION

It is interesting to note that a variation in color of the sol was obtained as the different ratios of silver nitrate to mulberry tree leave extract were used. The UV-Vis spectra were recorded and the data has been taken in the range 300–800 nm. The stabilized samples were further diluted for recording the UV-Vis data. The recorded data has been shown in Figure 2. The samples were calibrated using diluted mulberry tree leaves extract as reference.

Figure 2 clearly shows a well defined peak for the sample mixed in the ratio 2:1 (silver nitrate solution to mulberry tree leave extract). For different compositions, a shift in the maxima is observed which could be well correlated to the different sizes of the nanoparticles obtained for various compositions. A shift in peak is observed towards higher wavelengths as the amount of silver nitrate is increased in the solutions. The results have been tabulated in the Table 1.

The UV-Vis spectra of mulberry tree leaves extract was taken with water as reference and the shoulder at around 382 nm was observed. With increase in proportion of silver in the synthesis, the dominance of mulberry leaves extract around 380 nm is substituted by formation of nanoparticles whose signature starts appearing around 400 nm corresponding to the surface Plasmon resonance (SPR) peak. As the amount of silver nitrate is increased a shift in λ_{max} towards higher wavelengths is observed which could be attributed to an increase in particle size. The narrow nature of the SPR peak indicates monodispersed nanoparticles whereas a long tail in SPR peak points towards non-uniform size of nanoparticles.[7] Other studies show that as the diameter of the silver nanoparticles increases, the peak Plasmon resonance shifts to longer wavelengths and broadens.[6] Our results also show a shift in the peak to longer wavelengths is observed as the size of the silver nanoparticles increase. A well defined narrow peak in λ_{max} is observed for 2:1 composition which has shown the maximum stability and shelf life.

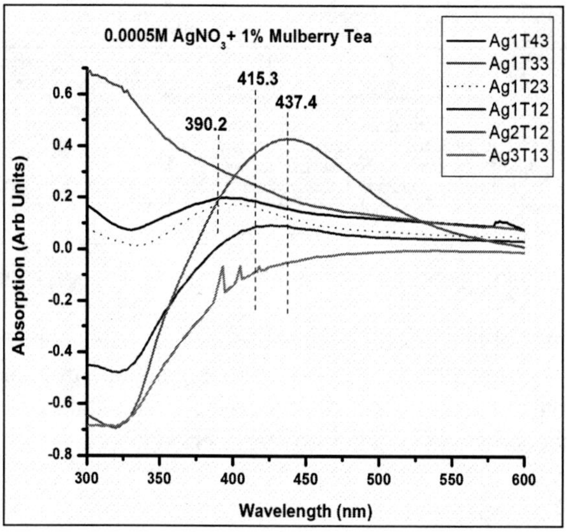

Fig. 2: The UV-Vis Spectrum of Samples of Silver Nanoparticles Synthesized Using Mulberry Tree Leaves Extract in Various Proportions. Samples are Ag: Tea (1:4) (1:3) (1:2) (1:1) (2:1) (3:1) Shown in Order in the Figure

Table 1: The Variation in the Peak Position Observed as a Function of Composition

Composition (silver nitrate solution to mulberry extract ratio)	*Position of λ_{max} (nm) as Seen from Figure 2*
1:2	390.2
1:1	415.3
2: 1	437.4
3:1	>437.4

An increase in particle size has been observed as the silver nitrate solution to mulberry extract ratio is increased. As shown by the Transmission Electron Microscopy of the samples 2:1 and 3:1 (Figure 3) as a representative result.

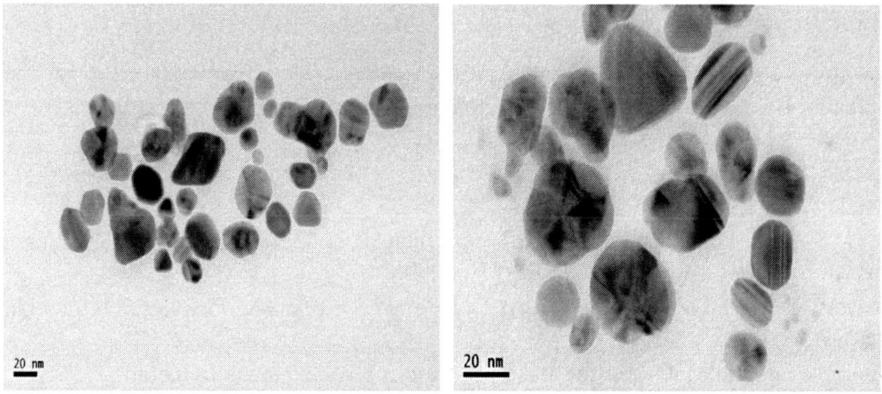

Fig. 3: Transmission Electron Micrographs of Two Samples Prepared in the Ratio 2:1 and 3:1 Silver Nitrate Solution (.0005 M) to Mulberry Tree Leaves Extract (1 gm) Stirred in 100 ml Deionized Water

The particle sizes are shown on a scale of 20 nm and are in the range (15–30 nm) and (30–40 nm) for 2:1 and 3:1 samples respectively.

The samples were further studied simultaneously for their stability and for antimicrobial studies under static conditions. It is observed that the samples having composition 3:1 yielded Ag nanoparticles settled at the bottom after 2–3 days. The growth of E-coli was monitored via optical density with and without nanoparticles. A decrease in optical density is observed as compared to the sample without nanoparticles. A combined study of the antimicrobial properties and the stability of these materials clearly indicates that 2:1 sample is better in contrast to the rest of samples which is approximately 30% inhibition of the E-coli and more detailed studies are underway.

CONCLUSIONS

The silver nanoparticles have been synthesized using mulberry leaves extract. These were prepared in various proportions and the color of the solution varied from pale yellow to brown yellow as the silver nitrate to mulberry tree extract ratio is changes (from 1:4 to 3:1). The UV-vis spectroscopy shows a shift in the λ_{max} to longer wavelengths as the silver nitrate proportion is increased in the final solution. This is correlated with the increase in particle size as evidenced from TEM. Our preliminary studies show that 2.1 sample has good shelf life (>1 month) and possesses antimicrobial property against growth of E-coli which is found to be 30% inhibition and more studies are in progress.

ACKNOWLEDGEMENTS

The work is supported by University of Delhi under the Innovation Projects in Colleges scheme. Authors Remy, Namrata and Rajeshwari are thankful for the financial support under the same scheme.

REFERENCES

[1] Huang, J., Li, Q., Sun, D., Lu, Y., Su, Y., *et al. Nanobiotechnol*, 18, 105104–105115, 2007.

[2] Elumalai, E.K., Prasad, T.N.V.K.V., Kambala, V., Nagajyothi, P.C. and David, E., *Archives of Applied Science Research*, 2(6): 76–81, 2010.

[3] Sharma, R., Sharma, A., Shono, T., Takasugi, M., Shirata, A. *et al. Biosci Biotechnol Biochem*, 65: 1402–5, 2001.

[4] Naowaboot, J., Pannangpetch. P., Kukongviriyapan, V., Kukongviriyapan, U., Nakmareong, S. *et al. Nutrition Research*, 29, 602–608, 2009.

[5] Anastas, P. and Warner, J.C., *Green Chemistry: Theory and Practice,* Oxford University Press, USA, 2005.

[6] Silver Nanoparticles: Properties and Applications www.sigmaaldrich.com

[7] Kulkarni, S.K., *Nanotechnology: Principle and Practices,* Capital Publishing, 2nd ed., 2010.

EFFICIENT PREPARATION AND SIZE EVALUATION OF GOLD NANOPARTICLES IN CHITOSAN SUSPENSION BY ONE-POT CHEMICAL REDUCTION METHOD

R. LAKSHMI NARAYANAN, M. SIVAKUMAR[1] AND K. VENKATESWARA RAO

Centre for Nano Science and Technology, Institute of Science and Technology, Jawaharlal Nehru Technological University, Hyderabad, A.P.
[1]Division of Nano Science and Technology, Anna University of Technology Tiruchirappalli, Tiruchirappalli, Tamil Nadu
E-mail: rlnarayanan10@gmail.com

ABSTRACT

Gold nanoparticles were synthesized in chitosan suspension by one-pot chemical reduction method without utilizing any chemical reducing reagent. In aqueous acidic medium, chitosan exhibits polycationic nature due to protonation of amine groups of each glucosamine repeating units. The adsorption of $AuCl_4^-$ ions occurs on chitosan through electrostatic attraction and further reduction of Au^{3+} to Au^0 proceeds through a three electron transfer mechanism by oxidation of protonated amine groups of chitosan chain. The change of color from dirty yellow to intense red indicates the formation of gold nanoparticles in chitosan suspension which acts as both reducing and stabilizing agent. Characterization techniques such as UV-VIS, FTIR, Zeta potential measurement, FESEM and TEM were done in order to examine the reduction and stabilization phenomena rendered by chitosan.

INTRODUCTION

Several linear polysaccharides such as chitosan possess metal chelating property which can able to bind metal ions in aqueous solution.[1] In recent years, research work was focused on

V. Rajendran, P. Prabu and K.E. Geckeler (eds.)
Nano Biomaterials, pp. 155–160 (2012).

simultaneous formation and conjugation of metal nanoparticles with chitosan under thermal treatment.[2] The metal nanoparticles that are formed strongly bound to chitosan and the resultant bio conjugates bear the advantage for various biological applications.[3]

Silver/chitosan nanocomposites were prepared by using basic chitosan suspension as a reductant in the absence of other chemicals. In Murugadoss and Chattopadhayay methodology, stable chitosan-Ag nanocomposites were prepared in the basic chitosan suspension at 95°C under reflux condition.[4] In Bhumkar *et al.* methodology, gold nanoparticles prepared with higher concentration (> 0.1% w/v) at pH below 5 were stable showing no signs of aggregation.[5] In our present work, efforts were made to prepare gold nanoparticles through one pot chemical reduction method in chitosan suspension with pH in the range of 5–6 by carrying out the entire reaction process at 100°C under reflux condition.

EXPERIMENTAL

Materials

Gold precursor stock solution was prepared by dissolving 10 mg of Hydroaurochloric acid (M.W. 393.83) in 100 ml Millipore water. The stock solution of medium molecular weight chitosan with 95% degree of deacetylation was obtained by dissolving 0.2 g of chitosan flakes in 50 ml diluted 1% acetic acid. The chitosan solution was allowed to stir overnight, sonicated for certain hours to dissolve chitosan completely and then filtered to obtain clear homogenous solution.

One Pot Preparation of Gold Nanoparticles in Chitosan Suspension

In this wet chemical reduction method, gold nanoparticles were prepared through reduction of gold precursors in chitosan suspension at 100°C with pH in the range of 5–6. In reaction vessel, 25 ml of 0.1 mg/ml gold precursor solution was added slowly in to 25 ml of 0.2% w/v chitosan solution under rapid stirring and refluxed for certain duration. Total volume of the final suspension was then centrifuged and washed with Millipore water for several times and obtained residue was suspended in Millipore water for analysis purpose.

Characterization Studies

UV-Visible absorption spectrum was observed using Perkin-Elmer Lambda-25 spectrometer in the wavelength range of 200–700 nm. FT-IR spectrum was recorded on Perkin- Elmer RX-1 model FTIR Spectrophotometer in the range of 400–4000 cm^{-1}. Particle size distribution was analyzed through DLS technique using Nanotrac instrument. Zeta potential measurements were recorded on Zetasizer Nano S, Malvern in the range of –100 mV to +100 mV. FESEM analysis was made using Ultra 55, Carl Zeiss instrument. TEM analysis was performed using JEOL Model 1200EX instrument operated at an accelerating voltage of 120 kV.

RESULTS AND DISCUSSION

Reduction and Stabilization of Gold Nanoparticles in Chitosan Suspension

During reaction process, there was a color change from dirty yellow to colorless and finally the incurrence of ruby red color indicates the formation of gold nanoparticles. Since chitosan was dissolved in aqueous acidic media, the amine groups of the D-glucosamine repeating units get protonated.[1,2] When diluted Hydroaurochloric acid was added in to chitosan solution, negatively charged gold (III) ions were adsorbed on various chains of chitosan due to electrostatic attraction with positively charged protonated amino groups.[3,5] Generally, the reduction of Au^{3+} (aq) to Au^0 follows three electron transfer mechanism.[6] But adsorbed gold (III) ions on various chitosan chains get reduced to neutral gold atoms through oxidation of three protonated amine groups without influence of external chemical reagent and further agglomerates to form gold nanoparticles. Thus, the overall reaction for the reduction mechanism can be expressed as follows:

$$AuCl_3\,(aq) + 3e^- + 3NH_3^+ \longrightarrow Au^0 + 3NH_2 + 3HCl$$

UV-Visible characterization can confirm the gold nanoparticles formation in chitosan suspension. In Figure 3a, curve 1 depicts the absorption spectrum of the Au Nps - chitosan suspension shows a strong and broad absorption peak at 521 nm which is the peculiar characteristics of surface plasmon resonance of gold nanoparticles. The curve 1 depicts the absorption spectrum for blank chitosan shows sharp absorption peak at 225 nm. The absence of chitosan absorption peak in curve 2 clearly states that all the amino groups of chitosan are actively involved in the reduction process leaving no traces of blank chitosan in the Au Nps-chitosan suspension.

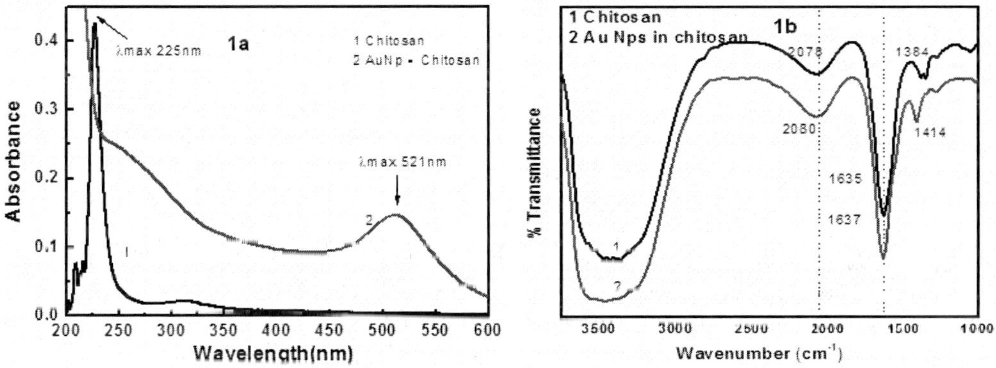

Fig. 1: Comparison of Blank Chitosan with Au Nps-Chitosan Suspension by UV-Visible Spectra (1a) and FTIR Spectra (1b)

After reduction, the formed gold nanoparticles strongly adsorbs to chitosan due to affinity binding of oxidized amine groups with gold surface. FT-IR spectra can reveal the specific interaction of amine group with gold nanoparticles through analysis of spectral shift.[7] In Figure 3b, curve 1 and 2 depicts the fingerprint pattern of FTIR spectra for blank chitosan and Au Nps-chitosan. In both curves there exists an overlapping band from 3300–3500 cm^{-1} which attributes to the stretching vibration of hydroxyl, amino and amide groups with slight widening for Au Nps-chitosan when compared to blank chitosan. For blank chitosan, a peak at 1635 cm^{-1} ascribed to bending vibration of $-NH_2$ group, shifted towards 1637 cm^{-1} for Au Nps- chitosan which clearly indicates that the formed gold nanoparticles are completely bound to chitosan through affinity interaction with amino group leaving no traces of blank gold nanoparticles in the suspension.

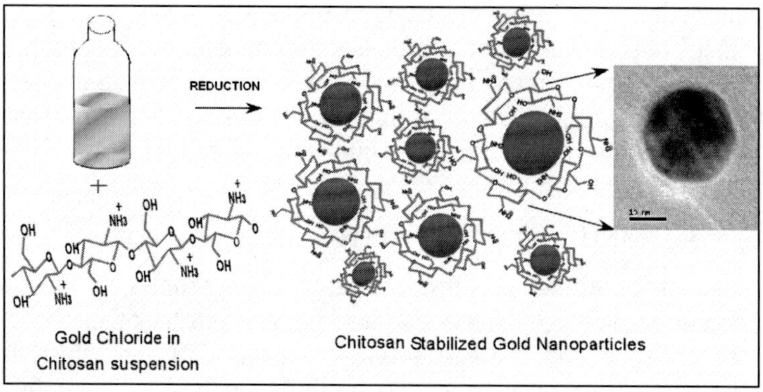

Fig. 2: Scheme for One-Pot Preparation of Gold Nanoparticles in Chitosan Suspension

The formed gold nanoparticles in chitosan suspension possess excellent stability and the particle aggregation is less likely to occur even after several months. The zeta potential of suspended gold nanoparticles was measured to determine the surface charge and stability of the nanoparticles in aqueous dispersion.[5] In colloids, usually particle aggregation is less likely to occur when the zeta potential is more than +30mV or less than –30 mV. The average zeta potential was measured about +33 mV clearly indicates that the stability arises due to electrostatic repulsion between positively charged amine groups of chitosan layer rendered on each gold nanoparticle surface. Thus the reduction and stabilization phenomenon-occurs in same suspension requiring no external chemical reagents and the methodology is simple which can be termed as one pot chemical reduction method which is depicted in Figure 2.

Morphological Studies of Chitosan Stabilized Gold Nanoparticles

In Figure 3a, particle size distribution of the chitosan stabilized gold nanoparticles shown with respect to intensity variation having wide range distribution from 10 to 90 nm. The Figure 3b depicts the FESEM secondary electron imaging of film casted Au Nps-chitosan suspension on glass substrate, gold nanoparticles possess uniform size distribution in the range of 5–50 nm. By closely examining the FESEM picture, gold nanoparticles on the top layer of chitosan appears bright than the gold nanoparticles lying on the bottom layer indicates that the formed nanoparticles are well dispersed with each other even though shrinkage occurs during film formation.

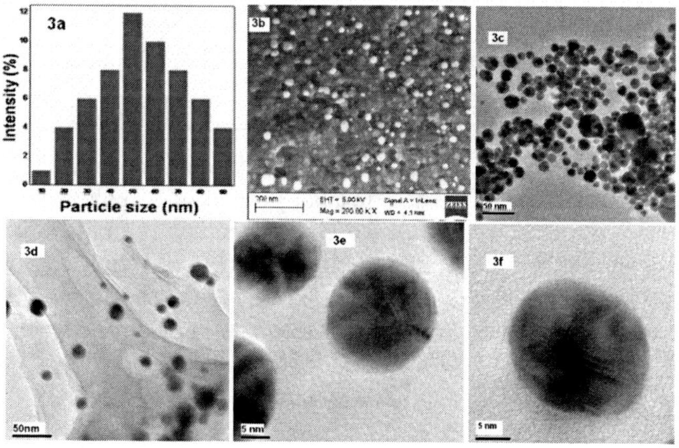

Fig. 3: Size and Morphological Studies of Chitosan Stabilized Gold Nanoparticles by DLS (3a), FESEM (3b) and TEM (3c, 3d, 3e, 3f)

At 120 KV, the resolution of TEM is 1 nm such that the gold nanoparticles formed in the size regime of 5 nm to 50 nm can be easily viewed and studied. In Figure 3c, the TEM image of drop casted Au NPs-chitosan suspension is shown at scale bar of 50 nm indicates that the gold nanoparticles are well dispersed having spherical morphology. Even though particles are closed arranged in different fashion there is no signs of agglomeration. Thus the electrostatic repulsion between NH_3^+ groups on gold nanoparticle makes it more stable in aqueous medium and the formed AuNps in chitosan suspension possess positive charge which is already confirmed through zeta potential analysis.

The size and morphology of the chitosan stabilized gold nanoparticles purely depends upon varying chain length of polymeric chitosan. When the chain length of chitosan is longer in compare to other chains, more number of gold ions adsorbed along the entire length which leads to bigger size nanoparticles and also slight variation in spherical geometry.[8] In Figure 3e, single gold nanoparticle is closely viewed at scale bar of 5 nm possess elongated spherical geometry with crystalline atomic lattice pattern in central region and amorphous in peripheral region.

Protonated amine groups of chitosan chain after reducing Au^{3+} ions gets conjugated to gold nanoparticle surface through Au-N bond formation which is already confirmed through FTIR studies. Chitosan is a biological macromolecule which is not easily observed in TEM studies due to its lower atomic number. But in Figure 3d, the polymeric chains of chitosan are observed along with adsorbed gold nanoparticles and it has been confirmed that the formed gold nanoparticles are completely bound to chitosan. In Figure 3e, the closer view of gold nanoparticles at 5 nm scale bar saying that the chitosan layer is conjugated around each gold nanoparticle surface. The layer formation is due to simultaneous adsorption and reduction of gold precursors mediated by protonated amine groups which subsequently stabilizes the gold nanoparticle through Au-N bond formation.[9] After stabilization the unreacted amine groups of conjugated chitosan gets exposed to the surrounding environment possessing smooth morphology of robust gold core and mild chitosan shell layer.

CONCLUSION

Chitosan stabilized gold nanoparticles is prepared by simple one pot chemical reduction method. The formation of gold nanoparticles in chitosan suspension is confirmed through existence of broad UV-Visible absorption peak at 521 nm. The protonated amine groups play a major role in the reduction of gold precursors and subsequent stabilization of gold nanoparticles through Au-N bond formation is analyzed through peak shift of FTIR spectrum. The FESEM image indicates the formation of well dispersed gold nanoparticles in the size range of 5–50 nm. The TEM image shows the formation of spherical nanoparticles with varying sizes. The gold nanoparticles are seen to be conjugated with chitosan showing no signs of agglomeration. It is evident that the chitosan layer is formed during subsequent reduction and stabilization and positive charge is rendered on nanoparticle surface due to presence of unreacted form of protonated amine groups which is measured through zeta potential analysis.

ACKNOWLEDGEMENTS

We acknowledge Dr. G.U. Kulkarni, DST Unit on Nanoscience, Jawaharlal Nehru Centre for Advanced Scientific Research, Bangalore, India for TEM studies. Our special thanks to TEM analyst Mrs. Usha Tumkurkar.

REFERENCES

[1] Dongwei Wei, Yongzhong Ye, Xueping Jia, Chao Yuan and Weiping Qian., Chitosan as an active support for assembly of metal nanoparticles and application of the resultant bioconjugates in catalysis, *Carbohydrate Research*, 345, 74–81, 2010.

[2] Huang, Haizhen and Yang, Xiurong, Synthesis of chitosan-stabilized gold nanoparticles in the absence/presence of tripolyphosphate, *Biomacromolecules*, 5, 2340–2346, 2004.

[3] Tan, Wee Beng and Zhang, Yong, Surface modification of gold and quantum dot nanoparticles with chitosan for bioapplications, Wiley InterScience, London, 2005.

[4] Murugadoss, A. and Chattopadhyay, Arun, A 'green' chitosan-silver nanoparticle composite as a heterogeneous as well as micro-heterogeneous catalyst, *Nanotechnology* 19, 015603, 2008.

[5] Bhumkar, Devika R., Joshi, Hrushikesh M., Sastry, Murali and Varsha, B., Chitosan Reduced Gold Nanoparticles as Novel Carriers for Transmucosal Delivery of Insulin, *Pharmaceutical Research,* 24, 8, 2007.

[6] Brust, M., Walker, M., Bethell, D., Schiffrin, D.J. and Whyman, R.J., Synthesis of thiol-derivatised gold nanoparticles in a two-phase Liquid–Liquid system, *J. Chem. Soc., Chem. Commun,* 801–802, 1994.

[7] Govindan, S., Nivethaa, E.A.K., Saravanan, R., Narayanan, V. and Stephen, A., Synthesis and characterization of chitosan-silver nanocomposite, *J. Appl Nanosci,* DOI 10.1007/s13204-012-0109-5.

[8] Sugunana, A., Thanachayanont, C., Dutta, J. and Hilborn, J.G., Heavy-metal ion sensors using chitosan-capped gold nanoparticles, *Science and Technology of Advanced Materials*, 6, 335–340, 2005.

[9] Bhattacharya, Santanu, Srivastava, Aasheesh, Synthesis of Gold Nanoparticles Stabilised by metal-chelator and the controlled formation of close-packed aggregates by them, *Proc. Indian Acad. Sci. (Chem. Sci.)*, 115, 5 and 6, 613–619, 2003.

EUCALYPTUS LEAF EXTRACT MEDIATED SYNTHESIS OF IRON OXIDE NANOPARTICLES

M. Balamurugan and S. Saravanan

Centre for Photonics and Nanotechnology, Department of Science, Sona College of Technology, Salem, Tamil Nadu
E-mail: chem.muruga@gmail.com

ABSTRACT

The primary goal of nanotechnology is the improvement of simple eco-friendly method for synthesis of nanoparticles. Bio synthesis of nanoparticles has been proposed as a cost effective environment friendly alternative to chemical and physical methods. The iron oxide nanoparticles have been synthesized by adding Eucalyptus globulus leaf extract into the aqueous ferric chloride solution. The synthesized nanoparticles were characterized by Ultra Violet-Visible spectrum (UV-Vis), Scanning Electron Microscopy (SEM), X-Ray diffraction (XRD) and Fourier Transform Infra Red spectroscopy (FT-IR). SEM image shows the nanostructures of prepared iron oxide nanoparticles.

INTRODUCTION

Oxides of iron are available widely in nature in different phases.[1] Additionally, iron oxides have attracted a great deal of attention among specialists because of their multivalent oxidation states,[2] a large set of possible polymorphisms,[3] and especially at the nano scale, their characteristic structural changes.[1] These materials have a considerable potential to be applied as sensors, in catalysis,[4] as high density magnetic recording media,[5] for targeted drug delivery in clinical trials,[6] and as substrates in cancer treatment methods.[7] There are various chemical and physical methods are available for synthesis of iron oxide nanoparticles. Those methods are used toxic or potentially hazardous as starting materials and more energy. The primary goal of nanotechnology is the improvement of simple eco-friendly method for

V. Rajendran, P. Prabu and K.E. Geckeler (eds.)
Nano Biomaterials, pp. 161–166 (2012).

synthesis of nanoparticles. By using the bio material such as microbes and plant part extract produced different types of nanoparticles and even nano rods. However, as some organisms are pathogens, it is risky to handle. Microorganisms need maintenance of cultures and controlled conditions such as temperature, pH and other factors for growth. Sometimes, the synthesis of nanoparticles utilizing plant parts could prove advantageous over other biological processes by eliminating the elaborate process of maintaining the microbial culture. Synthesis iron oxide nanoparticles by using plant extract has been quite limited and few works have been reported. The reports are available for synthesis of hexagonal metallic iron, amorphous iron, and α-Fe_2O_3 by using tea extract [8, 9] and iron nanoparticles by using Aqueous Sorghum Bran Extracts.[10] The tea extract mediated synthesized iron nanoparticles were found to be nontoxic when compared with iron nanoparticles prepared using conventional $NaBH_4$ reduction protocols. The concentration of the tea extract in the reaction mixture plays an important role in the size and crystallinity (hexagonal metallic iron, amorphous iron, and R-Fe_2O_3) of the synthesized iron nanoparticles.[8] In this study iron oxide nanoparticles were synthesized using a rapid, single step, and completely green biosynthetic method employing aqueous *Eucalyptus globulus* leaf (Family: Myrtaceae) extracts as both the reducing and capping agent. *Eucalyptus globulus* essential oil shows higher medicinal property and antimicrobial potential against the range of food spoiling microorganisms and it contain 1,8-cineole (45.4%), limonene (17.8%), p-cymene (9.5%), γ-terpinene (8.8%), α-pinene (4.2%) and α-terpineol (3.4%), while in the vapour, 1,8-cineole (34.6%), limonene (29.9%), p-cymene (10.5%), γ-terpinene (7.4%), α-pinene (4.0%) and α-phellandrene (2.4%).[11]

EXPERIMENTAL

Preparation of Eucalyptus Globulus Leaf Extract

The fresh *Eucalyptus globulus* leaf were collected and wash thoroughly and heated 100g/L fine piece of leafs at 80°C for 30 min followed by filtered through filter paper to separate out the broth.

Synthesis of Iron Oxide Nanoparticles

Iron oxide was prepared using green synthesis method. The *Eucalyptus globulus* leaf extract was used in order to reduce and cap the Fe ions. A solution of $FeCl_3$ (0.01M) was prepared by dissolving of solid $FeCl_3$ (Nice Chemicals Pvt. Ltd.) in double distilled deionized water. 20 ml of the above prepared *Eucalyptus globulus* leaf extract was dripped slowly into the aqueous solution of $FeCl_3$ with constant stirring at room temperature with normal atmosphere pressure. A visible color changes were observed, the yellow color aqueous solution of $FeCl_3$ turned to blackish brown (Figure 1).

RESULTS AND DISCUSSION

In the typical synthesis of Iron oxide nanoparticles a visible color change was observed after adding the eucalyptus leaf extract into aqueous Ferric chloride solution at room temperature.

Fig. 1: Digital Photograph of Visible Color Changes, after Adding *Eucalyptus Globulus* Leaf Extract into Aqueous Solution of Ferric Chloride

The intensity of color increased with time and dosage of plant extract (Figure 1). The color change is the most easy and commonly used indication of the formation of metal nanoparticles.[12]

The optical property of synthesized iron oxide nanoparticles is one of the important characteristics for evaluation of its optical and photocatalytic activity. UV-Visible absorption spectrum is the preliminary characterization to know the optical property. UV-Visible absorption spectrum was carried out at normal condition in the range of 300 to 800 nm (ELICO-BL 198). Figure 2 shows the UV-Vis spectrum of 100 ml of 0.01 M aqueous $FeCl_3$ and 20ml of *Eucalyptus globulus* leaf extract mixture.

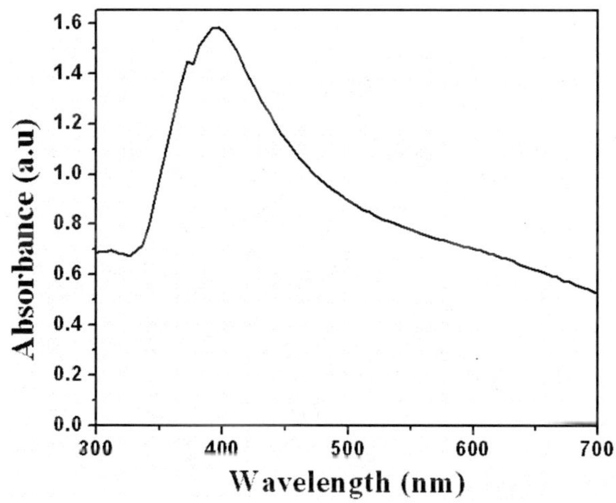

Fig. 2: UV-Visible Spectrum of 100 ml of 0.01 M aqueous $FeCl_3$ and 20 ml of *Eucalyptus Globulus* Leaf Extract Mixture

It shows an absorption peak around 402 nm indicating the formation of low dimensional β-Fe_2O_3. The absorption maximum of prepared nanoparticles is quite different to those observed earlier due to α-Fe_2O_3 and γ -Fe_2O_3 [13] but are similar to those reported for β-

Fe_2O_3.[1] The UV-Vis represents the formed nanoparticles are β-Fe_2O_3. The prepared iron oxide nanoparticles were analyzed by Scanning Electron Microscopy (SEM) with Energy-dispersive X-ray spectroscopy (EDX) (Quanta-200F SEM with AMETEK EDX) to know the morphology and atomic percentages. Figure 3 shows the SEM image of Iron oxide nanoparticles prepared from 100 ml of 0.01 M aqueous Ferric chloride solutions with 20 ml of eucalyptus leaf extract respectively. SEM images which constitute large number of iron oxide nanoparticles with well dispersed and clear morphology. The HR-SEM also reveals that nanoparticles are evenly spreaded and well separated. From the SEM image the size of particles are measured as about 200 nm. The EDX analysis (Figure 3c) of these particles indicates the presence of Fe and O in the pure form.

Fig. 3: (a, b) SEM, and (c) EDX Spectra of Iron Oxide Nanoparticles

Structural characterization was performed using Powder X-Ray diffraction (XRD) analysis. Figure 4(a) shows the XRD pattern (SEIFERT, XRD with Cu-Kα radiation, λ = 1.540598Å) of Iron Oxide nanoparticles prepared by using *Eucalyptus globule* leaf extract. The prepared colloidal solution was calcinated under 300°C and the obtained brown colour powder was used for the XRD pattern.

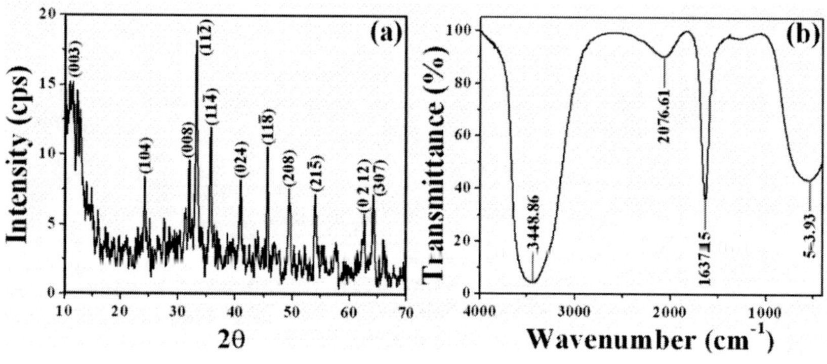

Fig. 4: (a) X-Ray Diffraction Pattern, and (b) Fourier Transform Infra Red Spectroscopy of Iron Oxide Nanoparticles

All the reflection peaks in this pattern were found to match with the β-Fe_2O_3 phase having hexagonal geometry (JCPDS 76–1821). The sample showed the major characteristic peaks for as grown crystalline iron oxide at 2θ values of 11.7 (003), 24.2 (104), 31.7 (008), 33.1 ($11\bar{2}$), 35.9 ($11\bar{4}$), 40.7 (024), 45.6 ($11\bar{8}$), 49.6 (208), 54.1 ($21\bar{5}$), 62.6 (0 2 12) and 64.8 (307) degrees. This indicates the crystallanity of the prepared nanoparticles are β-Fe_2O_3.

Figure 4(b) shows the FT-IR spectrum (Thermo Scientific, Nicolet is10 using KBr pellet technique) of prepared iron oxide nanoparticles. It displays three strong bands around 3449 cm^{-1} (br), 1637 cm^{-1} and 544 cm^{-1}. The observed bands are similar to those reported for β-Fe_2O_3.[14] The vibration bands are 544cm^{-1} (Fe-O str), 1637cm^{-1} (H_2O bending vibration) and a broad peak at 3448 cm^{-1} (H_2O str). Presents of organic molecule in the surface of iron oxide nanoparticles has the influence on the FT-IR peaks.[15] The broad peak observed around 544 cm^{-1} (Fe-O str), may due to the organic molecule which was from the leaf extract on the surface of iron oxide nanoparticles. The weak band at 2076 cm^{-1} may be due to the unsaturated Nitrogen (C≡N) compounds from the leaf extract.

CONCLUSION

In summery the present work provides a method to synthesize colloidal iron oxide by using *Eucalyptus globulus* leaf extract. From the UV-Vis and XRD we could conclude the prepared iron oxide nanoparticles were beta phase. The SEM image shows the nanostructure of β-Fe_2O_3 and size of particles are measured as about 200 nm. This technique has the ability of yielding crystalline β-Fe_2O_3.

REFERENCES

[1] Cornell, R.M. and Schwertmann, U., The Iron Oxides-Structure, Properties, Reactions, Occurrences and Uses, Wiley-VCH: Weinheim, Germany, 2003.

[2] Lisiecki, I., Billoudet, F. and Pileni, M.P., Control of the Shape and the Size of Copper Metallic Particles, *J. Phys. Chem.* 100, 4160–4166, 1996.

[3] Benz, M., Van der Kraan, A.M. and Prins, R., Reduction of aromatic nitrocompounds with hydrazine hydrate in the presence of an iron oxide hydroxide catalyst: II. Activity, X-ray diffraction and Mossbauer study of the iron oxide hydroxide catalyst, *J. Appl. Catal.*, A. 172, 149–157, 1998.

[4] Weiss, W. and Ranke, W., Surface chemistry and catalysis on well-defined epitaxial iron-oxide layers, *Prog. Surf. Sci.* 70, 1–151, 2002.

[5] Speliotis, D.E., Magnetic recording beyond the first 100 Years, *J. Magn. Magn. Mater.*, 193, 29–35, 1999.

[6] Kim, D.K., Zhang, Y., Voit, W., Rao, K.V., Kehr, J., Bjelke, B. and Mohamed, M., Superparamagnetic iron oxide nanoparticles for bio-medical applications, *Scr. Mater.*, 44, 1713–1717, 2001.

[7] Jordan, A., Scholz, R., Wust, P., Fahling, H. and Felix, R., Magnetic fluid hyperthermia (MFH): Cancer treatment with AC magnetic field induced excitation of biocompatible super paramagnetic nanoparticles, *J. Magn. Magn. Mater,* 201, 413–419, 1999.

[8] Nadagouda, M.N., Castle, A.B., Murdock, R.C., Hussain, S.M. and Varma, R.S., *In vitro* biocompatibility of nanoscale zerovalent iron particles (NZVI) synthesized using tea polyphenols, *Green Chem.*, 12, 114–122, 2010.

[9] Hoag, G.E., Collins, J.B., Holcomb, J.L., Hoag, J.R., Nadagouda, M.N. and Varma, R.S., Degradation of Bromothymol Blue by 'Greener' Nano-Scale Zerovalent Iron Synthesized Using Tea Polyphenols, *J. Mater. Chem.* 19, 8671–8677, 2009.

[10] Njagi, E.C., Huang, H., Stafford, L., Genuino, H., Galindo, H.M., Collins, J.B., Hoag, G.E. and Suib, S.L., Biosynthesis of Iron and Silver Nanoparticles at Room Temperature Using Aqueous Sorghum Bran Extracts, *Langmuir*, 27, 264–271, 2011.

[11] Tyagi, A.K. and Malik, A., Antimicrobial potential and chemical composition of Eucalyptus globules oil in liquid and vapour phase against food spoilage microorganisms. *Food Chemistry*, 126, 228–235, 2011.

[12] Toshima, N. and Yonezawa, T., Bimetallic nanoparticles-novel materials for chemical and physical applications, *New J. Chem.* 22, 1179–1201, 1998.

[13] Cherepy, N.J., Liston, D.B., Lovejoy, J.A., Deng, H. and Zhang, Z.J., Ultrafast Studies of Photoexcited Electron Dynamics in γ- and α-Fe_2O_3 Semiconductor Nanoparticles, *J. Phys. Chem. B.,* 102, 770–776, 1998.

[14] Kumar, A. and Singhal, A., Synthesis of colloidal β-Fe_2O_3 nanostructures-influence of addition of CO^{2+} on their morphology and magnetic behavior, *Nanotechnology*, 18, 475703 (7pp), 2007.

[15] Lee, J., Isobe, T. and Senna, M., Preparation of Ultrafine Fe_3O_4 Particles by precipitation in the presence of PVA at High pH, *J. Colloid Interface Sci.* 177, 490–494, 1996.

NOVEL RING LIKE METAL-ORGANIC ASSEMBLIES AS HOST FOR SMALL MOLECULE RECOGNITION

ROHITH P. JOHN, SUBASHIS PRADHAN, DOHYUN MOON[1] AND MYOUNG SOO LAH[2]

Department of Applied Chemistry, Indian School of Mines, Dhanbad, JH 826 004, India
[1]Pohang Accelorator Laboratory, Pohang, Kyung Buk 790 784
[2]AIRC, Ulsan National Institute of Science and Technology, Ulsan 689 798, South Korea
E-mail: rohithjohn@gmail.com

ABSTRACT

Four disc type metal-organic assemblies are reported. Two ligands used have different N-terminal residues that offer different sterric interaction to form octanuclear and dodeca nuclear metal-organic assemblies. The Mn^{3+} ions act as nodal points owing to their back to back bridging of pentadentate ligands. The octanuclear assembly can accommodate an ethanol molecule in its cavity, while dodecanuclear assembly is able to recognize cyclohexane, cyclooctane and adamntane in its hydrophobic cavity.

INTRODUCTION

Receptors for anions or small organic molecules assume significance owing to selective sensing. Therefore the design of receptors for anions involve, host systems that have suitably positioned NH groups for anionic recognition. For recognition of organic molecules several host systems are designed, some of which are derived from calixarene systems that encapsulate the guest molecules.[1] This recognition is driven by favourable interaction between host and the guest either via their functional groups or through a size selective process. The effective trapping of small organic molecules could be useful in trapping intermediates or in achieving catalysis with unusual product selectivity.[2,3] Similar observations are also made on metal-organic assemblies that have the advantage of tailoring cavity size.[4,5] In this

V. Rajendran, P. Prabu and K.E. Geckeler (eds.)
Nano Biomaterials, pp. 167–172 (2012).

context we have prepared cyclic metal-organic assemblies some of which can act as host that can effectively accommodate small organic molecules. We have reported in the past, disc like metal-organic assemblies that have hydrophobic groups directed to its central cavity.[6] Herein we report synthesis and structural characterization of an octanuclear manganese complex, $[MnL^1EtOH]_8 \supset EtOH$, and a dodecanuclear complex that encapsulates four different molecules, viz., $MnL^2MeOH]_{12} \supset C_6H_{12}$, $MnL^2MeOH]_{12} \supset C_8H_{16}$, $MnL^2MeOH]_{12} \supset C_{10}H_{16}$.

Experimental

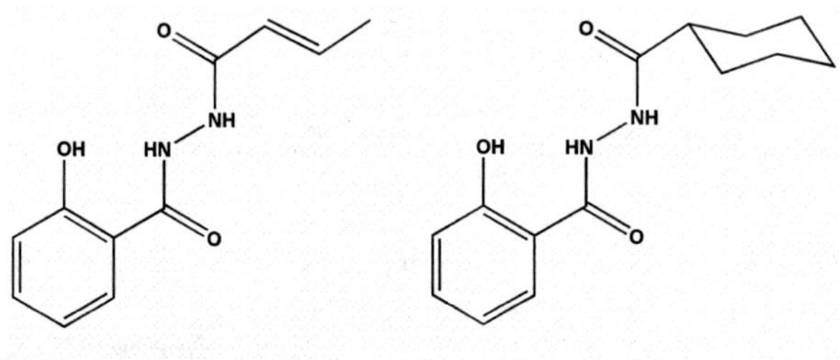

H3L1 H3L2

Fig. 1: Ligands H_3L^1 and H_3L^2 Used for Preparing Assemblies 1–4

MATERIALS AND METHODS

All reagents were obtained from commercial sources and were used as received. The ligands H_3L^1 and H_3L^2 were prepared as reported earlier.

Preparation of the Assemblies

$[MnL^1EtOH]_8 \supset C_2H_5OH$ (1): 0.1 mmol (22.0 mg) of the H_3L^1 was dissolved in 2 mL of DMF and was mixed with 6mL of ethanol and 2mL of methanol. To this solution 0.1mmol (24.5 mg) of manganese(II) acetate tetrahydrate was added. X-ray quality crystals were separated in 5-6 days. The product was filtered and washed with 8:2 mixture of ethanol and DMF, dried under vacuum. Elemental data: Found C, 45.52; H, 4.08; N, 9.41%. Calc. for $[Mn_8(L^1)_8(H_2O)_8]\cdot(EtOH)$: C, 45.64; H, 3.97; N, 9.47%.

$MnL^2MeOH]_{12} \supset C_6H_{12}$ (2): 0.1 mmol (26.3 mg) of the H_3L^2 was dissolved in 3 mL of DMF and was mixed with 7 mL of methanol. To this solution 0.1 mmol (24.5 mg) of manganese(II) acetate tetrahydrate was added followed by 15 μL of cyclohexane. X-ray quality crystals were separated in 3 days. The product was filtered and washed with 7:3 mixture of MeOH and DMF, dried under vacuum. Elemental data: Found C, 51.49; H, 6.21; N, 11.41%. Calc. for $[Mn_{12}(L^2)_{12}(dmf)_{12}]\cdot 3(dmf)\cdot 6(H_2O)$: C, 51.42; H, 6.02; N, 10.98%.

$[MnL^2MeOH]_{12} \supset C_8H_{16}$ (3): 0.1 mmol (26.3 mg) of the H_3L^2 was dissolved in 4 mL of DMF and was mixed with 6mL of methanol. To this solution 0.1 mmol (24.5 mg) of manganese(II) acetate tetrahydrate was added followed by 20 μL of cyclooctane. X-ray quality crystals were

separated in 3–4 days. The product was filtered and washed with 6:4 mixture of MeOH and DMF, dried under vacuum. Elemental data: Found C, 53.96; H, 6.08; N, 11.19%. Calc. for $[Mn_{12}(L^2)_{12}(dmf)_{12}]\cdot(dmf)\cdot(C_8H_{16})$,: C, 53.43; H, 5.99; N, 10.72%.

$[MnL^2MeOH]_{12} \supset C_{10}H_{16}$(4): 0.1 mmol (26.3 mg) of the H_3L^2 was dissolved in 4 mL of DMF and was mixed with 6 mL of methanol. To this solution 0.1mmol (24.5 mg) of manganese(II) acetate tetrahydrate was added followed by 1.5 mg of adamantane. X-ray quality crystals were separated in 3 days. The product was filtered and washed with 6:4 mixture of MeOH and DMF, dried under vacuum. Elemental data: Found C, 53.15; H, 5.72; N, 9.82%. Calc. for $[Mn_{12}(L^2)_{12}(H_2O)_8(dmf)_4]\cdot(C_{10}H_{16})$,: C, 53.16; H, 5.53; N, 9.82%.

X-ray Crystallography

Data collection for 2 and 3 were performed at –100°C with Mo Kα radiation (λ = 0.71073 Å) on a Bruker SMART X-ray diffractometer with a graphite crystal and incident-beam monochromator. The SMART and SAINT software packages[7] were used for data collection and integration, respectively. Collected data were corrected for absorbance using SADABS based upon Laue symmetry using equivalent reflections. The difference Fourier synthesis and the subsequent least square refinement revealed remaining structural solvents. The crystal structure of 2 contains at least three dmf sites and one water site with partial occupancies per asymmetric unit. A disordered cyclohexane site was observed at the inner cavity of the cyclic metal organic assembly. The crystal structure of 3 contained at least 3 dmf sites and one disordered water/methanol site. A cyclooctane molecule was observed at the inner cavity of metal-organic assembly. The free solvent molecules were then removed and the SQUEEZE routine[8] was performed to modify the disordered electron density structures of both 2 and 3. The cyclohexane/cyclooctane with prolate ellipsoidal thermal factors could be resolved using two disordered cyclohexane/cyclooctane molecules with half occupancy factors, where the 1–2 and 1–3 carbon distances of the cyclohexane/cyclooctane are restrained at 1.54 Å and 2.51 Å respectively.

For 1 and 4 diffraction data was measured using using a synchrotron radiation source 4AMXW ADSC Quantum –210 detector with a platinum coated silicon double crystal monochromator located at PAL, Korea. The HKL 2000 software was used for data collection, cell refinement, reduction and absorption correction. The structures were solved using direct methods and refined using either block diagonalised-matrix or full matrix least squares using SHELXTL-PLUS software.

RESULTS AND DISCUSSION

All the host systems assembled loses coordinated solvent molecules such as methanol and ethanol, upon applying vacuum. Their respective sites are subsequently replaced by water molecules when exposed to air. This results in discrepancies in the elemental analysis results. However most of the less volatile solvents such as dmf are retained in the sample unless it is washed excessively with methanol/ethanol.

Table 1: Crystallographic Data for 1–4

	1· 2EOH· 2MeOH	*2*	*3*	*4· 9dmf· 10w· 2.5MeOH*
Empirical formula	$Mn_8C_{112}H_{146}N_{16}O_{37}$	$Mn_{12}C_{186}H_{240}N_{24}O_{48}$	$Mn_{12}C_{188}H_{244}N_{24}O_{48}$	$Mn_{12}C_{221}H_{339}N_{33.50}O_{70}$
Formula weight	2747.97	4239.30	4267.35	5244.54
Temperature/*K*	90(2)	173(2)	173(2)	90(2)
Wavelength/Å	0.70000	0.71073	0.71073	0.70000
Crystal system	Tetragonal	Tetragonal	Tetragonal	Tetragonal
Space group	*P4/n*	$P4_2/n$	$P4_2/n$	$P4_2/n$
Unit cell dimensions				
a,b,c/Å	23.467(3), 23.467(3), 11.039(2)	27.189(3), 27.189(3), 18.904(4)	27.073(2), 27.073(2), 18.909(3)	27.048(4), 27.048(4), 18.895(4)
α, β, γ/°	α = β = γ = 90	α = β = γ = 90	α = β = γ = 90	α = β = γ = 90
Volume/$Å^3$	6079.4(17)	13974(3)	13860(3)	13823(4)
Z	2	2	2	2
μ mm^{-1}	0.890	0.580	0.586	0.607
Size/mm^3	0.40 × 0.25 × 0.18	0.80 × 0.50 × 0.30	1.25 × 0.88 × 0.65	0.30 × 0.25 × 0.10
θ/°	1.21 to 28.00	1.51 to 28.39	1.51 to 28.36	1.82 to 30.29
Reflections collected	27533	67612	73193	71970
Completeness to theta/%	96.0	91.7	97.3	98.5
Goodness-of-fit on F^2	1.038	1.057	1.061	1.024
R indices (all data)	R1 = 0.0531, wR2 = 0.1395	R1 = 0.1068, wR2 = 0.2049	R1 = 0.1174, wR2 = 0.2515	R1 = 0.0966, wR2 = 0.2473
Largest diff. peak and hole/e. $Å^{-3}$	0.834 and –0.722	0.796 and –0.617	0.873 and –0.711	1.380 and –0.903

X-ray quality single crystals were harvested from the mother liquor and coated with paratone oil before mounting over the goniometer head. The loss of volatile solvents affects the quality of the crystal upon exposure to air. All the elemental analyses were done on samples that were dried under vacuum after washing the harvested crystals using the same solvent combination as the mother liquor. It was observed that the trapped ethanol in assembly 1 was retained, while the coordinated ethanol molecules were replaced by water. This might be due to the rigid orientation of the N-terminal groups that prevent escape of the trapped ethanol. The trapped ethanol is disordered with the terminal methyl groups directed toward the hydrophobic tail units of the ligand, while the hydrophilic end is located at the center of the cavity (Figure 2).

The assembly 2 can also be prepared by addition of 15μL of cyclohexane into the solution once the crystals are formed. Assembly 3 and 4 can also be prepared in a similar fashion with the addition of respective guest molecules.

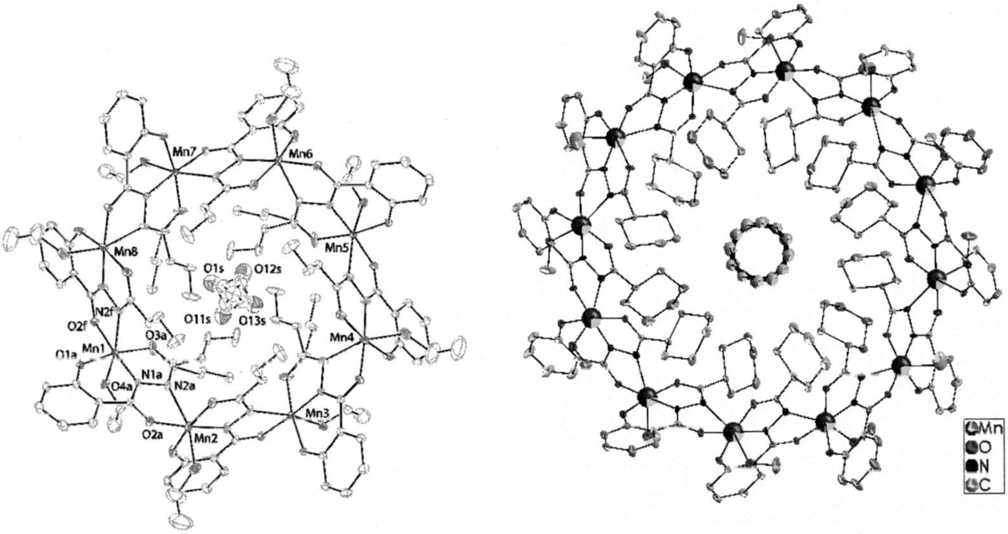

Fig. 2: An ORTEP Drawing of Assembly 1 with Disordered EtOH at the Cavity with 30% Ellipsoidal Probability and Diamond Graphics of Assembly 2 with Disordered Cyclohexane at the Cavity

It is also observed that a weakly bound guest such as cyclohexane or cyclooctane can be displaced from the cavity without disturbing the overall assembly. The above observation is an evidence for the preference of the cavity for hydrophobic molecules, and for a better fitting guest whenever available. The preference for adamantane over cyclohexane or cyclooctane may arise from maximizing of hydrophobic contact between the guest and the hydrophobic tails of the host assembly that form the flexible entry-exit port.

The assembly 2 can also be prepared by addition of 15 μL of cyclohexane into the solution once the crystals are formed. Assembly 3 and 4 can also be prepared in a similar fashion with the addition of respective guest molecules. It is also observed that a weakly bound guest such as cyclohexane or cyclooctane can be displaced from the cavity without disturbing the overall assembly. The above observation is an evidence for the preference of the cavity for hydrophobic molecules, and for a better fitting guest whenever available. The preference for adamantane over cyclohexane or cyclooctane may arise from maximizing of hydrophobic contact between the guest and the hydrophobic tails of the host assembly that form the flexible entry-exit port.

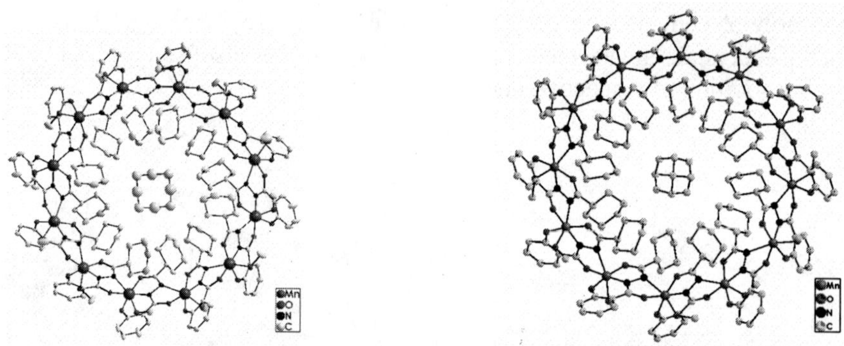

Fig. 3: A Diamond Graphic of 3 (left) with Cyclooctane in the Cavity and 4 (right) with Adamantane in the Cavity (disorder is omitted for clarity)

CONCLUSIONS

Four host-guest assemblies were prepared by self assembly method. The cavity shows a preference for small hydrophobic molecules such as cyclohexane, cylooctane and adamantane. Adamantane is preferred over lower cyclic hydrocarbons on account of better hydrophobic interaction with the host.

ACKNOWLEDGEMENTS

RJ thanks KRF for support grant and PAL for beamline use, SP thanks DST, SERC for a research fellowship.

REFERENCES

[1] Konrad Tiefenbacher and Julius Rebek, Jr., *J.Am. Chem. Soc.,* 2012, 134, 2914–2917.

[2] Berryman, O.B., Sather, A.C., Lledó, A. and Rebek Jr, J., *Angew. Chem. Int. Ed.*, 2011, 50, 9400–9403.

[3] Ams, M.R., Ajami, D., Craig, S.L., Yang, J.-S. and Rbeek, J., Jr., *J. Am. Chem. Soc.*, 2009, 131, 13190–13191.

[4] Murase, Takashi, Nishijima, Yuki and Fujita, Makoto, *J. Am. Chem. Soc.*, 2012, 134, 162–164.

[5] Ikemoto, Koki, Inokuma, Yasuhide and Fujita, Makoto, *J. Am. Chem. Soc.*, 2011, 133, 16806–16808.

[6] John, R.P., Lee, K., Kim, B.J., Suh, B.J., Rhee, H. and Lah, M.S., *Inorg. Chem.*, 2005, 44, 7109–7121.

[7] Sheldrick, G.M. Shelxtl-Plus, Crystal Structure Analysis Package; Bruker Analytical X-Ray, Madison, WI, USA, 1997.

[8] Spek, A.L., *Acta Crystallogr, Sect. A.,* 1990, 46, 194–201.

BIOGENIC SILVER NANOPARTICLES' *BOMBYX MORI* NUCLEAR POLYHEDROSIS VIRUS (BmNPV) INHIBITORY MECHANISM

S. TAMILSELVAN, T. ASHOKKUMAR, R. GEETHA, K. GOVINDARAJU[1], A. MOHAMED SADIQ[2] AND G. SINGARAVELU

Nanoscience Division, Department of Zoology, Thiruvalluvar University, Vellore
[1]Centre for Ocean Research, Sathyabama University, Chennai
[2]Department of Biochemistry, Aadhiparasakthi College of Arts and Science, Kalavai, Tamil Nadu
E-mail: gsvelu@gmail.com

ABSTRACT

Green chemistry approach on synthesizing noble metal nanomaterials is a powerful exciting technology to functionalize nanomaterials. Results addressed herein demonstrate interaction of biogenic silver nanoparticles on Bombyx mori nuclear polyhedrosis virus (BmNPV) inhibitory mechanism. SDS-PAGE analysis discloses that 35 kDa viral protein seems to be the target of Spirulina synthesized silver nanoparticles. HRSEM and TEM analysis clearly reveals that silver nanoparticles are accumulated on the outer membrane and also found to be penetrated and attached with DNA which has been confirmed with EDAX. AFM and Confoccal studies clearly indicate the denature of BmNPV. This might novel host protective epitope(s) and triggers insect innate immune systems. Present results highlight the possibility of nanoscience intervention on viral therapeutics.

INTRODUCTION

Nanotechnology is an emerging interdisciplinary cutting edge technology that has been booming in many areas including biology, chemistry and material science. Its profound societal impact has been considered at the huge momentum to usher in a second industrial revolution. Among

V. Rajendran, P. Prabu and K.E. Geckeler (eds.)
Nano Biomaterials, pp. 173–178 (2012).

noble metal nanoparticles silver nanoparticles has received considerable attention because of their unique electrical, optical as well as biological properties and are thus applied in catalysis,[1] biosensing,[2] imaging,[3] drug delivery,[4] nanodevice fabrication and medicine.[5,6]

Silver nanoparticles have a high specific area and a high fraction of surface atoms that lead to high antimicrobial activity compared to bulk silver metal.[7] Biological synthesis of silver nanoparticles at present seems to be of much interest due to the key advantage "ecofriendly nature". It has been demonstrated that by apt selection of reducing and stabilizing molecule it is possible to functionalize nanomaterials.[8] We have reported that silver nanoparticles synthesized using single cell protein *Spirulina platensis* found to possess antiviral activity against silkworm viral pathogen BmNPV [9] likewise *Psidium guavajava* synthesized gold nanoparticles found to possess antidiabetic activity which has been assessed by PTP 1B inhibitory activity.[8] Herein, we address on the inhibitory mechanism of BmNPV by *Spirulina* synthesized silver nanoparticles.

Bombyx mori nuclear polyhedrosis virus (BmNPV) is a circular double standard DNA virus belonging to the family baculoviridae, causing the deadly grasserie disease which is a scourge to silk industry.

MATERIALS AND METHODS

Synthesis of Silver Nanoparticles

Silver nanoparticles was synthesized using single cell protein *Spirulina platensis* as per our earlier report.[10] In brief, 500 mg of *S. platensis* biomass in a 250 mL Erlenmeyar flask with 100 mL of 10^{-3} M aqueous $AgNO_3$ was incubated at room temperature. The bioreduction of pure $AgNO_3$ was monitored using UV-vis spectrophotometer at regular intervals.

Isolation and Purification of BmNPV

BmNPV infected silkworm was collected from rearing from at lalapet, Vellore District, Tamilnadu. Haemolymph was collected from the diseased worms exhibiting typical symptoms, shortly before death by just cutting the front pair prolegs. The turbid milky haemolymph was stored in sterilized glass tubes and subjected to refrigeration for several days. Polyhedra settle at the bottom. After removing from the refrigerator, the haemolymph was diluted by adding distilled water and kept at 5 °C overnight. The haemolymph was filtered through two layers of cheese cloth. The filtrate was subjected to centrifugation at 15,000 rpm for 15 min using refrigerated centrifuge. The pellet was resuspended in half the quantity of original volume of sterile distilled water and centrifuged at 5000 rpm for 15 min. Repeated cycles of centrifugation at 15,000 rpm followed by 5000 rpm were done for 15 min each to obtain milky white amorphous sediment of highly purified nuclear polyhedra. Suspension was stored in distilled water at 4°C. [11,12]

SDS-PAGE Analysis

Interaction of silver nanoparticles with BmNPV was analyzed on SDS-PAGE according to Laemmli [13] to determine the target protein of viral inhibition.

Transmission Electron Microscopy (TEM) and High Resolution Electron Microscopy (HRSEM) with Energy Dispersive X-ray Analysis (EDAX) Studies

BmNPV was isolated from the infected silkworm and treated with silver nanoparticles and were mounted into copper stubs and coated with gold. The surface observation of the polyhedra interaction with silver nanoparticles was analyzed using TEM and HRSEM studies. The percentage of nanoparticles incorporated in the polyhedra was measured by Energy Dispersive X-ray analysis (EDAX).

Atomic Force Microscopy (AFM) and Confoccal Microscopy (CM) Studies

Atomic force microscopical and confocal microscopical analysis were made using Nanonics Multiview 1000 microscope equip with NWS software. In both AFM and CM, the samples were scanned along x- and y-axes, while the z-movement was effected by specially designed Z-Modulator.

Results and Discussion

Silk is one of the nature's gifts to mankind produced by silkworm. Among silkworm the most commercially exploited one are mulberry silkworm *B. mori* L. But it is unfortunate that the cultivation of silkworm has been seriously affected due to the outbreak of *Bombyx mori* nuclear polyhedrosis virus (BmNPV). The lack of effective therapeutics for BmNPV-caused viral diseases underscores the importance of searching for antiviral material.

Pharmaceuticals that can inactivate viruses and at the same time are low-toxic which represent a potential therapeutics with respect to both, treatment of the disease and vaccine production. Results addressed herein is such as a finding. Among the noble nanomaterials silver nanoparticles have received considerable attention due to their attractive physicochemical properties. The surface plasmon resonance and large effective scattering cross section of individual silver nanoparticles make them ideal candidates for molecular labelling,[14] where phenomena such as Surface Enhanced Raman Scattering (SERS) can be exploited. Present findings highlight the possibility of nanoscience intervention on viral therapeutics.

Result demonstrates the interaction of *Spirulina* synthesized silver nanoparticles with BmNPV. Silvernanotheraphy on the viral borne disease of silkworm *Bombyx mori* has been demonstrated by our research group.[9] In continuation, herein we report the inhibitory mechanism of BmNPV by silver nanoparticles synthesized using single cell protein *Spirulina.*

Interaction of silver nanoparticles with BmNPV was studied using SDS-PAGE, to identify the target viral protein of *Spirulina* synthesized silver nanoparticles. Results (Figure 1) show that eventhough different molecular weight proteins of BmNPV are identified a significant reduction was observed with 35 kDa protein when treated with *Spirulina* synthesized silver nanoparticles. Thus it is reasonable to infer that *Spirulina* synthesized silver nanoparticles directly binds with 35 kDa viral protein which leads to loss of stability.

The alteration of the structure and morphology of polyhedra was studied by TEM analysis. Figure 2, clearly shows that the silver nanoparticles were accumulated on the outer membrane, while some of them successfully penetrated into cell. In addition, leaking of interacellular substances and coagulation of silver nanoparticles at the polyhedra surface are also seen in the same TEM micrograph.

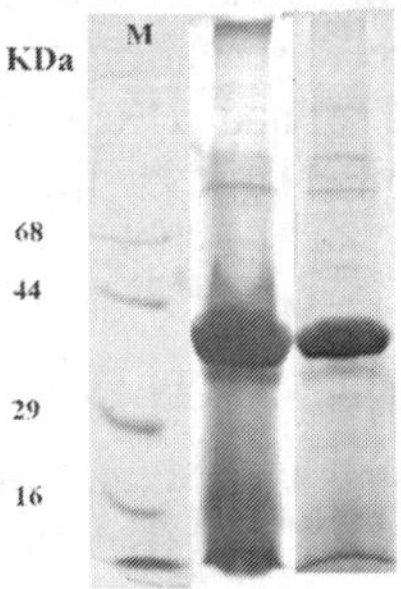

Fig. 1: SDS-PAGE Analysis Shows Polyhedra Treated with Silver nanoparticles. Lane 1-Standard marker, Lane 2-Proteins from pure polyhedra, Lane 3-silver nanoparticles treated polyhedra

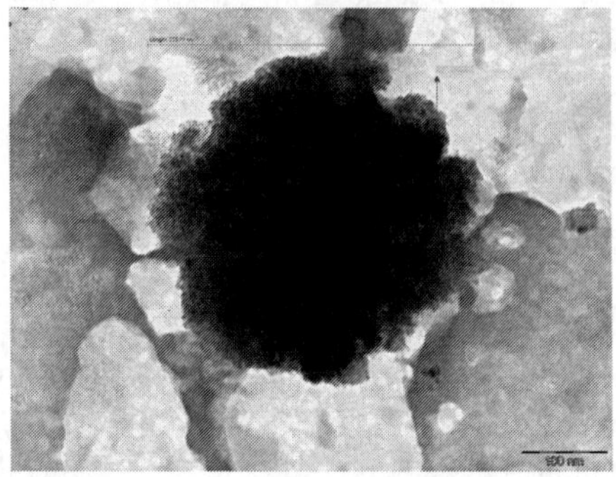

Fig. 2: Effect of Spirulina Synthesized Silver Nanoparticles Against BmNPV

SEM microscopy was used to identify the surface morphology of polyhedra treated with silver nanoparticles. The treated polyhedron shows the binding of silver nanoparticles to the BmNPV surface (Figure 3). The EDAX spectrum (Figure 4) showed the signal for silver nanoparticles which suggest that *Spirulina* synthesized silver nanoparticles were incorporated into the membrane of polyhedra and to the viral DNA.

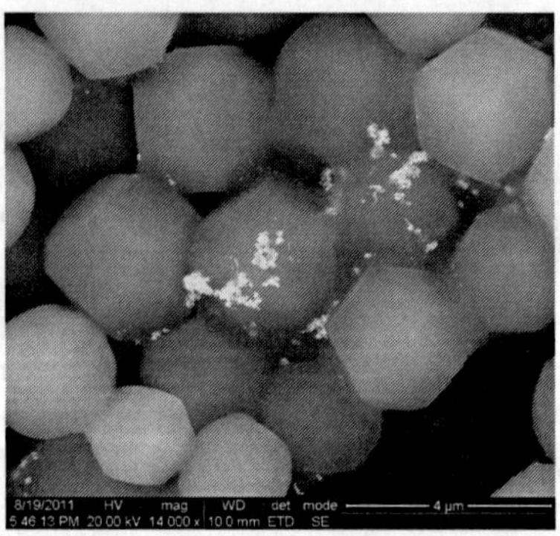

Fig. 3: SEM Image of BmNPV Treated with Silver Nanoparticles

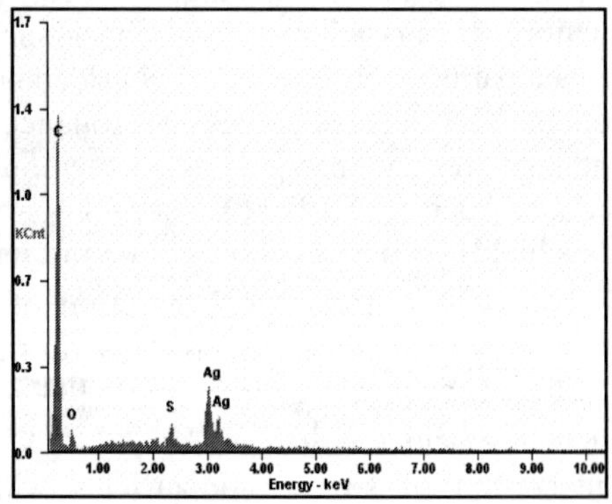

Fig. 4: EDAX Pattern of Silver Nanoparticles

Figure 5 the topography image of AFM (Figure 5) and CM (Figure 6) clearly show the damage in the surface of polyhedra and it might be due to the exposure of silver nanoparticles to polyhedra. This in turn, might influence innate immunological response in silkworm larvae and increases its life span. We have reported that *Spirulina platensis* synthesized silver nanoparticles are having potential antiviral activity against *Bombyx mori* nuclear polyhedrosis virus at 20 μL concentration against 2.22×10^6 polyhedral suspension inoculated silkworm larvae.[9] Silver nanoparticles subjected in the study is synthesized using *Spirulina platensis* which is a single cell protein, presumably that the silver nanoparticles with capping molecule protein may provide the property of antiviral activity. Indeed, the surface chemistry of the nanomaterials can modify their interactions with the external systems is accountable here. TEM, HR-SEM micrographs clearly reveals that the structure and morphology of polyhedra damaged and EDAX spectrum confirm the incorporation of silver nanoparticles in to the polyhedra which presumably inhibit BmNPV effectively. It was shown that the end group on the particle's surface is significant for the interaction between silver nanoparticles and the BmNPV.

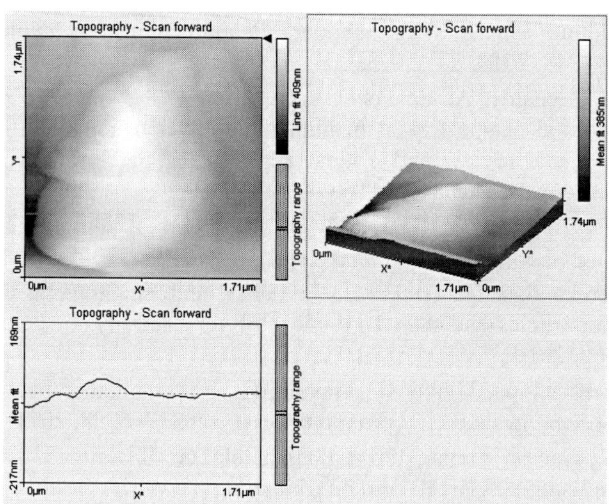

Fig. 5: AFM Depicts Ravage of Viral Structure and Its Surface

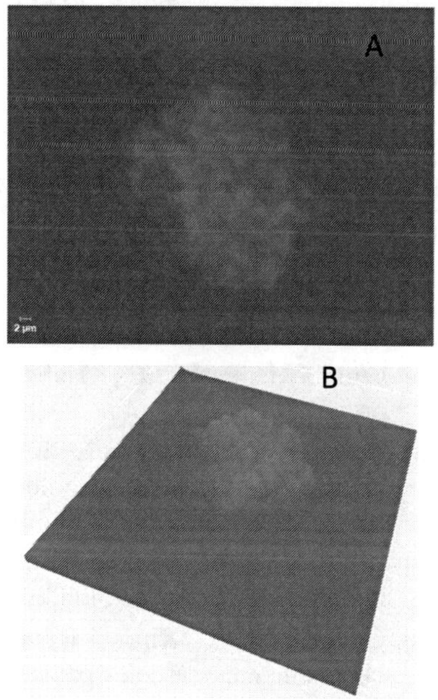

Fig. 6: A. Confocal Microscopy Shows Damaged Morphology of Polyhedra with Silver Nanoparticles B. 3D Image

Based on the results it is presumed that the silver nanoparticles could inhibit envelop glycoprotein knobs which might leads to failure on the interaction with host system. Results suggest *Spirulina* synthesized silver nanoparticles could be a good candidate as disinfectant as well as antiseptic to prevent the viral infection. Research on interaction of inorganic nanoparticles with biosystems has been started recently and its applications are going to be discovered. Based on the findings it is clear that ethnopharmacological approaches on the functionalization of silver nanoparticles may break through the limitations in infectious diseases.

ACKNOWLEDGMENT

GS thanks the Indian Council of Medical Research (ICMR), New Delhi for financial assistance.

REFERENCES

[1] Valden, M., Lai, X. and Goodman, D.W., Onset of catalytic activity gold clusters on titania with the appearance of nanometallic properties, *Science*, 281, 1647–1650, 1998.

[2] Mirkin, C.A., Letsinger, R.L., Mucic, R.C and Storhoff, J.J., A DNA – based method for rationally assembling nanoparticles into macroscopic materials, *Nature*, 382, 607–609, 1996a.

[3] Bielinska, A., Eichman, J.D., Lee, I., Baker, J.R and Balogh, L., Imaging {Au0-PAMAM} gold-dendrimer nanocomposites in cells, *J. Nanopart. Res.*, 4, 395–403, 2002.

[4] Niemeyer, C.M., Nanoparticles, proteins, and nucleic acids: biotechnology meets materials science, *Angew.Chem.Int.Ed.*,40, 4218–4158, 2001.

[5] Ahamed, M., Alshali, M.S and Siddiqui, M.K.J., Silver nanoparticle applications and human health, *Clin. Chim. Acta*,411,1841–1848, 2010.

[6] Tolaymat, T.M., Ei Badawy, A.M., Genaidy, A., Scheckel, K.G., Luxton, T.P and Suidan, M., An evidence – based environmental perspective of manufactured silver nanoparticle in syntheses and application: A systematic review and critical appraisal of peer- reviewed scientific papers, *Science of the total einvironment,*408, 999–1006, 2010.

[7] Cho, K.H., Park, J.E., Osaka, T and Park, S.G., The study of antimicrobial activity and preservative effects of nanosilver ingredient, *Electrochimica Acta*, 51, 956–960, 2005.

[8] Basha, K.S., Govindaraju, K. Manikandan, R., Ahn, J.S., Bae, E.Y and Singaravelu, G, Phytochemical mediated gold nanoparticles and their PTP 1B inhibitory activity. *Colloids Surf. B Biointerfaces,* 75(2),405–409, 2010.

[9] Govindaraju, K., Tamilselvan, S., Kiruthiga, V and Singaravelu, G., Silvernanotherapy on the viral borne disease of silkworm *Bombyx mori* L, *J. Nanopart Res*., 3, 6377–6388, 2011.

[10] Govindaraju, K., Khaleel Basha, S., Ganesh Kumar, V and Singaravelu, G., Silver, gold and bimetallic nanoparticles production using single cell protein (*Spirulina platensis*) Geitler. *J. Mater. Sci.,* 43,5115–5122, 2008.

[11] Kawarabata,T and Matsumoto, K., Isolation and structure of a nuclear polyhedrosis virus from polyhedra of the silkworm, *B. mori*, *Appl Ent Zool.,* 8, 227–233, 1973.

[12] Govindan, R., Narayanaswamy, T.K and Devaiah, M.C., Principles of silkworm pathology. *Seri Scientific Publication*, Bangalore, pp 64–65,1998.

[13] Laemmli, U.K., Cleavage of structural proteins during the assembly of the head bacteriophage T4, *Nature,* 227, 680–685, 1970.

[14] Sandhu, K.K., McIntosh, C.M., Simard, J.M., Smith, S.W and Rotello, V.M., Gold nanoparticle-mediated transfection of mammalian cells, *Bioconjug. Chem.* B, 13, 3–6, 2002.

BIOSYNTHESIS OF SILVER NANOPARTICLES USING THE EXTRACT OF *ALTERNANTHERA SESSILIS*

M. JANNATHUL FIRDHOUSE AND P. LALITHA

Department of Chemistry, Avinashilingam Institute for Home Science and Higher Education for Women University, Coimbatore, Tamil Nadu
E-mail: goldenlalitha@gmail.com

ABSTRACT

Green synthesis of silver nanoparticles was carried out using the aqueous extract of Alternanthera sessilis under various conditions. Alternanthera sessilis was found to exhibit strong potential for rapid reduction of silver ions. The synthesized silver nanoparticles were characterized with UV-visible absorption spectroscopy, FTIR, XRD and SEM. X-ray diffraction and SEM analysis showed that the average particle size is less than 100 nm. The present work of biosynthesis of silver nanoparticles using Alternanthera sessilis appears to be cost effective, eco-friendly and an alternative to conventional method of synthesis.

INTRODUCTION

Nanotechnology has gained ground in the twenty-first century and is rapidly growing due to the ability to manipulate and harness properties of assemblies that are at the nanosize scale of various biomolecules.[1] Nanoparticles exhibit completely new or improved properties based on specific characteristics such as size, distribution and morphology.[2] In the recent times, the advances in the field of nanosciences and nanotechnology has brought to fore the nanosized inorganic and organic particles which are finding increasing applications in industrial, medicine and therapeutics, synthetic textiles and food packaging products.[3]

Biosynthesis of nanoparticles using plant extracts is the favourite method of green, eco-friendly production of nanoparticles and exploited to a vast extent because the plants are

V. Rajendran, P. Prabu and K.E. Geckeler (eds.)
Nano Biomaterials, pp. 179–184 (2012).

widely distributed, easily available, safe to handle and with a range of metabolites.[4] The successful use of silver nanoparticles [Ag NPs] in diverse medical streams as antifungal and antibacterial agents has led to their applications in controlling phytopathogens.[5–6]

Alternanthera sessilis is a weed and occurs in both wet lands and uplands and can grow on a variety of soil types. It is a weed of rice throughout tropical regions and of other cereal crops, sugarcane and bananas. Although it is a weed, it has many utilities. In south-east Asia young shoots and leaves are ingested as vegetables. Previous phytochemical studies have reported the isolation of flavonols, triterpenoids, steroids and tannins; β-sitosterol, stigmasterol, campesterol, lupeol being few of its important constituents. The herb has been reported to have antipyretic, hepatoprotective, antiulcer, antibacterial, hematinic, diuretic activities.[7]

In the present work, we explored the green synthesis of silver nanoparticles using aqueous extract of *Alternanthera sessilis* as an alternative to chemical methods of synthesis. The synthesized silver nanoparticle was found to act as reducing as well as an capping agent.

EXPERIMENTAL

Fresh leaves of *Alternanthera sessilis* (20 g) were weighed and washed thrice with Millipore water. Then it is transferred into 500 ml Erlenmeyer flask and boiled for 5 mins with 100ml of Millipore water. The extract was filtered using Whatman filter paper and refrigerated for further use.

The aqueous extract of *Alternanthera sessilis* was treated with 3mM of silver nitrate solution under various conditions i.e room temperature (27–30°C), higher temperature (75°C) and sonication using ultrasonic bath (PCI Ultrasonics 1.5 L (H)). The reddish brown colour silver solution was centrifuged (Spectrofuge 7M) at 13000 rpm for 15 minutes. The silver nanoparticles were redispersed in water, centrifuged again and the supernatant solutions were analyzed.

UV-Visible absorption spectra were measured using Double beam spectrophotometer 2202- (SYSTRONICS). The size of synthesized silver nanoparticles was examined by X-ray diffraction analysis (SHIMADZU Lab X XRD-6000) with a Cu Kα radiation monochromatic filter in the range 10–80°. Morphology and size of silver nanoparticles were investigated by Scanning Electron Microscope using TESCAN instrument provided with Vega TC software. The FTIR spectrum (BRUKER FTIR Tensor-27) was also recorded for the synthesized nanoparticles.

RESULTS AND DISCUSSION

The detailed study on biosynthesis of silver nanoparticles by aqueous extract of *Alternanthera sessilis* with aqueous silver nitrate solution is reported in this work. The aqueous silver ions were reduced to silver nanoparticles when aqueous extract of *Alternanthera sessilis* was added. After 6 hrs, the yellow coloured solution was changed to reddish brown colour which indicates the formation of silver nanoparticles as shown in Figure 1. The formation of the silver nanoparticles was monitored by UV- visible spectrophotometric analysis. The UV-visible spectra showed the maximum absorbance at 420 nm corresponding to the surface Plasmon resonance of silver nanoparticles shown in Figure 2.

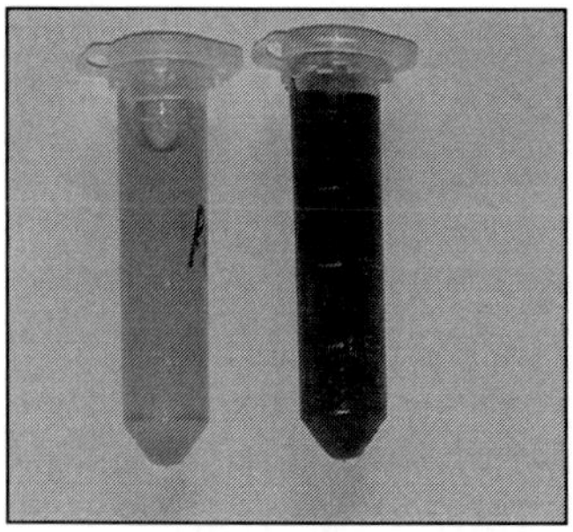

Fig. 1: Aqueous Extract of *Alternanthera sessilis* and Synthesized Silver Nanoparticles from Silver Nitrate Solution

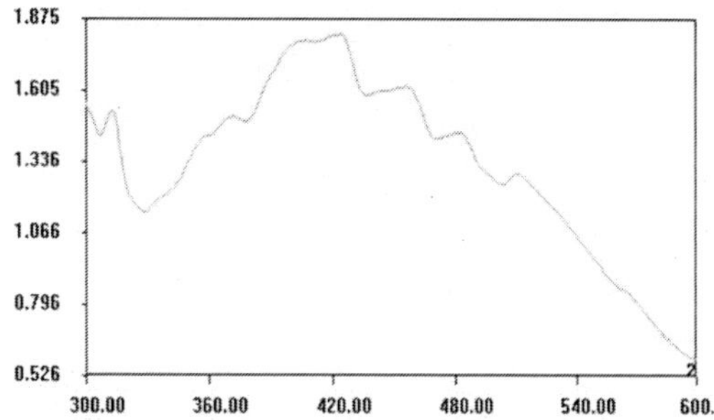

Fig. 2: UV-Visible Absorption Spectra of Silver Nanoparticles from Extract of *Alternanthera sessilis* and Silver Nitrate Solution

A comparative study on various experimental conditions was carried out to identify the effect of aqueous extract of *Alternanthera sessilis* and silver nitrate solution on the rate of bioreduction of silver ions. It is observed that the sonication method results in easy and rapid synthesis of silver nanoparticles as compared to other methods shown in Figure 3 because the ultrasound has ability to create clean, highly reactive surfaces on metals and thereby enhancing the rate of reactions.

Figure 4 shows the XRD patterns of drop coated silver nanoparticles synthesized using aqueous extract of *Alternanthera sessilis.* The XRD pattern shows a one intense peak of Bragg's reflection with 2θ values of 32.30° which may be indexed to the (101) based on the face centered cubic structre of silver nanoparticles. The particle size of the silver nanoparticles was calculated using Scherrer's equation is given in Table 1.

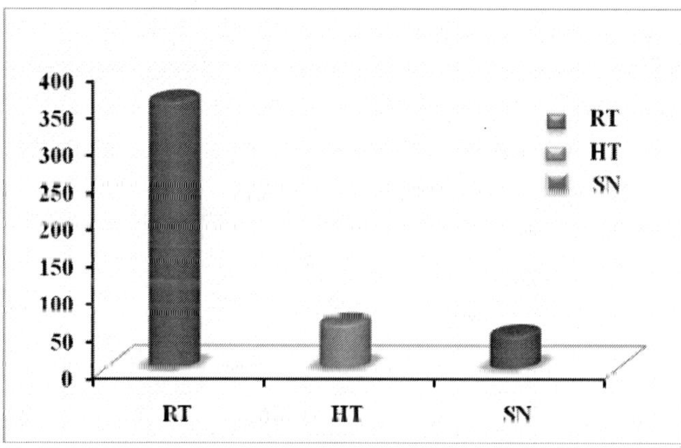

Fig. 3: Comparative Experimental Study on the Synthesis of Silver

Nanoparticles and its Variation with Time

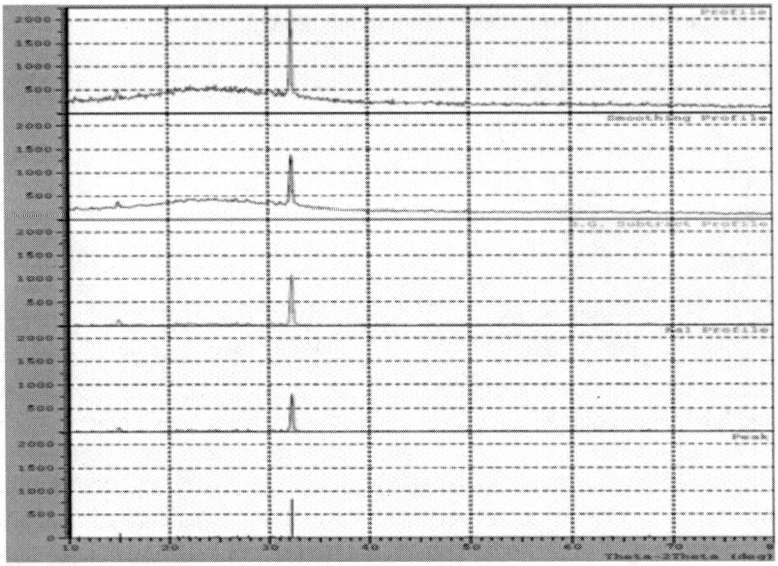

Fig. 4: XRD Patterns of Silver Nanoparticles Synthesized Using *Alternanthera sessilis*

Debye-Scherrer's Equation

The Debye-Scherrer's equation is commonly used to determine the crystalline size of nanoparticles.

$$D = k\,\lambda/\beta.\,\text{Cos}\theta$$

where,

D – Average crystalline size (nm)
k – Dimensionless shape factor (0.9)
λ – X ray wavelength (0.1541 nm)
β – Angular/line broadening at FWHM of the XRD peak at the diffraction angle
θ – Diffraction angle

Figure 5 represents the SEM image recorded from drop coated film of the silver nanoparticles synthesized using aqueous extract of *Alternanthera sessilis*. The SEM image showed spherical shape of silver nanoparticles formation with diameter range 30–50 nm.

Table 1: Determination of Crystalline Size of AgNP's Using Debye-Scherrer's Equation

S.No.	2θ	*FWHM*	$\beta = \pi$*FWHM/180	θ	$Cos\,\theta$	$D = k\,\lambda/\beta.\,Cos\,\theta$
1.	32.3005	0.25250	0.00440	16.15	0.96053	32.81

The particle size of silver nanoparticle is 32.81 nm.

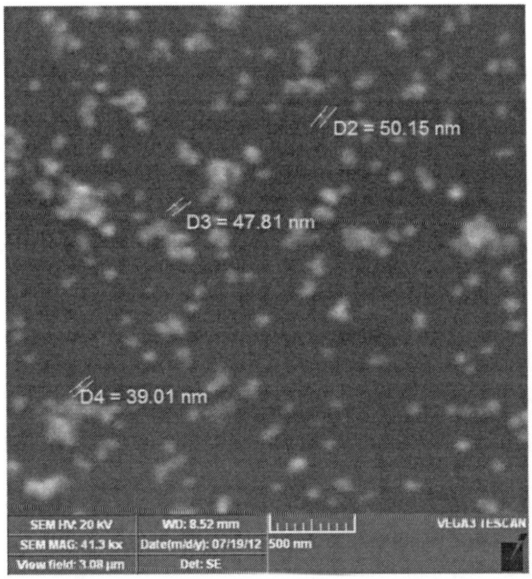

Fig. 5: SEM Images of Silver Nanoparticles Synthesized Using *Alternanthera sessilis*

Figure 6 shows the FTIR spectra of silver nanoparticles synthesized using aqueous extract of *Alternanthera sessilis*. The peaks located at 3253 cm^{-1}, 1634 cm^{-1} may be due to the presence of –NH or –OH group and carbonyl stretching in proteins. The other peaks at 2190 cm^{-1} and 2040 cm^{-1} are assigned to CC or CN triple bond respectively.

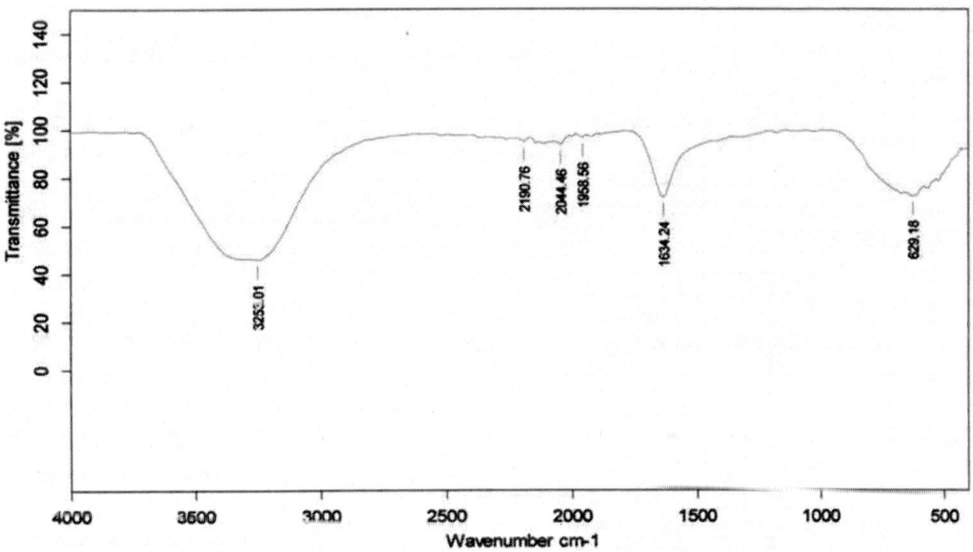

Fig. 6: FTIR Spectra of Silver Nanoparticles Synthesized Using *Alternanthera sessilis*

CONCLUSION

The biosynthesis of silver nanoparticle using the extract of *Alternanthera sessilis* was economical, non-toxic and environmentally benign. The formation of silver nanoparticles was characterized by UV-visible spectrophotometer. The synthesized silver nanoparticles were stable due to the reducing and capping nature of phytoconstituents present in the aqueous extract of *Alternanthera sessilis* analyzed by FTIR spectra. The particle size of the synthesized silver nanoparticles is less than 50 nm which was confirmed by XRD and SEM analysis.

ACKNOWLEDGEMENTS

The authors sincerely thank the Avinashilingam Institute for Home Science and Higher Education for Women University, Coimbatore, Tamilnadu, for providing research facilities, Karunya University for recording XRD and Periyar Maniammai University for recording SEM.

REFERENCES

[1] Panneerselvam, C., Ponarulselvam, S. and Murugan, K., Potential Anti-plasmodial Activity of Synthesized Silver Nanoparticle using *Andrographis paniculata* Nees (Acanthaceae), *Archives of Applied Science Research*, 3 (6), 208–217, 2011.

[2] Linga Rao, M. and Savithramma, N., Antimicrobial activity of silver nanoparticles synthesized by using stem extract of *Svensonia hyderobadensis* (Walp.) Mold—A rare medicinal plant, *Research in Biotechnology,* 3(3), 41–47, 2012.

[3] Ravishankar Rai, V. and Jamuna Bai, A., Nanoparticles and their potential application as antimicrobials Science against microbial pathogens, communicating current research and technological advances, A Mendez-Vilas (Ed)., *Formatex,* 2011.

[4] Kulkarni, A.P., Srivastava, A.A., Harpale, P.M. and Zunjarrao, R.S., Plant mediated synthesis of silver nanoparticles - Tapping the unexploited sources, *J. Nat. Prod. Plant Resour.,* 1 (4), 100–107, 2011.

[5] Panacek, A., Kolar, M., Vecerova, R., Prucek, R., Soukupova, J., Krystof, V., Hamal, P., Zboril, R. and Kvitek, L., Antifungal activity of silver nanoparticles against *Candida* spp, *Biometals,* 30, 6333–6340, 2009.

[6] Singh, M., Singh, S., Prasad, S. and Gambhir, I.S., Nanotechnology in medicine and Antibacterial effect of silver nanoparticles, *Digest J. Nanomater. Biostruct.,* 3, 115–122, 2008.

[7] Sahithi, B., Rajani, G. P., Sowjanya, K. and Gupta, D., Anti-inflammatory activity of Ethanolic and Aqueous extracts of *Alternanthera sessilis* lin, *Pharmacologyonline,* 1, 1039–1043, 2011.

SELF-ASSEMBLED Cu (II) MOFs WITH N, O DONORS MIXED LIGANDS

P. MOSAE SELVAKUMAR[1,2], E. SURESH[2] AND P.S. SUBRAMANIAN[2]

[1]*Department of Chemistry, Karunya University, Coimbatore, Tamil Nadu*
[2]*Discipline of Inorganic and materials and Catalysis, Central Salt and Marine Chemicals Research Institute (CSIR), Bhavnagar*
E-mail: pmosae@gmail.com; siva140@yahoo.co.in

ABSTRACT

Supramolecular coordination polymers, {Cu(bpy) (phd) $(H_2O)_2\}_n$ (1), {Cu(bpy) (phm) $(H_2O)\}_n$ (2), $\{Cu_2$ $(phm)_4(bpy)\}_n$ (4) and dimeric paddle wheel complex $[Cu_2$ $(phm)_4$ $(ACN)_2]$ (3) (where phd = phthalate dianion, phm = phthalate monoanion, bpy = 4, 4'-bipyridyl), have been synthesized by metal directed self-assembly. The single crystal X-ray structural investigation shows that all these complexes are self assembled through non-covalent interactions and resulted into a [1]D and [2]D supramolecular coordination polymer. Complex 1 and 2 were obtained as two and one-dimensional grid respectively. Complex 3 with M_2L_4 type paddle wheel complex upon interacting with 4, 4'-bipyridyl resulted into a linear coordination polymer 4.

INTRODUCTION

Materials based on the supramolecular assembly[1] of metal ions and organic ligands, often known as coordination polymers (networks) or metal–organic frameworks (MOFs) are metal-ligand compounds that extend into one, two or three dimensions (1D, 2D or 3D, respectively) *comprising* covalent metal-ligand bonding and non-covalent supramolecular interations.[2] Metallo supramolecular assemblies with extended structures due to the non-covalent bonds, such as coordination bonds, hydrogen bonds and $\pi \ldots \pi$ stacking interactions are currently attracted considerable attention.[1–9] In particular carboxylate group known for its multi-dentate nature capable to provide varieties of coordination mode such as *syn-syn*, *syn-anti*, *anti-anti*,

V. Rajendran, P. Prabu and K.E. Geckeler (eds.)
Nano Biomaterials, pp. 185–190 (2012).

monodentate, chelate etc. The metal carboxylates have emerged as an important family in the last few years.[10–14] The research in this area is rapidly developing owing to their intriguing structural motifs and unique properties such as porosity, supramolecular architectures for potential applications in material chemistry, separation technology, catalysis,[2,3,10,15–17] etc. Thus carboxylate gaining significant importance in the construction of varieties of supramolecular components such as MOF, CP, the present study reports syntheses, crystal structures of series of coordination polymers, $\{Cu(bpy)(phd)(H_2O)_2\}_n$ (1), $\{Cu(bpy)(phm)(H_2O)\}_n$ (2), $\{Cu_2(phm)_4(bpy)\}_n$ (4) and a dimeric paddle wheel complex $[Cu_2(phm)_4(ACN)_2]$ (3) (where phd = phthalate dianion, phm = phthalate monoanion, bpy = 4,4'-bipyridyl).

EXPERIMENTAL

All reagents and solvents employed were commercially available and used without further purification. Elemental analysis of the complexes were conducted a Perkin–Elmer PE 2400 series II CHNS/O elemental analyzer. Electronic spectra were recorded on a Shimadzu UV 3101PC spectrophotometer. IR spectra were recorded using KBr pellets on a Perkin–Elmer Spectrum GX FT-IR spectrometer.

Structural Determination

In each case of complex 1, 2, 3 and 4, a crystal of suitable size was selected and mounted on the tip of a glass fiber and cemented using epoxy resin. Intensity data for all crystals were collected using Mo-K_α (λ = 0.71073Å) radiation on a Bruker SMART APEX diffractometer equipped with CCD area detector at 100K. The data integration and reduction were processed with SAINT software. An empirical absorption correction was applied to the collected reflections with SADABS. The structures were solved by direct methods using SHELXTL and were refined on F^2 by the full-matrix least-squares method using the SHELXL-97 package.[18] Graphics are generated using PLATON and MERCURY 1.3. In all the complexes all non-hydrogen atoms were refined anisotropically till convergence is reached. Hydrogen atoms were fixed at idealized positions stereochemically.

RESULT AND DISCUSSION

Synthesis

Supramolecular coordination polymers, $\{Cu(bpy)(phd)(H_2O)_2\}_n$ (1), $\{Cu(bpy)(phm)(H_2O)\}_n$ (2), $\{Cu_2(phm)_4(bpy)\}_n$ (4) and dimeric paddle wheel complex $[Cu_2(phm)_4(ACN)_2]$ (3) (where phd = phthalate dianion, phm = phthalate monoanion, bpy = 4,4'-bipyridyl), have been synthesized by metal directed self-assembly as per scheme 1.

Complex 1 is self assembled into as two-dimensional rectangular network, while 2 resulted into one-dimensional straightforward chains. Interestingly, the phthalic carboxyl units in complex 1, occupying the transition position, both the carboxylate units in the phthalic dianions are involved in extending the two dimentional rectangular network. However in the case of 2, since one of the carboxylic group is methylated, it acts as monodentate and results into 1D linear polymeric network. Thus the prudent selection of bridging ligand such as methylated and non-methylated phthalate unit with appropriate diamine rigid spacer such as 4, 4'-bipy, one can control the formation of 1D and 2D network is spectated. Complex 3 with M_2L_4 type paddle wheel complex upon interacting with 4, 4'-bipyridyl resulted into a linear coordination polymer 4.

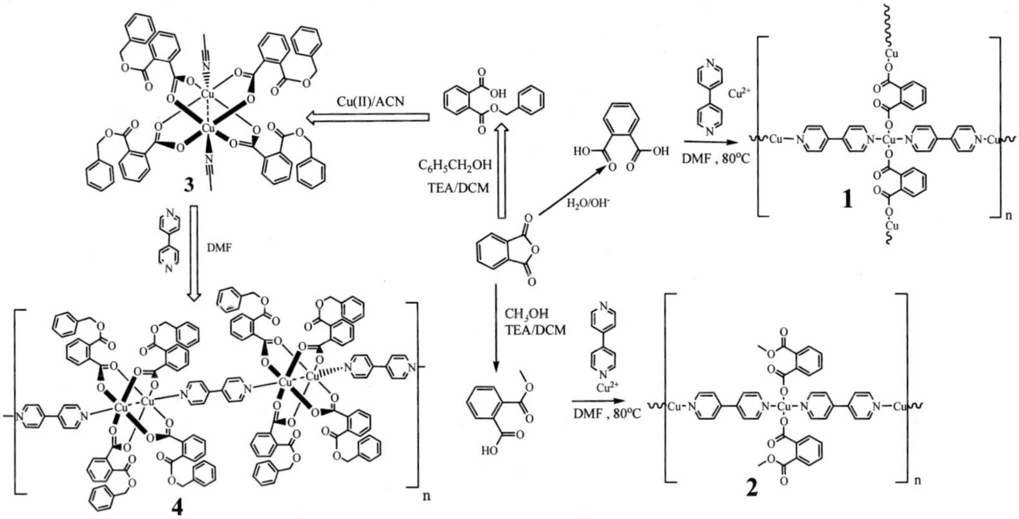

Scheme 1: Synthesis of 1–4 Using Metal Directed Self Assembly

Crystal Structures

Structure of Complex 1 and 2

The structure for title complex 1, was previously reported by Han Xu *et al.*[19] Figure 1(a) depict the ORTEP diagram with atom numbering scheme of complex 1, showing the coordination sphere around the Cu(11) centre. A view of the rectangular grid with micro pores of complex 1 is shown in Figure 1(b). It contains copper (II) centres bridged by 4, 4'-bipyridine ligands along the *c*-axis. Each Cu^{II} atom adopts a slightly distorted octahedral geometry, and possess coordination with two N atoms of 4, 4'-bipyridine, two O atoms of phthalate dianions and two O atoms of water molecules. The primary structure thus, engender chains of Cu^{II} centres bridged through bipyridine ligands along the *b*-axis, coordinated with two phthalate dianions, forming infinite linear cross linked 2D chain.

The title complex 2, crystallizes in the centrosymmetric monoclinic space group *P*21/*c*, contains linear 1D chain of copper(II) centres bridged through bipyridine ligands along the *b*-axis. Figure 1(c) depict the ORTEP diagram with atom numbering scheme of complex 2, and the coordination sphere around the Cu(11) centre and the linear 1D chain. Each Cu^{II} atom adopts a slightly distorted square pyramidal geometry, due to its coordination with two N atoms of *4,4′*-bipyridine, two O atoms of phthalate monoanions and one water molecule. The primary structure thus consisting chains of Cu^{II} centres bridged through bipyridine ligands along the *b*-axis, with two phthalate monoanions, generates an infinite linear one dimensional chain.

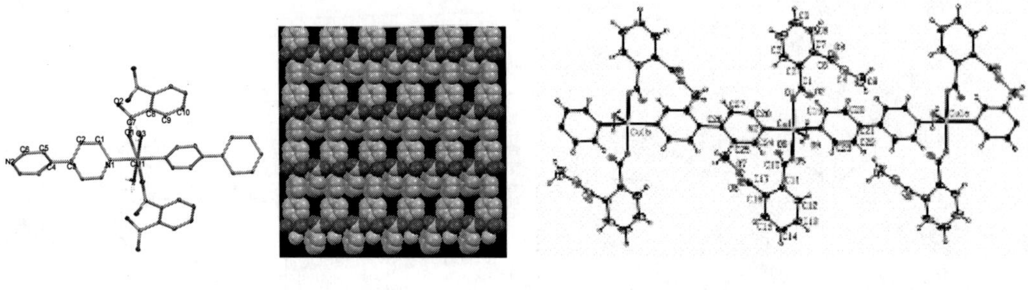

(a) (b) (c)

Fig. 1: (a) ORTEP Diagram of Complex 1 with Atom Numbering Scheme, (b) A View of the Rectangular Grid with Micro Pore of 1, (c) ORTEP Diagram of Complex 2 with Atom Numbering Scheme

Each metal centre is thus coordinated by two *trans* μ-*4,4′*-bipyridine ligands and two *trans* μ-phthalate ions in the equatorial positions [Cu-O = 2.413 (3) Å]. The axial coordination of one water molecules completes the slightly distorted square planar Cu^{II} coordination. The pyridyl rings of the bridging 4, 4'-bipyridine ligands are not twisted and almost planar. The layers are connected through H-bonds involving the coordinated solvent water molecules. Linear one dimensional chain in Figure 1 shows the water molecule coordinated with Cu are bonded alternative side.

Structure of Complex 3 and 4

The title complex 3, crystallizes in the centrosymmetric monoclinic space group P21/n, contains two copper (II) centres bridged by four carboxylate groups. Figure 2 shows Mercury diagram of dimeric paddle wheel complex 3 and its polymeric complex 4. Each Cu^{II} atom adopts a slightly distorted square pyramidal geometry, with Cu(II) center coordinating two N atoms of ACN solvent, eight O atoms of phthalate monoanions.

The title Complex 4, crystallizes in the centrosymmetric monoclinic space group P21/n, and engender a linear 1D chain of copper(II) centres bridged through 4, 4′-bipyridine ligands along the *c*-axis. Each Cu^{II} atom adopts a slightly distorted square pyramidal geometry, with the corners of the coordination occupied by one N atoms of 4, 4'-bipyridine and eight O atoms of phthalate monoanions. The primary structure consists of chains of dimeric Cu^{II} paddle wheel type units bridged by bipyridine ligands along the *c*-axis, forming infinite linear one dimensional chain. This structure provides and interesting evidence that the axial solvent

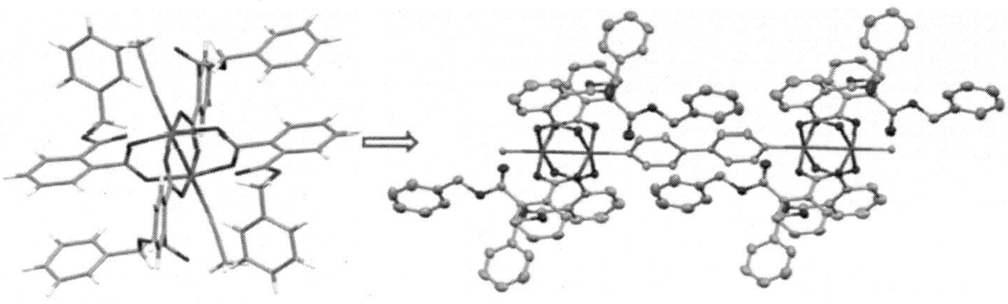

Fig. 2: Mercury Diagram of Dimeric Paddle Wheel Complex 3 and its Polymeric Complex 4

molecules in complex 3, were selectively replaced by judicious selection of 4, 4'-bipy ligand without disturbing the paddle wheel geometry. Thus the simple replacement provides an aesthetically appealing 1D linear polymeric network 4 retaining the paddle wheel geometry is wonderful molecular engineering technique for the synthesis of single pot synthesis of monomer to 1D polymeric CP.

CONCLUSION

In conclusion, we have successfully synthesized a series of coordination polymers, $\{Cu(bpy)(phd)(H_2O)_2\}_n$ (1), $\{Cu(bpy)(phm)(H_2O)\}_n$ (2),$\{Cu_2\ (phm)_4(bpy)\}_n$ (4) and dimeric paddle wheel complex $[Cu_2\ (phm)_4\ (ACN)_2]$ (3) (where phd = phthalate dianion, phm = phthalate monoanion, bpy = 4,4'-bipyridyl) by metal directed self-assembly. All these complexes are structurally characterized by single crystal X-ray diffraction analyses. The resultant structures of these complexes 1 and 2 clearly illustrates the formation of 2D and 1D coordination polymeric chains, the Complex 3 with M_2L_4 type paddle wheel complex upon interacting with 4,4'-bipyridyl resulted into a linear coordination polymer. The simple incorporation of 4, 4'-bipyridine ligand with its appropriate molar ratio, selectively replace the axial solvents in complex 3 and results into an interesting 1D polymeric chain is illustrated a simple route for the synthesis of such chains.

ACKNOWLEDGEMENTS

We acknowledge Department of Science and Technology, Council of Scientific and Industrial Research, India and Karunya University for support.

REFERENCES

[1] Lehn, J.-M., Supramolecular Chemistry Concepts and Perspectives; VCH: New York, 1995.

[2] Janiak, C., Engineering coordination polymers towards applications, *Dalton Trans.*, 2781–2804, 2003.

[3] Rosi, N.L., Eddaoudi, M., Kim, J., O'Keeffe, M. and Yaghi, O.M., Advances in the Chemistry of Metal-Organic Frameworks, *Cryst Eng Comm.*, 4, 401–404, 2002.

[4] Subramanian, P.S, Suresh, E. and Casella, L., Supramolecular helical architectures dictated by folded and extended conformations of the amino acid in ternary CuII-diamine-racemic amino acid complexes, *Eur. J. Inorg. Chem.,* 1654, 2007.

[5] Selvakumar, P.M., Suresh, E. and Subramanian, P.S., Synthesis, Spectral Characterization and Structural Investigation on some 4-Aminoantipyrine Containing Schiff base Cu(II) complexes and their molecular association, *Polyhedron,* 26, 74, 2007.

[6] Selvakumar, P.M., Suresh, E. and Subramanian, P.S., Hydrogen bonded helices: synthesis, crystal structure and self assembled microtubes, *J. Mol. Str.,* 919, 72–78, 2009.

[7] Selvakumar, P.M., Suresh, E. and Subramanian, P.S., Single Stranded Helical Supramolecular Architecture with Left handed Helical Water Chain in Ternary Copper(II) Tryptophan/Diamine Complexes, *Polyhedron.,* 28, 245–252, 2009.

[8] Cherian, A., Bisht, K.M., Subramanian, P.S. and Suresh, E., Structural Diversity in Two Dimensional Chiral Coordination Polymers Involving 4, 4'-Bipyridine and L-Cysteate as Bridging Ligands with Zn and Cd Metal Centres: Synthesis, Characterization and X-ray Crystallographic Studies, *Inorg. Chim. Acta.*, 365, 363–370, 2011.

[9] Kalyanaraman, S., Krishnakumar, V., Selvakumar, P.M., Subramanian, P.S., Ganesan. K. Synthesis, Structure and Spectral Investigation of Hydrogen bonded Helical Chain in Mixed ligand Cu(II) Pyrimidone complexes, *Spectrochimca Acta Part A, Molecular and Biomol Spect.,* 71, 609–615, 2008.

[10] Rao, C.N.R., Natarajan, S. and Vaidhyanathan, R., Metal Carboxylates with Open Architectures *Angew. Chem. Int. Ed.*, 43, 1466, 2004.

[11] Selvakumar, P.M., Suresh, E. and Subramanian, P.S., New Macrocyclic Cu(II) –bischelates with Paddle Wheel Cu2- acetate cage, *Inorg. Chim. Acta.,* 1503–1509, 2008.

[12] Selvakumar, P.M., Studies on Supramolecular Helical Compounds: Synthesis, Structural Investigation and Functional Application, *Ph.D Thesis, Bhavnagar University*, 2011.

[13] Selvakumar, P.M., Neha. C., Suresh, E. and Subramanian, P.S., Synthesis and Crystal Structure for some macrocyclic ring Compounds, *Syn. Comm.,* 41, 1309–1319, 2011.

[14] Selvakumar, P.M., Suresh, E., Waghmode, S. and Subramanian, P.S., Synthesis, Structure, EPR and DFT calculation on dinucelar Paddle Wheel Cu(II) Complexes with Bis-chelate rings, *J. Coord. Chem.,* 56, 3495–3509, 2011.

[15] Selvakumar, P.M., Nadella, S., Prathap, K.J., Kureshy, R.I., Suresh, E. and Subramanian, P.S., Synthesis, Crystal Structure and Catalytic Studies on dinuclear Copper(II) Mesocates, *Inorg. Chim Acta.,* 375, 106–113, 2011.

[16] Selvakumar, P.M, Jebaraj, P.Y., Sahoo, J., Suresh, E., Prathap, K.J., Kureshy, R.I. and Subramanian, P.S., The first bromide ion directed double helicate and its role in catalysis, *RSC Adv.*, 2, 7689–7692, 2012.

[17] Selvakumar, P.M., Nadella, S., Fröhlich, R., Albrecht, M. and Subramanian, P.S., A new class of solvatochromic material: Geometrically unsaturated Ni (II) complexes, *Dyes and Pigments*, 95, 563–571, 2012.

[18] Spek, A.L., *Acta Cryst.*, D65, 148–155, 2009.

[19] Xu, H., Sun, R., Y.-Z. Li, Y.-Z. and Bai, J.-F., Poly[[diaqua-μ-1,2-benzene-1,2-dicarboxylato-μ-4, 4'-bipyridine-copper(II)] dihydrate]: A two-dimensional copper(II) polymer, *Acta Cryst.* E62, m1156-m1158, 2006.

GELATIN MICRO PARTICLE LOADED WITH *EPHEDRA GERARDIANA* EXTRACT FOR LUNG TISSUE REGENERATION

DEEPTI SINGH, DOLLY SINGH, SOON MO CHOI AND SUNG SOO HAN

Department of Nano, Medical and Polymer Materials, Yeungnam University 214-1 Daedong, Gyeongsan 712-749, Korea
Polymer Gel Research Cluster Center, Yeungnam University, Gyeongsan 712-749, Korea
E-mail: sshan@yu.ac.kr

ABSTRACT

A major challenge in tissue engineering and regenerative medicine is attaining local and efficient release of growth factor to guide cell function. Gelatin is known polymer with RGD sequence that enhances the cellular attachment and can be easily degraded through enzymatic action. In this study, we have synthesized gelatin microspheres with diameters in the range of 1–6 μm by using an oil-water emulsification process and were stabilized by crosslinking with glutaraldehyde. The degree of crosslinking was altered by controlling the incubation time in cross linking solution. Drug loading rate studies and a model protein bovine serum albumin (BSA), showed quick protein uptake over initial 24h, followed by a lag phase and then a further increase was noted as the microsphere swelled up in approximately 3 days. Gelatin particle were loaded with solvent-free ethyl acetate extract of Ephedra gerardiana and seeded with human lung epithelial cells. The cell proliferation and metabolic state was checked periodically. The MTT assay performed indicated the increased metabolic activity of cells on gelatin particle with solvent free extract of Ephedra gerardiana in comparison to particles and 2D culture of cells without extracts. These results exhibit that gelatin particles can be used as deliver vehicle for growth factors or drugs locally to cells for directing their function.

V. Rajendran, P. Prabu and K.E. Geckeler (eds.)
Nano Biomaterials, pp. 191–196 (2012).

INTRODUCTION

Delivery of various growth factors and anti-microbial drugs is impetus in achieving tissue regeneration.[1] A wide range of the drug carriers have been fabricated using natural and synthetic polymeric materials [2–4] as these system offers advantage such as controlled release of drug, provides a high surface area which allows efficient drug loading and releasing from these polymeric matrix [5] along with attaining local site specific drug delivery.[6] Choosing the polymer for the fabrication is as important as the drug or a bioactive molecule that is loaded onto these matrices. Gelatin a known hydrolyzed collagen is extensively used in biomedical and food industries.[7] It is often chosen in tissue engineering due to RGD motifs that acts as cell binding receptors and is easily biodegradable in body simulating fluid undergoing hydrolytic cleavage.[8] Micro particles prepared using gelatin offers superior advantage as it is easily soluble in water and thus making it simpler for loading of bioactive compounds and drugs. The herbal plant *Ephedra gerardiana* is an ayurvedic medicine used for treating bronchitis and asthma along with various other lung related disorders. Traditionally it contains 50% ephedrine and rest 50% constitutes of other alkaloids like pseudoephedrine and is consumed as herbal tea to relieve bronchitis and blood pressure.[9]

MATERIALS AND METHODS

Micro Particle Fabrication

Gelatin particles were prepared by conventional oil-water emulsion technique. Briefly 5% working solution of porcine gelatin was prepared in aqueous solution and warmed at 37°C and added drop-by-drop in emulsification bath containing grape seed oil set at 2000 rpm. Gelatin was allowed to emulsify for 30 min and later cooled below the critical temperature in acetone for obtaining solid particles which were later washed using surfactant and centrifuged at 1200 rpm to remove all the traces of oil followed by lyophilization until further used.

Plant Extraction and Loading on Microparticles

Ethyl acetate extract of *Ephedra gerardiana* was prepared by soxhlet extraction method.[10] 300 μl of solvent free extract at 100 μg/ml concentration was loaded on gelatin particles and left overnight to be saturated with the extract. After overnight incubation at room temperature, particles were washed with PBS to remove any unbound drug on surface.

Drug Release Profile Studies and *in-vitro* Cell Proliferation

For cell compatibility testing microparticles loaded with extract were placed in 24 well plate and seeded with human epithelial lung cells (1.7×10^6) and 500 μl of fresh DMEM containing 10% FBS was added to each well. For drug release profile media was collected in hourly manner for first 10 h followed by 24 hr for period of 6 days. Media was collected for burst release and cumulative release for 6 days respectively. All experiments were set up in triplicates for concurrent readings.

Imaging of Microparticles and Cells

Cell seeded particles were viewed using fluorescent microscopy to check for cellular proliferation and attachment to the particles. Live-Dead staining kit obtained from sigma was used to examine the cell-particle interaction. Ethidium bromide stock (2 mM) was diluted in 10 ml distilled PBS (tissue culture grade) along with 4 µm of calcein AM. The working solution of the stain 200 µl was added to the test well and incubated for 45 min and stained cells were observed under Nikon fluorescent microscope.

RESULTS AND DISCUSSION

Characterization of Microparticles

The gelatin microparticles were prepared in different polymer concentration and post fabrications were cross-linked using glutaraldehyde (Figure 1). The oil-water emulsion technology was used for fabrication of gelatin particles in spherical shape. Various parameters were used to optimize microparticles with average diameter of 3.5 µm with range of 1–6 µm. The size of the particles was controlled by the speed of oil-gelatin slurry mixing. Post fabrication cross-linking was performed under constant stirring to stabilize the particles. Microparticles formed and cross-linked were lyophilized. Generally post fabrication and cross-linking microparticles tends to aggregate resulting in major hindrance [9] during application however use of porcein skin gelatin at 5% concentration and precipitation using acetone resulted in formation of gradient of the particles. The smallest nanoparticles were floating upon centrifugation and microparticle formed sediments at the bottom of the tube. No aggregations of the particles were observed during cross-linking when performed under constant stirring.

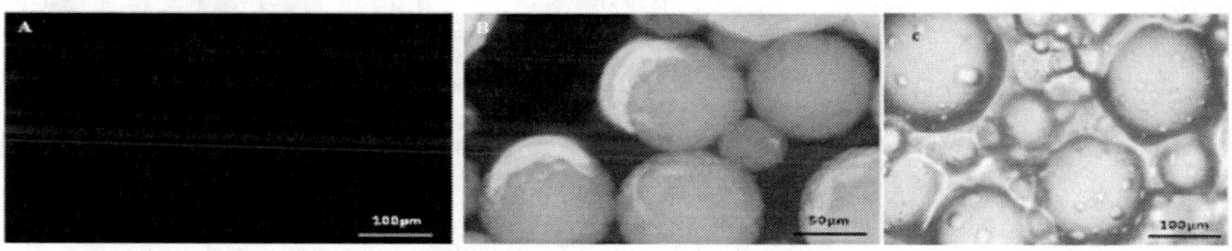

Fig. 1: Fluorescent (A) and Bright Field (B) Microscopy Image of Cross-Linked Gelatin Particles. Image (C) Non-cross Linked Gelatin Microparticle

Drug Release and Cell Compatibility Profile

The rate and degree of protein and drug loading into cross-linked microparticle was examined by quantification of the amount taken up by the particle over the period of incubation. There was no burst release from the particle as observed by spectrophotometric study and over the period of 6 days it was observed that the protein and extract release from the particle was steady and controlled (Figure 2A). MTT assay performed to check the metabolic activity of the cells; initially all the three systems didn't show any significant difference however on day 3 cells cultured on gelatin microparticles with and without the extract where almost in par with each other. On day 5th higher cellular proliferation in extract incorporated microparticles was observed and cells were more active in comparison to cells on gelatin particles or 2D without drugs (Figure 2B). Human epithelial lung cells cultured on gelatin microparticles (Figure 3A) was found to maintain the normal cell phenotype and cells cultured in presence of *Ephedra gerardiana* (Figure 3B) were found to be metabolically more active in comparison to the 2D system. The cells were observed to be in contact with gelatin micro-particle over 2 weeks culture time however particles were found to degrade over the time (Figure 4).

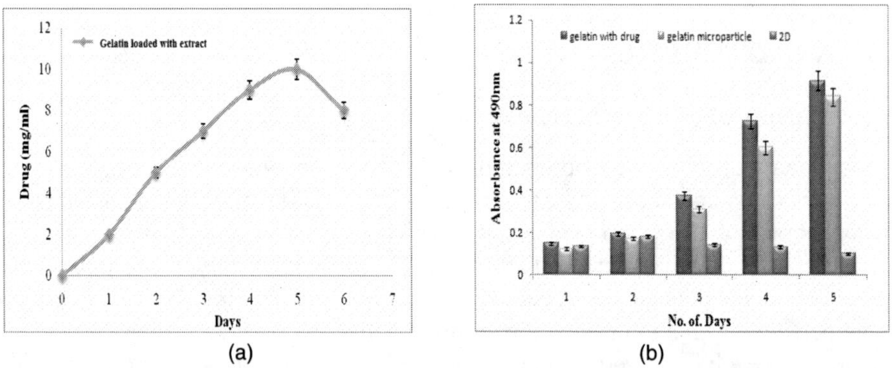

Fig. 2: Drug Release Profile of Gelatin Microparticle (A) and Lung Cell Proliferation Rate Analyzed by MTT (B) on Gelatin Microparticle Loaded with Herbal Extract in Comparison with Particle and 2D System

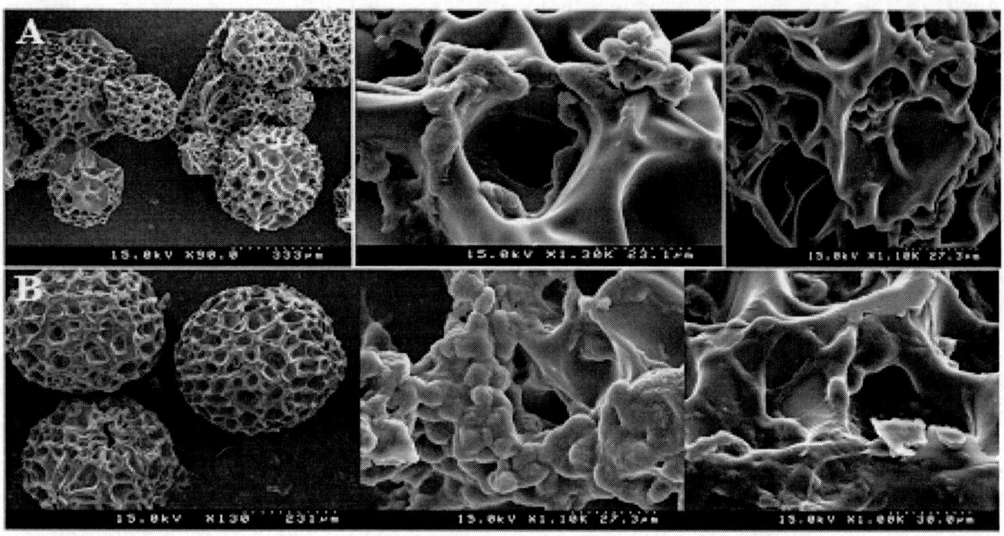

Fig. 3: Scanning electron micrography image of gelatin microparticle (A) and particles loaded with *Ephedra gerardiana* (B) seeded with lung cells. Particles with *Ephedra gerardiana* extract showed higher cellular attachment and proliferation in comparison with cells on pure gelatin microparticle

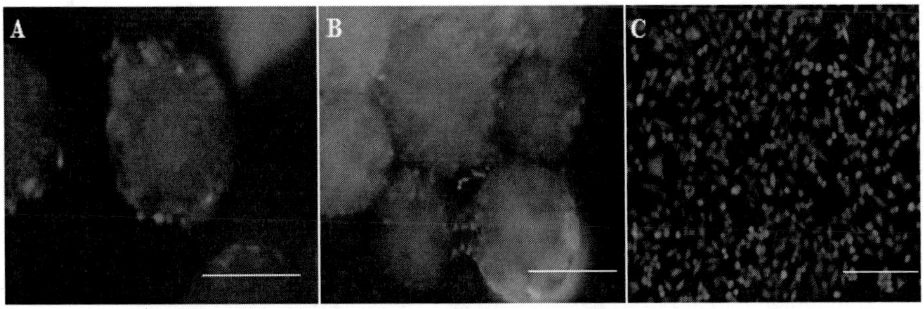

Fig. 4: Live/Dead Staining of Cells Cultured on Gelatin Micro Particle (A) Loaded with Herbal Drug (B) and 2D (C) on Day 1

CONCLUSION

Fabrication of gelatin (porcine skin) particles using oil-water emulsion is simple and robust and post synthesis cross-linking allows stabilization of the particle. The degree of cross-linking is easily modulated by varying the incubation time with glutaraldehyde. The *Ephedra gerardiana* is known for helping in treating the lungs related disorder proved to be potential bioactive component for lung regeneration. The particles incorporated with solvent free extracts showed higher cellular proliferation in comparison to other systems. Gelatin based microparticles could be simplest delivery vehicle for *Ephedra gerardiana* as the release was found to be controlled and significant amount of the drug was released over one week time and rate of drug release was in accordance to the cellular proliferation profile indicating the components from the plants could play a significant role in the higher cellular metabolic rate and can be explored as potential drug for lung tissue regeneration especially in case of emphysema where loss of tissue could lead to morbidity. Microparticle could be used for delivery of various bioactive components or can be used as part of scaffold to provide ideal condition to control cell functions.

ACKNOWLEDGMENT

This work was supported by grant No. RTI04-01-04 from the Regional Technology Innovation Program of the Ministry of Knowledge Economy (MKE).

REFERENCES

[1] Adhirajan, N., *et al.*, Gelatin microspheres cross-linked with EDC as a drug delivery system for doxycyline: development and characterization, *J. Microencapsula,* 24, 647–659, 2007.

[2] Cruz, D.M, *et al.*, Chitosan microparticles as injectable scaffolds for tissue engineering, *J. Tissue Eng Regen Med.,* 2, 378–380, 2008.

[3] Habraken, W.J. *et al.*, *In vitro* growth factor release from injectable calcium phosphate cements containing gelatin microspheres, *J. Biomed Mater Res A.*, 91, 614–622, 2009.

[4] McGuigan, A.P. and Sefton, M.V. Modular tissue engineering: Fabrication of a gelatin-based construct, *J. Tissue Eng Regen Med.*, 1, 136–145, 2007.

[5] Silva, G.A., *et al.*, Materials in particulate form for tissue engineering 1. Basic concepts, *J. Tissue Eng Regen Med.,* 1, 4–24, 2007.

[6] Yamamoto, M, *et al.*, Controlled release of growth factors based on biodegradation of gelatin hydrogel, *J Biomater Sci Polym Ed*., 12, 77–88, 2001.

[7] Young, S., *et al.*, Gelatin as a delivery vehicle for the controlled release of bioactive molecules, *J. Control Release*., 109, 256–274, 2005.

[8] Patel, Z.S., *et al.*, Biodegradable gelatin microparticles as delivery systems for the controlled release of bone morphogenetic protein-2, *Acta Biomater*, 4, 1126–1138, 2008.

[9] Kawai, K., Accelerated tissue regeneration through incorporation of basic fibroblast growth factor-impregnated gelatin microspheres into artificial dermis, *Biomaterials,* 21, 489–499, 2000.

[10] Singh, D. *et al.*, Biochemical characterization of triphala extracts for developing potential herbal drug formulation for ocular diseases, *Int J. Pharm Pharm Sci.,* 3, 516–523, 2011.

SYNTHESIS AND STRUCTURAL PROPERTIES OF TiO_2 NANOCAPSULES PREPARED BY HYDROTHERMAL METHOD: GROWTH MECHANISM FOR MASS PRODUCTION

M.A. SHAH, S. RUBAB AND S.A. AKHOON

Electron Microscopy Centre, Department of Physics, Faculty of Science,
National Institute of Technology, Srinagar
E-mail: mashahnit@gmail.com; sashraf@kau.edu.sa

ABSTRACT

Carbon nanotubes were not the only materials to raise red flags. Nanostructures including particles, rods, wires, belts, cages, walls, flowers and rings have been reported to be harmful. Their unique physical characteristics raise concerns that tiny clamps of metals, ceramics and organics could prove uniquely toxic as well. Nanomaterials made from the urea, polyethylene glycol, polyvinylpyrrolidone, polytertrafluoethylene or PTFE, showed even more dramatic effects.[1] To address this burning issue, we have made a small breakthrough in preparation of nanomaterials including the present one in discussion.

Growth of titanium oxide (TiO_2) nanocapsules of 20 ± 5 nms in width and ~ 60nm in length has been synthesized through an economic and harmless route. The approach is based on a simple reaction of titanium powder and water at ~110°C without use of any additives. Powder X-ray diffraction (XRD) studies reveal that the as prepared nanocapsules are mixture of pure rattle and anatase in phase whereas the morphological investigations reveal an interesting and exciting shape of capsules. The energy dispersive x-ray spectrometry (EDX) shows that the as grown product is of titanium (Ti) and oxygen (O_2) only. The prospects of the process are bright and promising and can be readily used to prepare other metal and metal oxides.

V. Rajendran, P. Prabu and K.E. Geckeler (eds.)
Nano Biomaterials, pp. 197–202 (2012).

INTRODUCTION

Titanium dioxide (TiO_2) minerals are found in three different structures; rutile (tetragonal), anatase (tetragonal) and brookite (orthorhombic). Only rutile and anatase phases are stable whereas brookite transforms into rutile at quite low temperature. TiO_2 enjoys a high popularity in surface science and especially the rutile (100) has evolved as a model system in the surface science of metal oxides.[1] Rutile is thermodynamically most stable phase and the (110) face has the lowest surface energy. The bulk terminated (1x1) surface is shown in Figure 1.

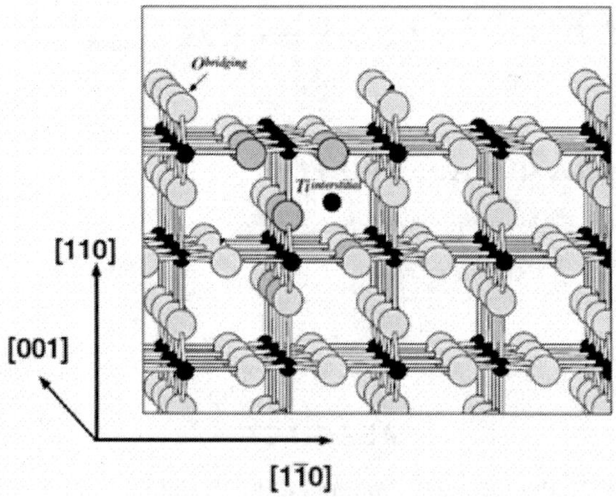

Fig. 1: Ball and Stick Model of Rutile (110) Surface. Large Balls; Oxygen and Small Balls; Titanium

TiO_2 finds use in wide range of important technological applications where a more detailed understanding of the fundamental aspects of TiO_2 surfaces or interfaces would potentially have a positive impact. It is used in heterogeneous catalysis, as a photocatalyst; in solar cells for production of hydrogen and electric energy; as a gas sensor; as a white pigment (e.g in paints and in cosmetic products); as a corrosive protective coating; as an optical coating, in ceramics, in electric devices such as varistors, plays a role in biocompatibility of bone implant, finds applications in Li- based batteries and electrochemical devices.[2] The rutile phase has been widely used for pigments of materials because of its chemical stability. However, an anatase phase has been widely used photodegration due to its high photoactivity. The majority of the applications of TiO_2 are strongly influenced by the crystalline phase, crystallite size and shape.

In order to obtain high crystalline TiO_2 at low temperature a long time is necessary to get pure anatase, or mixture of anatase or rutile phase. TiO_2 have been generated by physical methods, chemical approaches have proved to be more effective and efficient as they provide better control over the size and shape, which is one of the essential requirements for the preparation of nanomaterials. Chemical synthesis of nanomaterials has been reviewed by few authors but innumerable improvements and better methods are being reported continually in the last few years.[3]

Cassaignon *et al.* (2007) reported selective synthesis of brookite, anatase and rutile nanoparticles by the thermo-hydrolysis of TiO_4 in concentrated aqueous nitric acid, wherein the selectivity was strongly dependent the medium acidity.[4] Similarly, using aqueous solutions of TiO_4 and acidic additives, Yin *et al.* (2002) and Franklyn *et al.* (2007) confirmed the selective formation of nanoparticulate anatase and rutile nanorods.[5,6] In addition, Jiu *et al.* (2006) reported the synthesis of a pure, highly crystalline anatase phase through hydrolyzing titanium isopropoxide rather than a $TiCl_4$ precursor in presence of ethylenediamine basic catalyst and cetyltrimethylammonium bromide (CTAB) surfactant.[7] Li *et al.* (2006) synthesized TiO_2 nanorods by adding $Ti(OBu)_4$ into the mixed solution of linoleic acid, triethylamine and cyclohexane and then sealing the solution for two days.[8] All of the aforementioned investigations have used organics, surfactants and catalysts during synthesis at very high temperatures. Moreover, most of the pathways suggested for the synthesis of nanocrystals involve environmentally malignant chemicals which are toxic and not easily degraded in the environment.[9] Environmental friendly chemical synthesis requires alternative solvents such as ionic liquids, liquid and water.[10]

Recently, an interesting and encouraging result obtained by the reaction of titanium metal with water by Shah and Fahad (2011) provides motivation to synthesize TiO_2 nanostructures without additives or organics at low temperatures.[11] In this study, an interestingly, nanocapsules of 20±5 nms in width and ~ 60nms in length has been obtained by a simple reaction of titanium powder/foil and de-ionized water at very low temperature of 110°C. The reported method besides being organics free is economical, fast, environmentally benign and free of pollution, which will make it suitable for large scale production. Variety of morphologies is obtained by changing the initial conditions like temperature, time and the precursor.

EXPERIMENTAL

Materials

Titanium powder or foil (titanium powder, Ranbaxy, > 3 μm) was used as a source of titanium and was cleaned by ultra-sonication in acetone and water for 20 minutes in each solvent. The de-ionized water used was prepared in the laboratory.

Syntheses of TiO_2 Nanoparticles

A closed cylindrical Teflon lined stainless steel chamber of 50 ml capacity was used for the preparation of TiO_2 nanocapsules. In a typical preparation process, 2 mg of titanium powder was added to 20 ml of distilled water in a glass vial. The reaction mixture was sonicated for about 10 minutes in a glass veil, transferred into a stainless steel Teflon lined metallic bomb of 50 ml capacity and sealed under inert conditions. The closed chamber was then placed inside a preheated box furnace and the mixture was heated slowly to 110°C and maintained at this temperature for 10 hours. The furnace is allowed to cool after the desired time and the resulting suspension was centrifuged to retrieve the product, washed and then finally vacuum dried for few hours. The samples can even be air dried.

Characterization of TiO_2 Nanoparticles

The phase composition of the as synthesized samples were identified by X-ray diffraction on a Siemens D 5005 diffractometer using Cu Kα (λ =0.15141 nm) radiation operated at 30kV and 30 mA. The morphology and the size of the products was carried out using high resolution FE-SEM (FEI NOVA NANOSEM-600) coupled with energy dispersive x-ray spectrometer (EDX).

RESULTS AND DISCUSSION

Structural information of the final product could be given by powder X-ray diffraction (XRD) analysis. In the XRD pattern of sample (Figure 2), all the observed peaks can be indexed to a pure tetragonal rutile and anatase phase (JCPDS card, 21–1272) with lattice constants a = 3.785 Å and c = 9.513 Å. No peak of other phases was observed, which indicates that the product is mixture of rutile and anatase and well crystallized.

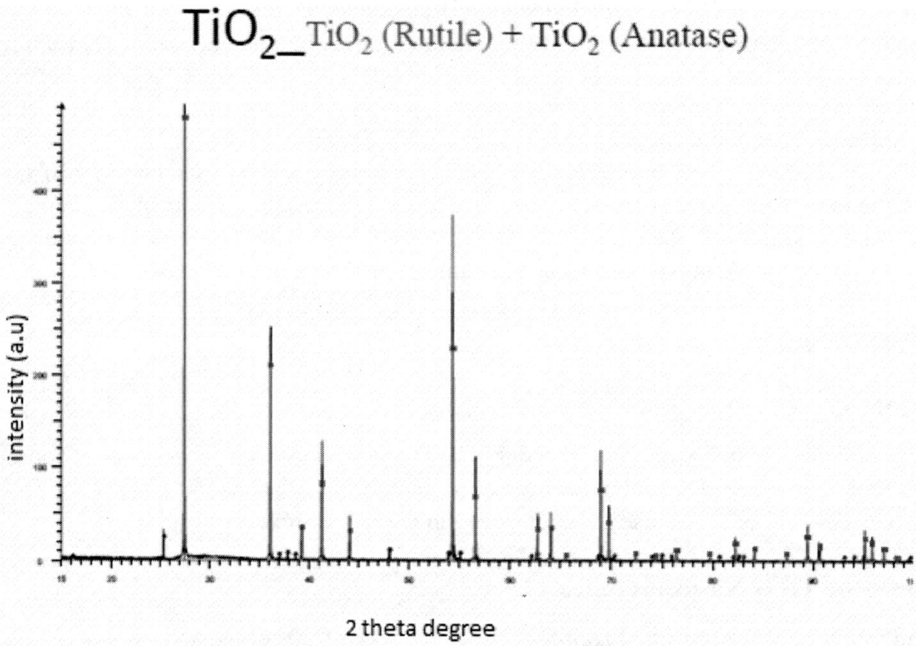

Fig. 2: The XRD Pattern of the Sample

By the morphological investigations using FESEM, it was observed that the grown products are nanocapsules with an average diameter of 20 ± 5 nms in width and ~ 60 nms in length. Figure 3 (a) and (b) show the low and high magnification FESEM images of the nanocapsules and confirms that the products are grown in a very high density having shape of capsule. EDX analysis was performed. It is confirmed from the EDX analysis that the grown nanoparticles are composed of titanium and oxygen only.

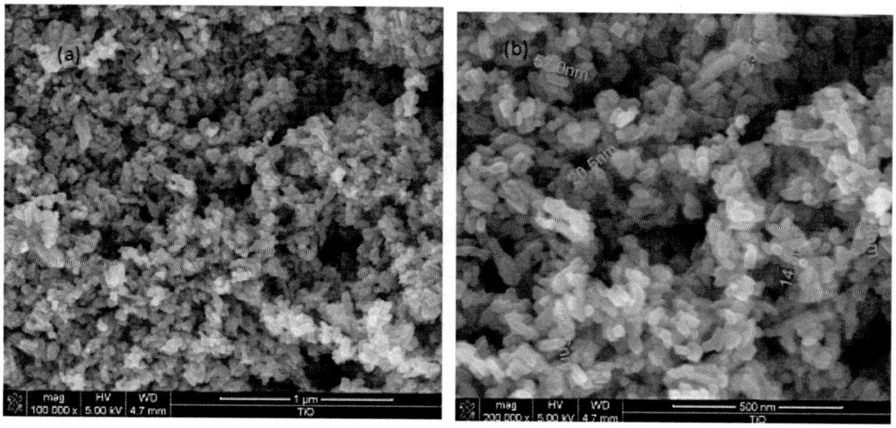

Fig. 3: (a) Low and (b) High-Resolution FESEM Images of TiO_2 Nanocapsules

As initially, titanium does not react with water molecules but at 110°C and under pressure in Teflon-lined stainless chamber, the powder reacted with water and forms a protective titanium hydroxide ($Ti(OH)_2$) layer with dissolved hydroxide ions onto the surfaces of the titanium particles according to the following reaction mechanism:

As the concentration of the Ti^{2+} and OH^- ions exceeds a critical value, the precipitation of TiO_2 nuclei starts. The $Ti(OH)_2$ can be transformed into the TiO_2 crystals via the simple chemical reactions mentioned below:

$$Ti(OH)_2 \xrightarrow{\Delta} TiO_2 + H_2O$$

The precipitates of $Ti(OH)_2$ are more soluble as compared to the TiO_2 precipitates, therefore, the formed $Ti(OH)_2$ precipitates tend to continuously produce Ti^{2+} and OH^- ions which form the TiO_2 nuclei. Titanium hydroxide ($Ti(OH)_2$) layer with dissolved hydroxide ions onto the surfaces of the titanium particles as shown in Figure 4 (a & b).

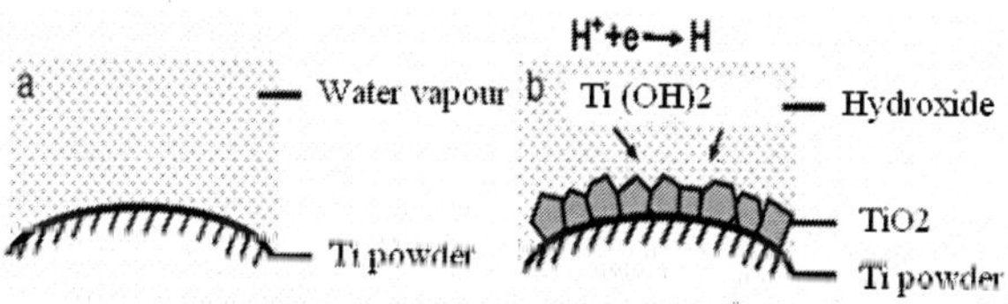

Fig. 4: A Schematic Illustration of the Growth (a) Ti Particles and Water Vapour co-Exists. (b) Layer of $Ti(OH)_2$ Formed on the Particle Surface and Subsequent Decomposition to TiO_2

Moreover, water at elevated temperatures plays an essential role in the precursor material transformation because the vapor pressure is much higher and the state of water at elevated temperatures is different from that at the room temperature. The solubility and the reactivity of the reactants also change at high pressures and high temperatures and high pressure is favorable for crystallizations.[12]

CONCLUSIONS

Growth of titanium oxide (TiO_2) nanocapsules of 20±5 nms in width and ~ 60 nms in length has been synthesized through an economic and safe route. The proposed single source and catalyst free method is simple, soft, economic and environmentally benign which will make it suitable for various applications. Furthermore, it is well expected that such a technique would be extended to prepare many other important metal oxide nanostructures.

ACKNOWLEDGEMENT

We are pleased to acknowledge the support by KAUST for XRD measurements.

REFERENCES

[1] Diesord, U., Structure and properties of TiO_2 surfaces; a brief review, *Applied Phys. A*, 76, 1–7 (2002).

[2] Chen, X. and Mao, S.S., Titanium dioxide Nanomaterials: Synthesis, properties, modifications and applications, *Chem. Rev.,* 107 (2007), 2891–2959.

[3] Shah, M.A. and Towkeer, A., Principles of Nanoscience and Nanotechnology, Naroosa Publishing House, New Delhi (2010).

[4] Cassagnon, S., Koelsch, M. and Jolivet, J.P, Selective synthesis of brookite, anatase and rutile nanoparticles: Thermolysis of $TiCl_4$ in aqueous nitric acid, *J. Mat. Sci.,* 42 (2007), 6689–6695.

[5] Yin, H.B., Wada, Y., Kitamura, T., Sumida, T., Hasegawa, Y. and Yanagida, S., Novel synthesis of phase pure nanoparticulate anatase and rutile TiO_2 using $TiCl_4$ aqueous solution, *J. Mater. Chem*., 12 (2002), 378–383.

[6] Franklyn, P.J., Levendis, D.C., Demetrius, C., Coville, N.J. and Maaza, M., Phase transformation of hydrothermally synthesized nanoparticles of TiO_2: from anatase to rutile nanorods, *S.Afr. J. Chem*., 60 (2007) 71–75.

[7] Jiu, J.T., Isoda, S., Wang, F.M. and Adachi, M., Dye sensitized solar cells based on a single crystalline TiO_2 nanorod film, *J.Phys. Chem.,* B 110 (2006), 2087–2092.

[8] Li, X.L., Peng, Q., Yi, J.X., Wang, X., Li, Y.D., Near monodisperse TiO_2 nanoparticles and nanorods, *Chem Eur. J* 12 (2006) 2383–2391.

[9] Zhang, J., Xiao, X. and Nan, J., Hydrothermal–hydrolysis synthesis and photocatalytic properties of nano TiO_2 with an adjustable crystalline size, J. of Hazardous Materials ,176 (2010), 617–622.

[10] Verianshyah, B., Park, T.J., Lim, J.S. and Lee, Y.W., *J. Solid Supercritical fluids*, 34 (2005), 51–61.

[11] Shah, M.A. and Al-Marzouki, F.M., TiO_2 nanoparticles prepared without harmful organics; A biosafe and economical approach, Scientia Irancia, Transctions F; Nanotechnology (2011) dio: 10.1016/j.Scient. 2011.06.006.

[12] Chen, X.B., Mao, S.S., Synthesis of titanium dioxide nanomaterials, *J. Nanosci. Nanotechnol.,* 6 (2006), 906–925.

CYTOTOXIC RESPONSE TO CARBON NANOTUBES BY LYMPHOCYTE CELLS

SEEMA MANCHANDA, MADHURI SHARON[1],
MAHESHWAR SHARON[1] AND AMOL KAKADE

Birla College of Arts, Science and Commerce, Kalyan, Maharashtra
[1]SICES College, Ambarnath, Maharashtra
E-mail: seema231106@gmail.com

ABSTRACT

A large number of studies have been performed in the past several years to explore the potential toxic effects of carbon nanotubes. Recent research showed that unfunctionalized, long MWCNTs may pose a carcinogenic risk in mice. It is thus critical and urgent to clarify the toxicity issue of carbon nanotubes. The current status is that toxicity appears to be dependent on the material preparation, especially geometry and surface functionalization.

In present paper, Lymphocyte Proliferation assay was done to test the Cytotoxicity of CNT. Blood samples were taken from 7 healthy patients who were not on medication nor had a history of any illness in the past 2 years. Cell growth, viability, apoptosis and metabolic activity were evaluated in proliferating human peripheral blood lymphocytes. The viability of freshly isolated lymphocyte suspension assessed by trypan blue staining was 96 ± 6%.

INTRODUCTION

Safety is the first requirement of any material used in medicine. A large number of studies have been performed in the past several years to explore the potential toxic effects of carbon nanotubes. The conclusions of these reports varied drastically, showing a large dependence on the type of nanotube materials as well as functionalization approaches. Cell culture experiments and *in vivo* pilot studies conducted by various groups observed no obvious toxicity of properly functionalized carbon nanotubes (Kam *et al.*, 2004, Dumortier *et al.*,

V. Rajendran, P. Prabu and K.E. Geckeler (eds.)
Nano Biomaterials, pp. 203–210 (2012).

2006, Schipper *et al.,* 2008 and Wu *et al.,* 2008). On the other hand, raw carbon nanotubes were shown to be toxic to mice after inhalation into the lung (Lam *et al.,* 2004, Warheit *et al.,* 2004, Shvedova *et al.,* 2005 and Muller *et al.,* 2005). Recent research showed that unfunctionalized, long MWCNTs may pose a carcinogenic risk in mice (Poland *et al.,* 2008). As a result of the wide variety of reports, both the public and research communities are currently concerned about using carbon nanotubes for biomedical applications. It is thus critical and urgent to clarify the toxicity issue of carbon nanotubes. The current status is that toxicity appears to be dependent on the material preparation, especially geometry and surface functionalization. Well functionalized CNTs with biocompatible surface coatings have been shown to be nontoxic *In vitro* to cells and *in vivo* in mice.

In vitro Toxicity of Carbon Nanotubes

Even in cell culture experiments, the issue of toxicity of carbon nanotubes is still controversial. While inhibition of HEK 293 cell proliferation after exposure to SWCNTs was reported by Cui *et al.* (2005). Ding *et al.* (2005) observed that MWCNTs induce cell cycle arrest and increase apoptosis/necrosis of human skin fibroblasts. However, neither of these studies used functionalized carbon nanotubes. Apoptosis of T lymphocytes induced by oxidized MWCNTs was observed by Bottini *et al.* (2006). Sayes *et al.* (2006) further reported that the toxicity of CNTs was dependent on the density of functionalization, with minimal toxicity for those heavily functionalized with the highest density of phenyl-SO3X groups. These results are understandable because CNTs without proper functionalization have a highly hydrophobic surface, and thus may aggregate in the cell culture and interact with cells by binding to various biological species, including proteins, via hydrophobic interactions, and induce certain cell responses such as cell toxicity. Other factors may also contribute to the observed toxicity of CNT samples *In vitro*. Surfactants, present in excess in the CNT suspensions, are known to be highly toxic to cells (Dong *et al.,* 2008). The metal catalyst content in CNTs should also be considered when the toxicity of carbon nanotubes is investigated (Plata *et al.,* 2008). Moreover, proper assays must be employed in toxicity tests to avoid interference of carbon nanotubes with the assay reagents (Casey *et al.,* 2007, Worle-Knirsch *et al.,* 2006). For these reasons, *In vitro* toxicity assays of carbon nanotubes should be carefully designed and performed with well prepared and characterized materials, as well as suitable assay methods. Cells exposed to SWCNTs, PEGylated by various PL PEG amphiphiles, exhibited neither enhanced apoptosis/neurosis, nor reduced proliferation of various cell lines *In vitro* [Liu *et al.,* 2007& Kam *et al.,* 2005). Carbon nanotubes covalently functionalized by 1,3-dipolar cycloaddition developed by Prato *et al.,* also appeared to be safe to the tested cell lines, including primary immune cells (Wu *et al.,* 2005 and Dumortier *et al.,* 2006). Carbon nanotubes with a biomimetic coating engineered by Bertozzi *et al.* were also nontoxic to cells (Chen *et al.,* 2004 and Chen *et al.,* 2006). Several other independent groups also reported that CNTs coated by DNA, amphiphilic helical peptides and serum proteins were not toxic to cells (Heller *et al.,* 2005, Chin *et al.,* 2007 and Yehia 2007). In very recent work, Jin *et al.* discovered that SWCNTs taken up by cells via endocytosis exited cells through exocytosis without affecting the viability of cells (Jin *et al.,* 2008). It appears that raw CNTs and CNTs without serum-stable functionalization show toxicity to cells at moderate dosage, while serum-stable, functionalized CNTs show little toxicity even at high dosages.

MATERIALS AND MEHODS

Cyto-Toxicity Response to Carbon Nano Tubes by Lymphocyte Cells

MWCNT: These MWCNTs were synthesized by CVD method using Fe/Ni as catalyst having 95% purity. It has a diameter of 30–40 nm and length of 5–50 μm. It has < 2% amorphous carbon and < 0.2% ash. The surface area of MWCNT was $300m^2/g$, where as its thermal conductivity was 1812+300 W/m.K. It was procured from Monad Nanotech Pvt Ltd. It was further purified by acid treatment.

Source of Lymphocytes

Blood samples were not taken from 7 healthy patients who were not on medication nor had a history of any illness in the past 2 years.

Medium Used for Culturing Lymphocytes was Ficoll Hapaque, RPMI-1640 Medium, Foetal Bovine Serum (FBS) (from Sigma-Aldrich) and Gentamycine (from Nicholas Piramal).

Reagent tried for Lymphocyte Proliferation assay was Alamar Blue (AB) (from Biosource).

Lymphocyte Proliferation Assay was done to test the Cytotoxicity of CNT blood were collected in Heparin tubes. Peripheral blood (PB) T lymphocyte was isolated by Ficoll h-paque separation method. Cells were then culture in RPMI- 1640 media supplemented with FBS and gentamycine making the concentration of cells 1 million/ml. Incubation of cells was done by adding 20 μl of nanomaterials to 180 μl of cell suspension at 37°C in 5% CO_2 atmosphere for 72 hours. The final concentration of added nanomaterials was 10, 25, 75, 100 mg/ml. After 72 hours of treatment with MWCNT, 20 μl of Alamar Blue was added in each well and incubated for 24 hours. Reduction in Alamar Blue showed the percentage proliferation of lymphocyte. The proliferation of lymphocyte was calculated by formula,

$$\text{Percentage reduction in Alamar blue} = \frac{(O2 \times A1) - (O1 \times A2) \times 100}{(R1 \times N2) - (R2 \times N1)}$$

Where:

O1 = molar extinction coefficient (E) of oxidized Alamar Blue (Blue) at 540nm
O2= E of oxidized Alamar Blue at 630 nm
R1 = E of reduced Alamar Blue at 540 nm
R2– E of reduced Alamar Blue at 630 nm
A1 = absorbance of test wells at 540 nm
A2 = absorbance of test wells at 630 nm
N1 = absorbance of negative control well (media plus Alamar Blue but no cells) at 540 nm
N2 = absorbance of negative control well (media plus Alamar Blue but no cells) at 630 nm.

Effect of MWCNT on Lymphocyte Proteins by SDS PAGE

Pellet of the suspension of cells by centrifugation at 2500 x g for 10 minutes. Supernatant was discarded. 1 ml of Mammalian Protein Extraction Reagent (M-PER) (Thermo scientific) was added to each pellet. After shaking for 10 minutes debris were removed by centrifugation at 14000 x g for 15 minutes. The supernatant were then analyzed on 8% SDS PAGE.

Cell growth, viability, apoptosis and metabolic activity were evaluated in proliferating human peripheral blood lymphocytes.

Data Analysis

Total IgE serum level was analyzed by Mann-Whitney Test. Cell growth data were analyzed by applying the paired sample t-test. In all cases, P value lower than 0.01 were considered as statistically significant. The statistical tests were performed by SPSS 15 software.

RESULTS

The total IgE serum level was 184.57 ± 31.61.

The viability of freshly isolated lymphocyte suspension assessed by trypan blue staining was 96 ± 6%.

Figure 1 demonstrates the dynamics of reduced AB accumulation in PB lymphocytes (from 1, 2.5, 7.5, 10 μg/ml of MWCNT. It shows that there is a significant decrease in all PB lymphocytes with increasing concentration of MWCNT as compare to control.

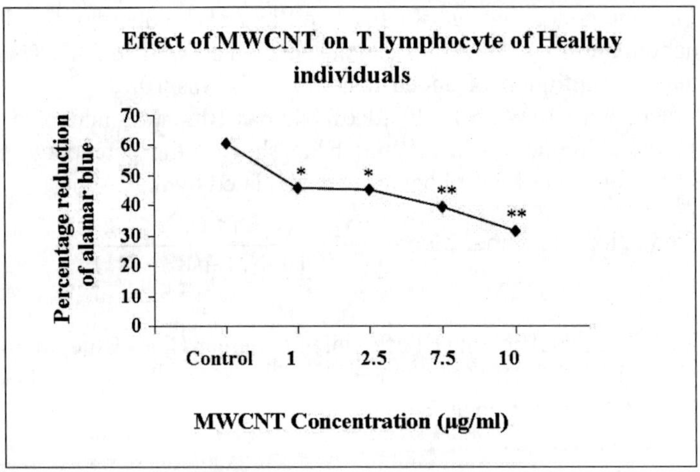

*P<0.05, **P<0.01; pair sample t-test.

Fig. 1: Effect of different concentration of MWCNT on the proliferation of T lymphocyte as indicated by reduction of Alamar Blue. Data presented is mean of 7 readings obtained from 7 healthy donors. Significant difference for treated cells with MWCNT vs. untreated cells (control) group

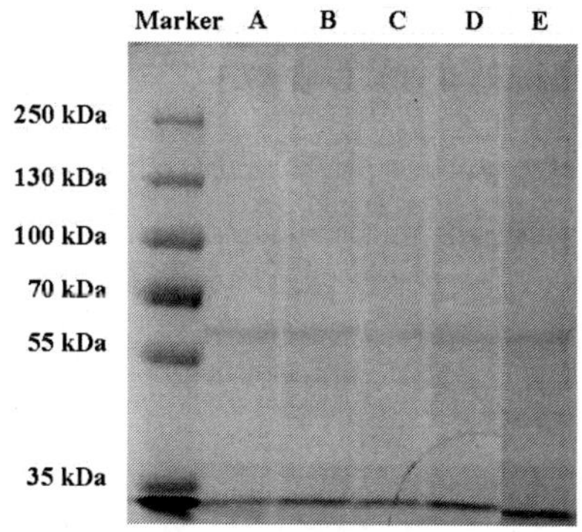

Fig. 2: Effect of Different Concentration of MWCNT on the Protein of the T Lymphocyte (A) 10 µg/ml MWCNT (B) 7.5 µg/ml MWCNT (C) 2.5 µg/ml MWCNT (D) 1 µg/ml MWCNT and (E) Control (no MWCNT)

Lymphocytic protein are more to be affected by the exposure to pathological protein hence it was decided to assessed lymphocytic protein by SDS-PAGE analysis of lymphocyte treated with different doses of MWCNT(5).

DISCUSSIONS

Time dependent and dose dependent toxicity of SWCNT has been shown by (Cui *et al.,* 2005) on proliferation of kidney epithelial cell where as similar findings were obtained by MWCNT on skin epithelial cells (Monteiro-Riviere *et al.,* 2005). Zengl *et al.* (2008) have shown that SWCNT concentration of upto 10 µg/ml filled to induced DNA damage but higher concentration like 25 and 50 µg/ml caused decreased cell growth in cultured human blood cells.

In the interesting finding by Ryan *et al.* (2007) have shown that human mast cells and pheripheral basophil which are involved in allergic response where inhibited by fullerene and Carbon nano materials.

Since our efforts have been studying the in vitro toxicity of MWCNT, we feel that to address the possible side effects of CNTs on human health and our environment, researchers have investigated the toxicology of CNTs in animal models. Non-functionalized raw CNTs have been intra-tracheally (IT) instilled into animals, showing obvious pulmonary toxicity including unusual inflammation and fibrotic reactions due to the aggregation of hydrophobic raw CNTs in the lung airways (Lam *et al.,* 2004, Warheit *et al.,* 2004, Shvedova *et al.,* 2005 and Muller *et al.,* 2005). Those results suggest that aerosol exposure of raw CNTs in the workplace should be avoided to protect human health. Nevertheless, toxicities observed by intratracheal instillation of large amounts of raw CNTs may have little relevance to the toxicology profile of functionalized soluble CNTs for biomedical applications, especially when they are administered through other routes such as intraperitoneal (IP) and intravenous (IV) injections, by which lung airways are not exposed to CNTs.

Furthermore, length dependent pathogenicity was observed, as no obvious toxic effect was observed for shorter and smaller MWCNTs (length 1– 20 μm, diameter 10–14 nm), indicating that the toxicology profiles of CNTs may significantly differ between CNTs of various sizes (diameter and length). It is worth noting that functionalized SWCNTs used in typical biomedical research have length 50–300 nm and diameter 1–2 nm, which are entirely different from the geometry of MWCNTs used by Poland *et al.*

To fully address the toxicity concern of CNTs, further investigations including animal models other than mice and on larger scales, are still required. Moreover, the interactions between administered CNTs and the immune complement system, whose activation is an important first line of defense against foreign species, especially microbes, require more attention (Salvador-Morales *et al.,* 2006). Moreover, increased efforts are needed not only from the chemical aspect, i.e. further optimizing CNT surface chemistry and geometry for improved biocompatibility, but also from those with biological expertise, to systematically study the complete CNT toxicology profile in different animal models with different routes of administration.

REFERENCES

[1] Kam, N.W.S., Jessop, T.C., Wender, P.A. and Dai, H.J. Nanotube molecular transporters: Internalization of carbon nanotube-protein conjugates into mammalian cells. *J. Am. Chem. Soc.* 2004, 126, 6850–6851.

[2] Dumortier, H., Lacotte, S., Pastorin, G., Marega, R., Wu, W., Bonifazi, D., Briand, J. P., Prato, M., Muller, S. and Bianco, A., Functionalized carbon nanotubes are noncytotoxic and preserve the functionality of primary immune cells. *Nano Lett.,* 2006, 6, 1522–1528.

[3] Schipper, M.L., Nakayama-Ratchford, N., Davis, C.R., Kam, N.W.S., Chu, P., Liu, Z., Sun, X., Dai, H. and Gambhir, S.S., A pilot toxicology study of single-walled carbon nanotubes in a small sample of mice. *Nat. Nanotech.,* 2008, 3, 216– 221.

[4] Wu, P., Chen, X., Hu, N., Tam, U. C., Blixt, O., Zettl, A. and Bertozzi, C.R., Biocompatible carbon nanotubes generated by functionalization with glycodendrimers. *Angew. Chem. Int. Ed.,* 2008, 47, 5022–5025.

[5] Lam, C.W., James, J.T., McCluskey, R. and Hunter, R.L., Pulmonary toxicity of single-wall carbon nanotubes in mice 7 and 90 days after intratracheal instillation. *Toxicol. Lett.,* 2004, 77, 126–134.

[6] Warheit, D.B., Laurence, B.R., Reed, K.L., Roach, D.H., Reynolds, G.A.M. and Webb, T.R. Comparative pulmonary toxicity assessment of single-wall carbon nanotubes in rats. *Toxicol. Lett.,* 2004, 77, 117–125.

[7] Shvedova, A.A., Kisin, E.R., Mercer, R., Murray, A.R., Johnson, V.J., Potapovich, A.I., Tyurina, Y.Y., Gorelik, O., Arepalli, S., Schwegler-Berry, D. *et al.,* Unusual inflammatory and fibrogenic pulmonary responses to single-walled carbon nanotubes in mice. *Am. J. Phys. Lung Cell. Mol. Physiol.,* 2005, 289, L698 L708.

[8] Muller, J., Huaux, F., Moreau, N., Misson, P., Heilier, J. F., Delos, M., Arras, M., Fonseca, A., Nagy, J.B. and Lison, D., Respiratory toxicity of multi-wall carbon nanotubes. *Toxicol. Appl. Pharmacol.* 2005, 207, 221–231.

[9] Poland, C. A., Duffin, R., Kinloch, I., Maynard, A., Wallace, W. A. H., Seaton, A., Stone, V., Brown, S., MacNee, W. and Donaldson, K., Carbon nanotubes introduced into the abdominal cavity of mice show asbestos-like pathogenicity in a pilot study. *Nat. Nanotech.* 2008, 3, 423 428.

[10] Cui, D.X., Tian, F.R., Ozkan, C.S., Wang, M. and Gao, H.J., Effect of single wall carbon nanotubes on human HEK293 cells., *Toxicol. Lett.,* 2005, 155, 73–85.

[11] Ding, L.H., Stilwell, J., Zhang, T.T., Elboudwarej, O., Jiang, H.J., Selegue, J.P., Cooke, P.A., Gray, J.W., Chen, F.Q.F., Molecular characterization of the cytotoxic mechanism of multiwall carbon nanotubes and nano onions on human skin fibroblast. *Nano Lett.,* 2005, 5, 2448–2464.

[12] Bottini, M., Bruckner, S., Nika, K., Bottini, N., Bellucci, S., Magrini, A., Bergamaschi, A. and Mustelin, T., Multi-walled carbon nanotubes induce T lymphocyte apoptosis. *Toxicol. Lett.,* 2006, 160, 121–126.

[13] Sayes, C.M., Liang, F., Hudson, J.L., Mendez, J., Guo, W.H., Beach, J.M., Moore, V.C., Doyle, C.D., West, J.L., Billups, W.E., *et al.* Functionalization density dependence of single-walled carbon nanotubes cytotoxicity *In vitro*. *Toxicol. Lett.,* 2006, 161, 135–142.

[14] Dong, L., Joseph, K.L., Witkowski, C.M. and Craig, M.M., Cytotoxicity of single-walled carbon nanotubes suspended in various surfactants. *Nanotechnology,* 2008, 19, 255702.

[15] Plata, D.L., Gschwend, P.M. and Reddy, C.M., Industrially synthesized single-walled carbon nanotubes: Compositional data for users, environmental risk assessments, and source apportionment. *Nanotechnology,* 2008, 19, 185706.

[16] Casey, A., Herzog, E., Davoren, M., Lyng, F.M. and Byrne, H.J., Chambers, G. Spectroscopic analysis confi rms the interactions between single walled carbon nanotubes and various dyes commonly used to assess cytotoxicity. *Carbon,* 2007, 45, 1425–1432.

[17] Worle-Knirsch, J.M., Pulskamp, K., Krug, H.F., Oops they did it again! Carbon nanotubes hoax scientists in viability assays. *Nano Lett.,* 2006, 6, 1261–1268.

[18] Liu, Z., Sun, X., Nakayama, N. and Dai, H., Supramolecular chemistry on water- soluble carbon nanotubes for drug loading and delivery. *ACS Nano,* 2007, 1, 50–56.

[19] Kam, N.W.S., O'Connell, M., Wisdom, J.A. and Dai, H. Carbon nanotubes as multifunctional biological transporters and near-infrared agents for selective cancer cell destruction. *Proc. Natl. Acad. Sci. USA* 2005, 102, 11600 11605.

[20] Wu, W., Wieckowski, S., Pastorin, G., Benincasa, M., Klumpp, C., Briand, J.P., Gennaro, R., Prato, M. and Bianco, A., Targeted delivery of amphotericin B to cells by using functionalized carbon nanotubes. *Angew. Chem. Int. Ed.* 2005, 44, 6358 6362.

[21] Dumortier, H., Lacotte, S., Pastorin, G., Marega, R., Wu, W., Bonifazi, D., Briand, J.P., Prato, M., Muller, S., Bianco, A., Functionalized carbon nanotubes are noncytotoxic and preserve the functionality of primary immune cells. *Nano Lett.,* 2006, 6, 1522–1528.

[22] Chen, X., Lee, G.S., Zettl, A. and Bertozzi, C.R., Biomimetic engineering of carbon nanotubes by using cell surface mucin mimics. *Angew. Chem. Int. Ed.* 2004, 43, 6111–6116.

[23] Chen, X., Tam, U.C., Czlapinski, J.L., Lee, G.S., Rabuka D. and Zettl, A. and Bertozzi, C.R., Interfacing carbon nanotubes with living cells, *J. Am. Chem. Soc.,* 2006, 128, 6292–6293.

[24] Heller, D.A., Baik, S., Eurell, T.E. and Strano, M.S., Singlewalled carbon nanotube spectroscopy in live cells: Towards long-term labels and optical sensors, *Adv. Mater.,* 2005, 17, 2793–2799.

[25] Chin, S.F., Baughman, R.H., Dalton, A.B., Dieckmann, G.R., Draper, R.K., Mikoryak, C., Musselman, I.H., Poenitzsch, V.Z., Xie, H. and Pantano, P., Amphiphilic helical peptide enhances the uptake of single-walled carbon nanotubcs by living cells. *Exper. Biol. Med.* 2007, 232, 1236 1244.

[26] Yehia, H.N., Draper, R.K., Mikoryak, C., Walker, E.K., Bajaj, P., Musselman, I.H., Daigrepont, M.C., Dieckmann, G.R. and Pantano, P., Single-walled carbon nanotube interactions with HeLa cells, *J. Nanobiotech,* 2007, 5, 8.

[27] Jin, H., Heller, D.A. and Strano, M.S., Single-particle tracking of endocytosis and exocytosis of single-walled carbon nanotubes in NIH-3T3 cells. *Nano Lett.,* 2008, 8, 1577–1585.

[28] Cui, D.X., Tian, F.R., Ozkan, C.S., Wang, M. and Gao, H.J., Effect of single wall carbon nanotubes on human HEK293 cells, *Toxicol. Lett.,* 2005, 155, 73–85.

[29] Monteiro-Riviere, N.A., Nemanich, R.J., Inman, A.O., Wang, Y.Y. and Riviere, J.E., Muti-walled carbon nanotube interactions with human epidermal keratinocytes. *Toxicol. Lett.*, 2005 155, 377–384, 2005.

[30] Zeng, L., Alemany, L.B., Edwards, C.L. and Barron, A.R., Demonstration of covalent functionalization of SWCNT by NMR spectroscopy: side chain length dependence on the observations of the sidewall sp3 Carbons. *Nano Res.*, 2008, 1, 72–88.

TEMPERATURE AND pH RESPONSIVE *IN SITU* GEL FORMING NANODISPERSIONS FOR EFFICIENT OCULAR DELIVERY OF IBUPROFEN

SUBRAMANIAN NATESAN, VINOTH JEEVANESAN, VANITHA KUPPUSAMY AND SIVASANKARI GOMATHINAYAGAM

Laboratory for Lipid Based Systems, Department of Pharmaceutical Technology, Anna University of Technology Tiruchirappalli, Tiruchirappalli
E-mail: natesansubbu@gmail.com

ABSTRACT

The maintenance of therapeutic drug concentration over the prolong period of time is required for the effective management of various pathological condition of eye. The various precorneal clearance factors limit the ocular absorption of drug and in turn resulted in frequent dose administration. The stimuli responsive in situ gel forming systems have been investigated recently as an ocular drug delivery system due to its ability to offer extended precorneal residence time and enhanced local ocular bioavailability. In this study, we have developed and characterized the dual stimuli responsive system consist of pH responsive Ibuprofen loaded cellulose acetate phthalate (CAP) nanoparticles dispersed in temperature responsive gel forming solution containing Pluronic F127 and F68. The nanoparticles dispersed in Pluronics solution (21% F127 and 10% F68) showed good phase transition behaviour at ocular temperature and moderate viscosity makes the system suitable for the ocular administration. The particle sizes of the nanoparticles were ranged between 159 to 356 nm. The in vitro drug release study showed about 70% of drug release in 8 hours. The prolonged retention of gel consistency and controlled corneal permeability of the drug (180 µg in 2 hours) from the system in response to the pH and temperature was observed. Hence, the developed temperature and pH responsive in situ gelling topical nanodispersion may act as a better alternative to conventional topical ocular formulations.

V. Rajendran, P. Prabu and K.E. Geckeler (eds.)
Nano Biomaterials, pp. 211–218 (2012).

INTRODUCTION

Most of the ocular diseases like inflammation, glaucoma, dry eye syndrome, conjunctivitis, corneal ulcer and uveitis etc., require frequent administration of drug to improve/manage the pathological condition. However, due to precorneal clearance factors such as blinking reflex, lacrimation, tear turnover, limited cul-de-sac volume, rapid drainage, metabolism/degradation of drug in ocular environment and poor corneal permeability of the drugs, only a small fraction of the instilled dose (about 1%) is locally absorbed with the conventional ophthalmic drug delivery systems. So, the development of an ideal topical ocular formulation which offers good ocular bioavailability remains as a challenge task to be resolved satisfactorily. The temperature and pH responsive polymers are extensively used in ocular environment responsive *in situ* gel forming (sol-gel) ophthalmic formulations. These systems exhibit extended precorneal residence time, enhanced local ocular bioavailability by providing sustained drug release, increased drug stability, decreased dosing frequency and better patient compliance.

The mixture of Pluronics was reported as an *in situ* gelling vehicle to enhance the ocular bioavailability of drugs.[1] Mishra *et al.*[2] demonstrated that the higher drug transport across corneal membrane and increased ocular retention time of the drug from their developed temperature and ionic responsive *in situ* gelling ophthalmic vehicle composed of Pluronic F-127 and chitosan. Gurny *et al.*[3] formulated the cellulose acetate phthalate (CAP) latex as a pH triggered ophthalmic drug delivery system, in which the CAP get collapse and release the loaded drug when the pH of the environment changes from its native pH 4.5 to pH 7. If the CAP particles are surrounded by ocular temperature responsive gel forming Pluronic solution, the lachrymal fluid (pH-7) expected to slowly diffuse through the Pluronic gel matrix to solubilize the CAP nanoparticles and releases the drug slowly from the system to ocular environment.

Hence in the present study physiological dual stimuli responsive *in situ* gelling system consists of Ibuprofen loaded CAP nanoparticles dispersed in thermo-responsive Pluronics (Pluronic F127 and F68) solution were developed for spontaneous gelation upon instillation to eye, increased corneal residence time, prolonged pH responsive drug release and enhanced ocular bioavailability. The developed particles were characterized for its size, polydispersity index, zeta potential, drug content, encapsulation efficiency, loading capacity, phase transition behaviour and evaluated for *in vitro* drug release and *in vitro* ocular permeability of the drug from the system.

EXPERIMENTAL PROCEDURES

Materials

Ibuprofen and CAP were obtained as generous gift from Fourrts (India) laboratories Ltd., Chennai and Apex laboratories Ltd., Chennai respectively. Pluronic 407 and 188 were procured from BASF, Mumbai. All other chemicals used are analytical grade and purchased from Merck India, Mumbai.

Formulation and Characterization of Ibuprofen Nanoparticles

Four different batches of Ibuprofen loaded nanoparticles (NP50, NP100, NP150 and NP200) were formulated with varied concentration of polymer (CAP) by emulsification and solvent evaporation method. Accurately weighed Ibuprofen and CAP were dissolved in the organic phase containing acetone and ethanol (1:2) and added slowly drop-wise into the aqueous phase consists of Tween 80 water (1:100), under constant stirring at 2000 rpm upto 24 hours. After complete evaporation of organic solvent, the nanoparticles were centrifuged at 17000 rpm for 10 minutes and repeatedly washed with distilled water. The resulted Ibuprofen nanoparticles were collected, dried and stored.

An accurately weight quantity of formulated nanoparticles were dissolved in phosphate buffer pH 7.4 and the absorbance was measured using UV spectrophotometer (Shimadzu Corporation, Japan) at 221 nm and the drug content was calculated. Encapsulation efficiency and loading capacity of the formulated nanoparticles were calculated by the following formulas.

Encapsulation efficiency (%) = [Loaded drug/Total drug] × 100

Loading capacity (%) = [Loaded drug/Nanoparticle weight] × 100

The mean particle size, polydispersity index and zeta potential of the nanoparticles were determined using Zetasizer (Malvern, UK) based on the dynamic light scattering and laser Doppler anemometer techniques.

DEVELOPMENT AND EVALUATION OF TEMPERATURE RESPONSIVE *IN SITU* GELLING SYSTEM

Combination of Pluronic F127 and Pluronic F68 at different percentage of 18 and 13%, 21 and 10% and 24 and 7% were prepared by dissolving the appropriate amount of Pluronics in phosphate buffer pH 4.5. Phase transition behaviour of the various combinations of Pluronic solutions were analysed by optical cloud point determination [4] and test tube-inversion method.[5] Briefly, Pluronic solutions were taken in a test tube and immersed into gradually heating water bath. The temperature at which the opaqueness appeared was noted as cloud point. The thermo-responsive sol-gel phase transition was studied by test tube-inversion method, in which the polymeric solution in the test tube was continuously tilted at 90° angle in a slowly raising temperature water bath. The temperature at which the gelling occurred was noted. The both studies were done in triplicate and average time was obtained from the three recorded results.

FORMULATION AND CHARACTERIZATION OF DUAL STIMULI RESPONSIVE OCULAR TOPICAL NANODISPERSION

The known amount of four batches of Ibuprofen-CAP nanoparticles was dispersed in Pluronics solution (21% of Pluronic F127 and 10% of Pluronic F68 in phosphate buffer pH 4.5) to get 0.5% Ibuprofen nanodispersion of F50, F100, F150 and F200 respectively. The nanoparticles dispersion in solution was ensured by sonication and stored at the temperature of 20°C.

The *in vitro* drug release was carried out by a known quantity of nanodispersed formulations were placed in a dialysis membrane bag and suspended in phosphate buffer solution pH 7.4. The entire system was maintained at 37±0.5°C with continuous constant stirring at 100 rpm. At predetermined time intervals, the sample was withdrawn and simultaneously the same volume of fresh medium was replaced to maintain the sink condition. The amount of Ibuprofen released in to the release medium was analysed by UV spectrophotometer at 221 nm using phosphate buffer solution pH 7.4 as blank. Release kinetics of optimized formulation was determined by fitting the drug relcase data with the zero order, first order, Higuchi equation and Korsmeyer-Peppas equation. *In vitro* transcorneal permeation of Ibuprofen nanodispersions across the goat cornea was performed using modified side-by-side diffusion cell.[6] The nanoparticles containing low (F50) and high (F200) concentration of CAP were subjected to *in vitro* corneal permeability. The goat cornea was carefully removed along with 2–4 mm of surrounding scleral tissue from the eye within 30 minutes of slaughtering of the animal and washed with cold saline. The cornea was placed between the donor and receptor cells, in such a way that the epidermal surface of the cornea facing the donor side and clamped. The corneal area available for diffusion was 0.78 cm^2. The cells were filled with phosphate buffer pH 7.4 solution and both the cells were stirred using magnetic stirring bar at 150 rpm, maintained at 37±0.5°C. Nanodispersion was dispersed in donor cell and samples were withdrawn from receptor cell at predetermined time intervals and replaced with fresh buffer solution to maintain sink condition. The samples were analysed for the amount of drug permeated across the cornea using UV spectrophotometer at 221 nm.

RESULTS AND DISCUSSION

Characterization of Ibuprofen-CAP Nanoparticles

The drug content, encapsulation efficiency, loading capacity, particle size and polydispersity index and zeta potential of the formulated nanoparticles are given in the Table. 1.

Table 1: Characterization of Ibuprofen-CAP Nanoparticles

Code	*Ibuprofen Content (mg)*	*Encapsulation Efficiency (%)*	*Loading Capacity (%)*	*Average Particle Size (nm)*	*Poly Dispersity Index*	*Zeta Potential (mV)*
NP50	34.2	68.4	60.2	159	0.812	–3.38
NP100	46.1	92.2	45.5	227	0.617	–3.20
NP150	48.3	96.6	35.4	246	0.615	–6.80
NP200	49.0	98.0	25.6	326	0.902	–1.20

The drug content and encapsulation efficiency was increased with the increase in the polymer concentration, which indicates the efficiency of drug incorporation and encapsulation was directly proportional to the concentration of CAP in the formulation. Whereas the loading capacity of the nanoparticles decreased with increase of polymeric concentration, indicates the more amount of polymer in formulation dramatically affects the loading of drug into the nanoparticles. The particle size of the nanoparticles also increased in respond to the increase in polymer concentration. The polydispersity index of all the nanoparticles were laid below the value 1 indicates its mono-disperse in nature. The zeta potential of the nanoparticles was observed around –1 mV to –6 mV. Theoretically, these values denote the instability of these particles in dispersion, but when the particles were suspended into the higher concentration of viscous Pluronic solution in the final nanodispersed formulation, the zeta potential was found to be –21 mV. Hence, higher concentration of Pluronic might contribute the dispersion stability of these nanoparticles by adhering on the surface of the nanoparticles.

Development and Evaluation of Temperature Responsive *In situ* Gelling Systems

The cloud point and phase transition temperature of the various combinations of Pluronic F127 and Pluronic F68 is given in Table. 2. The Pluronic solution having 21% of Pluronic F127 and 10% of Pluronic F68 showed cloud point and phase transition to gel at 37°C. So this combination of Pluronics was selected for the formulation.

Table 2: Evaluation of Temperature Responsive *In situ* Gelling System

Pluronic Concentration (%)		*Cloud Point Temperature (°C)*	*Sol-gel Transition Temperature (°C)*
F127	*F68*		
18	13	39	39
21	10	37	37
24	7	35	35

Characterization of Stimuli-responsive *In situ* Gel Forming Ibuprofen Nanodispersions

The cumulative percentage of Ibuprofen released from *in situ* gel forming ocular nanodispersion is shown in Figure 1. The results indicate that the F50 formulation had initial slow release for the first two hours followed by rapid release of drug. The other formulations were followed the controlled release of drug of about 30% within eight hours. This indicated that the CAP nanoparticles slowly erode/dissolves and release the drug on pH change. Moreover, the *in situ* gel formation by Pluronic in 37°C allows the dissolution media to diffuse at very slow rate to contact the CAP nanoparticles to dissolve it and also limits the drug diffusion through gel matrix.

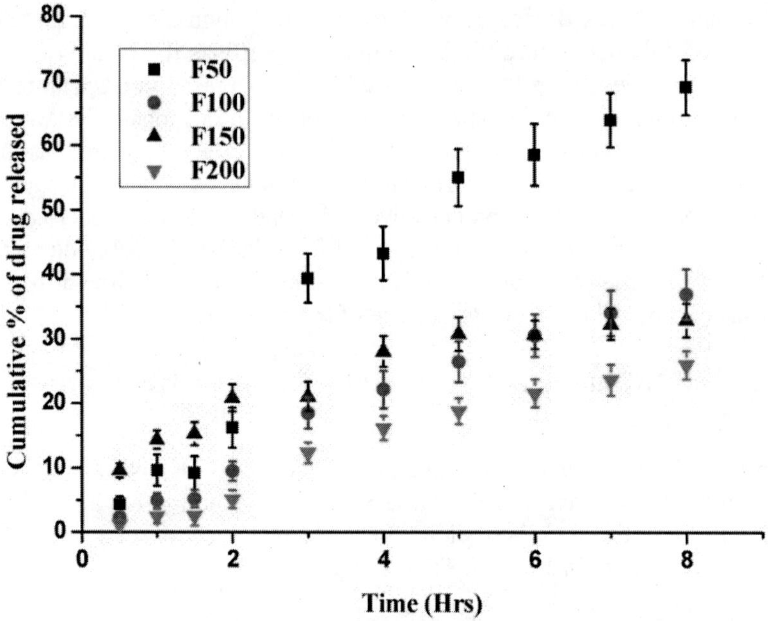

Fig. 1: In vitro Drug Release of in situ Gel Forming Ocular Nanodispersion of Ibuprofen

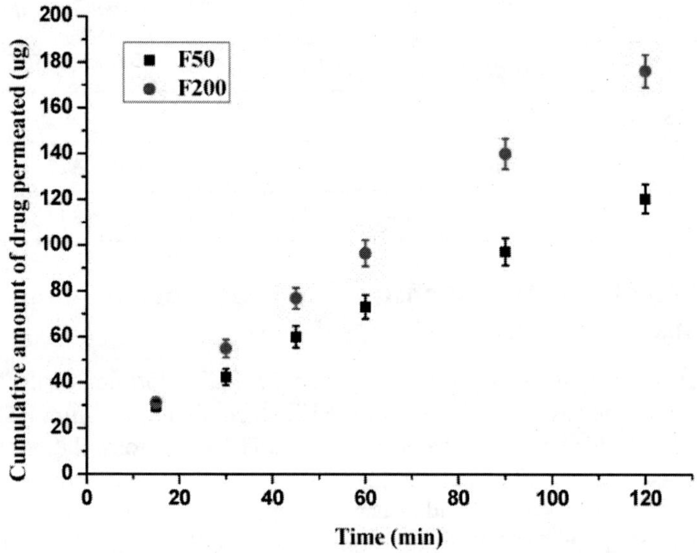

Fig. 2: In vitro Corneal Permeability of Ibuprofen from Stimuli-Responsive in situ Gel Forming Ocular Dispersion

The release kinetics study of the optimized formulation (F200) obeys the first order kinetics and denotes the drug release was dependent of its concentration. The Higuchi plot confirmed the drug released through diffusion from polymeric matrix. In the korsmeyer-Peppas plot, the 'n' values of the formulation was found more than 0.5, confirmed that the release followed the non-Fickian diffusion or anomalous transport.

The corneal permeability study indicates the Ibuprofen permeated from F200 formulation was higher than F50 formulation in respond to the change in pH of the environment (Figure 2). This might be due to the higher concentration of drug encapsulated in the F200 formulation might undergo rapid saturation solubility which facilitate more amount of drug release and permeation in a controlled prolonged manner.

CONCLUSION

The developed dual stimuli responsive nanodispersion was found to have the promising *in vitro* characteristics for the temperature responsive gel phase transition and pH responsive controlled drug release, which indicates the suitability of the formulation for topical ocular administration to overcome the problems existing in the conventional formulations. Hence, we postulate that the developed temperature and pH responsive *in situ* gelling topical nanodispersion may act as a better alternative to conventional topical ocular formulations.

ACKNOWLEGEMENT

The authors are very thankful to the Department of Biotechnology, Ministry of Science and Technology, Government of India, for their support to carry out this research work.

REFERENCES

[1] Lin, H.R., Sung, K.C., Vong, W.J., In situ gelling of alginate/pluronic solutions for ophthalmic delivery of pilocarpine. *Biomacromolecules*, 5, 2358–65, 2004.

[2] Gupta, H., Jain, S., Mathur, R., Mishra, P., Mishra, A.K. and Velpandian, T., Sustained ocular drug delivery from a temperature and pH triggered novel in situ gel system. *Drug Deliv.*, 14, 507–15, 2007.

[3] Gurny, R., Boye, T. and Ibrahim, H., Ocular therapy with nanoparticulate systems for controlled drug delivery. *J Control Release*, 2, 353–61, 1985.

[4] Rafie, F., Javadzadeh, Y., Javadzadeh, A.R., Ghavidel, L.A., Jafari, B., Moogooee, M. and Davaran, S., In vivo evaluation of novel nanoparticles containing dexamethasone for ocular drug delivery on rabbit eye. *Curr Eye Res*. 35, 1081–9, 2010.

[5] Kim, I.Y., Yoo, M.K., Kim, B.C., Park, I.Y., Lee, H.C. and Cho, C.S. Thermogelling behaviors of poly(caprolactone-b-ethylene glycol-b-caprolactone) triblock copolymer in the presence of hyaluronic acid. *J Polym Sci A: Polym Chem.* 46, 3629–37, 2008.

[6] Malhotra, M., Majumdar, D.K., In vitro transcorneal permeation of ketorolac tromethamine from buffered and unbuffered aqueous ocular drops. *Indian J Exp Biol,* 35, 941–7, 1997.

ENHANCEMENT OF MODULUS, STRENGTH AND CONDUCTING PROPERTIES IN REDUCED GRAPHENE OXIDE-GRAFT-POLY (METHYL METHACRYLATE)/POLY VINYL CHLORIDE NANOCOMPOSITE

RAMA K. LAYEK, ANIRUDDHA KUNDU AND ARUN K. NANDI

Polymer Science Unit, Indian Association for the Cultivation of Science, Jadavpur, Kolkata
E mail: psuakn@iacs.res.in

ABSTRACT

The highly soluble nanocomposites composed of reduced graphene oxide (rGO) and poly (vinyl chloride) (PVC) has been successfully fabricated by dissolving poly(methyl methacrylate) (PMMA) modified reduced graphene oxide (MG) and PVC in tetrahydofuran (THF) and casting from THF solution. The solubility of MG/PVC composite increases remarkably due to a polymer/solvent synergy. For the increase of compatibility of graphene in PVC matrix, we have synthesized PMMA functionalized rGO hybrid material obtained by a simple approach via in situ atom transfer radical polymerization (ATRP) of monomer with rGO dispersion in 1,2-dichlorobenzene (ODCB) which is highly soluble in THF. The synthesized Graphene oxide (GO) and MG has been characterized by AFM, TEM, and FTIR. It is clear from the TEM and image that the polymer brush has grown from rGO surface. Finally this MG sheets are used to fabricate graphene/PVC composite from THF solution. Mechanical parameters such as the tensile stress at break and young's modulus of the composite films increases 167.5% and 224% respectively compare to pure polymer for 8%(w/w) MG content. Not only the mechanical property, but also the conductivity of the composite increases dramatically over PVC.

V. Rajendran, P. Prabu and K.E. Geckeler (eds.)
Nano Biomaterials, pp. 219–226 (2012).

INTRODUCTION

Graphene has generated a great deal of interest owing to their unique physical properties such as large aspect ratio, excellent mechanical, high electrical and thermal conductivity.[1–3] Therefore graphene is an attractive candidate for wide range of application including electrochemical, field-emission and electronic devices.[4–6] Among the various applications, perhaps the most promising and potential application is associated with mechanical property improvement. For example Young's modulus of single-layer graphene can reach upto 1 TPa while strength has the upper limit 130 GPa, is the strongest material ever measured.[1] Therefore the judicious incorporation of graphene in the polymer matrix by homogeneous dispersion and fine interface control may lead to the formation of high performance composite. However, it is very difficult to use graphene as successive nano filler in a polymer matrix, because the insolubility in solvents makes them difficult to achieve efficient dispersion. In addition, intrinsic van der Waals attraction among the graphene layers in combination with their high surface area and high aspect ratio, often leads to significant agglomeration, thus preventing efficient transfer of their superior properties to the matrix. Moreover, the non-interactive graphene surface results weak graphene-polymer interactions and poor interfacial adhesion, which may lead to graphene aggregation within the matrix. Therefore, to prepare high performance graphene polymer nanocomposite, fundamental criteria is the improvement of solubility of graphene and the graphene-polymer interactions. For the stable dispersion and uniform filler loading into polymer matrix interfacial engineering is required to maximize the load transfer from graphene sheets to polymer matrix.

Several methods have been reported to produce graphene polymer composite with good dispersion and good load transfer between the polymer and graphene sheets such as melt casting,[7] insitu polymerization [8] and aid of surfactant to aid dispersion.[9] Many authors have used the functionalized graphene sheets to improve the mechanical property of graphene polymer nanocomposites in recent literature.[10–12] Among them to achieve stable dispersions of graphene and optimize the microstructure of nanocomposites, polymer functionalization of the graphene surface is often necessary.[10] Polyvinyl chloride (PVC) is the third extensively produced plastic, after polyethylene and polypropylene. The poly(vinyl chloride) (PVC) is a technologically important polymer and a large number of research efforts have been given to various aspects of its preparation, processing, and properties. It is extensively used in construction as it is cheaper than more durable longer lasting alternatives. Again it can be made flexible by the addition of plasticizers for clothing and upholstery, electrical cable insulation, inflatable products and in many other applications in which it replaces rubber.[13] The fabrication of mechanically strong and conducting PVC nano composite may be very much beneficial for the industrial application. In very recent times the graphene/PVC nano composite have been fabricated from sonicated graphite flakes [14] and covalently modified graphene via solvent cast method.[15]

In this work, we reported the preparation of highly soluble nanocomposites composed of graphene and poly (vinyl chloride) (PVC) by mixing PMMA modified reduced graphene oxide with PVC dissolved in tetrahydofuran (THF). For the increase in compatibility of graphene in PVC matrix, we have synthesized PMMA functionalized reduced graphene oxide hybrid material obtained by a simple approach via in situ ATRP of monomer [16] with rGO dispersion in 1,2–dichlorobenzene (ODCB) which is highly soluble in THF. Finally this functionalized graphene sheets are used to fabricate graphene PVC composite from THF solution. Mechanical parameters such as the young's modulus and tensile strength of the composite films significantly increase compare to pure polymer. Not only the mechanical property but also the conductivity of the composite increases dramatically compare to pure PVC.

EXPERIMENTAL SECTION

Samples

Graphite (Aldrich, USA), sodium nitrate and potassium permanganate (Merck, Mumbai) are used as received. Double distilled water was used. Hydrazine hydrate solution (99%, synthetic grade, Merck, Mumbai) and N, N, N, N, N pentamethyl diethylene triamine (PMDETA, Aldrich) are used as received. THF (Renkem, New Delhi, analytical grade) and methyl methacrylate (MMA, Loba chemicals) are distilled in vacuum before use. CuCl (Aldrich, USA) is purified by washing with 10% HCl followed by methanol and diethyl ether(Renkem, New Delhi, analytical grade) in a schlenk tube under nitrogen atmosphere.

Synthesis and Purification of Graphene Oxide (GO)

GO was prepared according to modified Hummers method.[17] In this procedure, 46 ml concentrated sulfuric acid was taken a conical flask in an ice bath and stirred with a bar magnet. Then 0.1 gm sodium nitrate was added into the flask under constant stirring. 2 gm Graphite powder was added very slowly into the conical flask with vigorous stirring at the mention temperature. After the complete addition of graphite powder, 6 gm potassium permanganate was very slowly added into the conical flask maintaining the temperature at 0–5°C. Then the reaction mixture was allowed to stir at room temperature forming a thick paste for 6 hour. After that 92ml double distilled water was poured into the flask under a constant stirring. The temperature of the solution was raised to about 90°C and the mixture was allowed to stir for another 30 minutes. Finally 280 ml of water was added followed by slow addition of 3 ml H_2O_2 (30%) and the colour of the solution changed from dark brown to yellow. The solution was centrifuged to collect the product at bottom and supernatant liquid was thrown out. The product was re-dispersed with water and centrifuged until the pH of the solution become neutral. Finally the brown mass was collected and dried at 60°C in vacuum to obtain graphene oxide (GO).

Synthesis of Reduced Graphene Oxide (rGO)

rGO was synthesized according to Stankovich method.[18] In This typical procedure, at first GO (100 mg) was dispersed in 100 mL water in 250 mL round bottom flask to obtain inhomogeneous solution. The sonication of the solution was performed until clear homogeneous solution of brown colour appears. After that 1.00 mL hydrazine hydrate solution was then added into the solution and the solution was allowed to reflux in an oil bath at 100°C. Reduced GO gradually precipitated out from the solution in the form of black solid. This product was collected by repeated re-dispersion in water and centrifugation. The black solid was dried in vacuum at 60°C to obtain rGO.

Synthesis of rGO-Graft-Poly (methyl methacrylate) (MG)

250 mg of rGO was sonicated in 15 mL of *o*-dichlorobenzene (ODCB) in a dry tube at room temperature for 1 h followed by purging dry nitrogen gas. After sonication, 10 mg of CuCl was added. After the mixture was degassed three times, the tube was sealed with rubber septum. 4 mL of MMA was purged with nitrogen gas and injected into the reaction mixture. 15 μL of PMDETA was added with slow stirring and it was kept in an oil bath at 60°C for 24 h.[16] At the end of the reaction viscosity of the reaction mixture was increased dramatically. The reaction mixture was diluted with $CHCl_3$ and was precipitated in methanol and filtered through PTFE membrane. The residue was re-dispersed in $CHCl_3$ by sonication, and then filtered for several times to remove ungrafted PMMA. The black mass was dried in vacuum at 60°C to obtain RGO-PMMA. Gravimetric analysis suggests that MG samples contain 25% rGO.

Preparation of PVC/MG Films

The MG reinforced PVC films were prepared by solvent cast method. In this procedure, 250 mg PVC was dissolved in 5 ml THF and in another tube MG was dispersed very well in 5 ml THF and mixed with the PVC solution so that MG concentration becomes 8% (w/w) with respect to PVC. The solution was sonicated for 30 minutes to make homogeneous solution. The solution was then poured in a petri disc for evaporation at open air to obtain the nanocomposite film and it is designated as MG8. Pure PVC film was also prepared in same manner.

RESULT AND DISCUSSION

Functionalization of rGO with PMMA

Atomic force microscopy (AFM) gives the information about the thickness and length of the graphene sheets. For AFM measurement the 0.01 mg GO was dispersed in 1ml in water by ultrasonication and then cast a drop on a freshly cleaved mica surface. After drying the sample it was observed that the thickness of the graphene oxide sheets and presented in Figure 1 which is almost similar to the previous report, i.e. 1 nm.[19] For detail morphological investigation TEM images of the GO and the MG nanosheets are shown in Figure 2a and 2b. Figure 2a shows clearly shows that the sheet like morphology of GO where the edges tend to fold and roll. The TEM image of the MG sample shows some dark spots which is absent in pristine GO (Figure 2). The presence of such spots may be due to the formation of polymer brushes by surface functionalization which very similar reported to other polymers.[10,11]

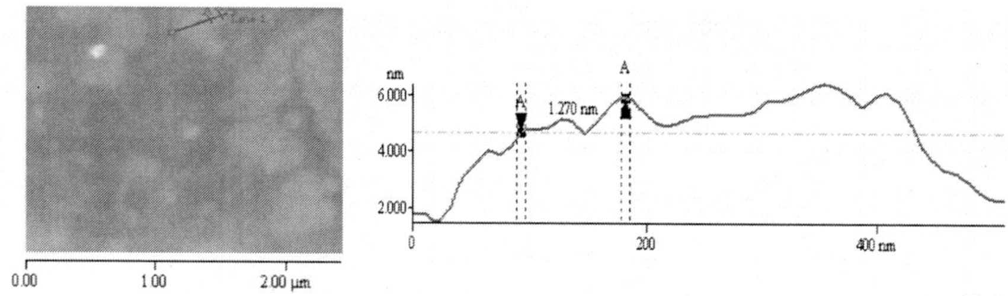

Fig. 1: AFM Image of GO

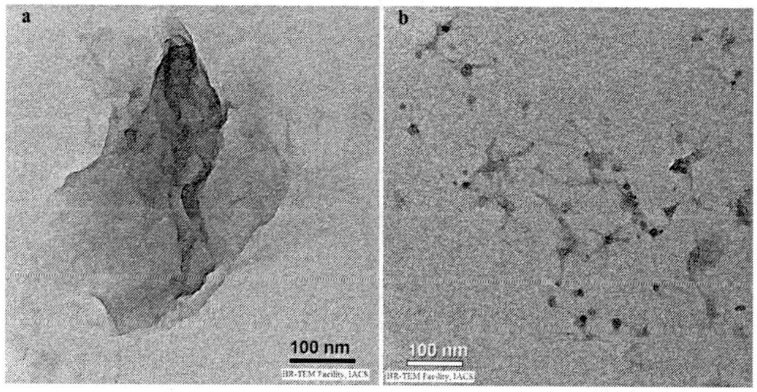

Fig. 2: TEM Images of (a) GO, and (b) MG

To elucidate the formation of GO, rGO and MG, the FTIR spectra of the sample are analyzed and presented in Figure 3. GO shows characteristic peak due to many oxygen containing moiety such as at 1728 cm^{-1} corresponds to C=O stretching vibration, 3418 cm^{-1} for O-H stretching vibration, 1052 cm^{-1} for C-O stretching, and 1612 cm^{-1} for skeletal vibration from

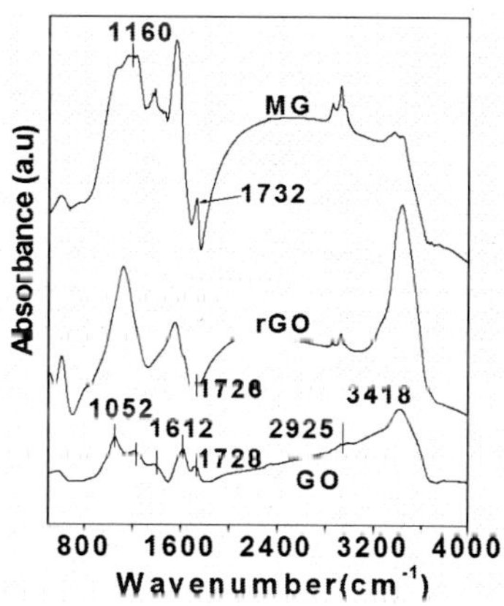

Fig. 3: FTIR Spectra of GO, rGO and MG

unoxidized graphite domain. Beside this, the peak at 2925/2855 cm^{-1} for sp^3 C-H stretching tells that there are some CH_2 or CH groups which exist in the GO samples.[19] In case of rGO samples the peak of epoxide group disappears due to hydrazine hydrate treatment. But the peak of hydroxyl group reduces but still remains at 3430 cm^{-1} after hydrazine hydrate treatment. The FTIR spectra of rGO show that hydrazine hydrate treatment can not restore all sp^2-CH but remain some sp^3-CH.[19] The FTIR spectra of MG shows characteristic peaks at 1732 cm^{-1} and 1160 cm^{-1} for the ester group of PMMA and 2936 cm^{-1} for sp^3 C-H stretching which is more intense than that of GO and rGO.[12]

Nanocomposite Films of PVC Prepared with MG Fillers (PMMA/GPMMA)

Mechanical Property

The cross sectional FESEM image (Figure 4) of the MG8 shows the good dispersion of MG in PVC matrix The mechanical behaviour relating to Young's modulus, tensile strength of PVC and MG/PVC composite films has been studied and typical mechanical property enhancement of MG8 over PVC has presented. The MG8 sample shows 167.5% increase in tensile stress and 224% increase of Young,s modulus than that of pure PVC films. The FESEM images of the fracture surface of the MG8 samples show the dispersion of MG throughout the PVC matrix and we assumed that the mechanical properties of the graphene/polymer composite are highly dependent on graphene dispersion, which straightforwardly control sheet to sheet and sheet to polymer interactions in the composite materials. Such molecular level dispersion may play significant role in graphene to polymer load transfer and interfacial bonding between graphene sheets and polymer matrix.

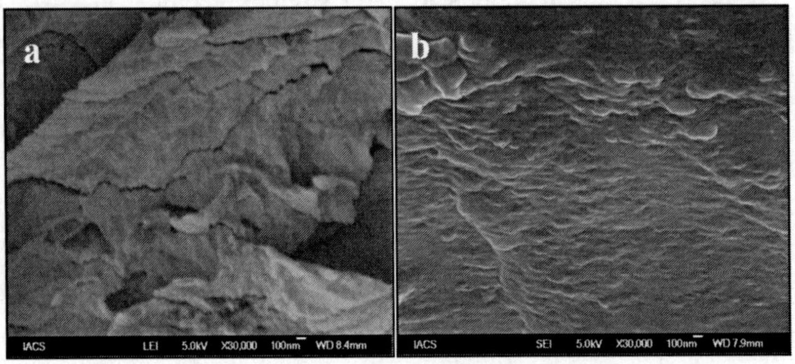

Fig. 4: FESEM Image of (a) PVC and (b) MG8

CONDUCTIVITY

Pure PVC having dc conductivity of 5×10^{-12} S/m is an insulator in nature. For MG loading (8%) the conductivity reaches at 7×10^{-2} S/m. The dramatic rise of conductivity is due to the homogeneous distribution of graphene sheets through out the PVC matrix.

CONCLUSION

In summary, high performance electrically conductive and mechanically strong PVC composite have been successfully fabricated by using PVC as matrix and MG as nano filler in solvent cast method. FESEM image is the evidence for the homogeneous dispersion of MG sheets in PVC matrix. The MG8 sample shows increase of tensile strength and young modulus 167.5% and 224% respectively. These composite films may have potential applications in structural materials with high strength building material and many other applications in which it replaces rubber nano composite.

REFERENCES

[1] Lee, C., Wei, X., Kysar, J. W., Hone, J., Measurement of the Elastic Properties and Intrinsic Strength of Monolayer Graphene, *Science*, 321, 385–388,2008.

[2] Du, X., Skachko, I., Barker, A., Andrei, E. Y., Approaching ballistic transport in suspended graphene, *Nature Nanotechnol.,* 3, 491–495, 2008.

[3] Balandin, A. A., Ghosh, S., Bao, W., Calizo, I., Teweldebrhan, D., Miao, F., Lau, C. N., Superior Thermal Conductivity of Single-Layer Graphene, *Nano Lett.,* 8, 902–907, 2008.

[4] Liu, N., Luo, F., Wu, H., Liu, Y., Zhang, C., Chen, J., One-Step Ionic-Liquid-Assisted Electrochemical Synthesis of Ionic-Liquid-Functionalized Graphene Sheets Directly from Graphite, *Adv. Funct. Mater.* 18, 1518–1525, 2008.

[5] Ye, D., Moussa, S., Ferguson, J. D., Baski, A. A., El-Shall, M. S., Highly Efficient Electron Field Emission from Graphene Oxide Sheets Supported by Nickel Nanotip Arrays, *Nano Lett.,*12, 1265–1268, 2012.

[6] Avouris, P., Graphene: Electronic and Photonic Properties and Devices, *Nano Lett.*,10, 4285–4294, 2010.

[7] Kim, H., Miura, Y., Macosko, C. W., Graphene/Polyurethane Nanocomposites for Improved Gas Barrier and Electrical Conductivity, *Chem. Mater. 22*, 3441–3450, 2010.

[8] Sahoo, S., Karthikeyan, G., Nayak, G. C., Das, C. K., Electrochemical characterization of in situ polypyrrole coated graphene nanocomposites, *Synthetic Metals,* 161, 1713–1719, 2011.

[9] Lee, H. B.; Raghu, A. V.; Yoon K. S.; Jeong, H. M., Preparation and Characterization of Poly(ethylene oxide)/Graphene Nanocomposites from an Aqueous Medium, *J Macromol Sci Part B Phys,* 49, 802–809, 2010.

[10] Fang, M., Wang, K., Lu, H., Yang, Y., Nutt. S., Covalent polymer functionalization of graphene nanosheets and mechanicalpropertics of composites, *J. Mater. Chem.*, 19, 7098–7105, 2009.

[11] Goncalves, G.; Marques, P.A.A.P.; Barros-Timmons, A. I. Bdkin, M.K. Singh, N. Emami, J. Gracio Graphene oxide modified with PMMA via ATRP as a reinforcement filler, *J. Mater. Chem,*. 20, 9927–9934, 2010.

[12] Layek, R. K., Samanta, S., Chatterjee, D.P. and Nandi, A.K., Physical and mechanical properties of poly(methyl methacrylate)—functionalized graphene/poly(vinylidine fluoride) composites: Piezoelectric β polymorph formation, *Polymer,* 51, 5846–5856, 2010.

[13] Titow, W.V., PVC technology. Springer, pp. 6– ISBN 978-0-85334-249-6, 31 December 1984.

[14] Vadukumpully, S., Paul, J., Mahanta, N. and Valiyaveettil, S., Flexible conductive graphene/ poly(vinyl chloride) composite thin films with high mechanical strength and thermal stability, *Carbon,* 49, 198–205, 2011.

[15] Salavagione, H.J. and Martinez, G., Importance of Covalent Linkages in the Preparation of Effective Reduced Graphene Oxide Poly(vinyl chloride) Nanocomposites, *Macromolecules* 2011, 44, 2685–2692.

[16] Liu, M., Zhu, T., Li, Z. and Liu, Z., One-Step in Situ Synthesis of Poly(methyl methacrylate)-Grafted Single-Walled Carbon Nanotube Composites, *J. Phys. Chem. C*, 113, 9670–9675, 2009.

[17] Hummer, W.S. and Offeman, R., Preparation of Graphitic Oxide, *J Am Chem Soc* 1958, 80, 1339.

[18] Stankovich, S., Dikin, D.A., Piner, R.D., Kohlhaas, K.A., Kleinhammes, A., Jia, Y., Wu, Y., Nguyen, S.T. and Ruoff, R.S., Synthesis of graphene-based nanosheets via chemical reduction of exfoliated graphite oxide, *Carbon*, 45, 1558–1565, 2007.

[19] Stankovich, S., Dikin, D.A., Dommett, G.H.B., Kohlhaas, K.M., Zimney, E.J., Stach, E.A., Piner, R.D., Nguyen, S.T. and Ruoff, R.S., Graphene-based composite materials, *Nature Nature*, 442, 282–286, 2006.

[20] Ren, P., Yan, D., Ji,, X., Chen, T. and Li, Z., Temperature dependence of graphene oxide reduced by hydrazine hydrate, *Nanotechnology*, 22, 055705 (8pp), 2011.

[21] Kong, H., Gao, C. and Yan, D., Controlled Functionalization of Multiwalled Carbon Nanotubes by in Situ Atom Transfer Radical Polymerization, *J Am Chem Soc*, 126, 412–413, 2004.

BIOSYNTHESIS AND CHARACTERIZATION OF SILVER NANOPARTICLES USING *ASPERGILLUS NIGER SP.*

A. Sivasamy, K. Jagan, S. Nethaji and Sugania Josphine

Chemical Engineering Area, CSIR-Central Leather Research Institute, Adyar, Chennai
E-mail: arumugamsivasamy@yahoo.co.in

ABSTRACT

In this study, AgNP's were biosynthesized using Aspergillus niger isolated from contaminated soil. The pure culture was obtained from the sub-cultured culture. The pure culture was inoculated in the fungal broth respectively and left to see for its growth under the incubation of 30°C and mycelial mat formation was observed after seven days and further it was filtered. To the extra cellular matrix solution appropriate amount of substrate silver precursor solution is added and the particle formation was observed with respective time. In course of reaction the colour changed from transparent to yellowish-brown indicating the formation of Ag in the solution. The kinetics for the synthesis of nanoparticle were monitored and found that the reaction time was very less and after 24 h. From FT-IR, SEM and TEM analyses it was that the formed NP's were very smooth and spherical in shape with an average diameter of 35 nm. XRD results were comparable with previous literature, and confirm that the prepared Ag nanoparticle has FCC lattice arrangement and the sharpening of peaks confirm that the sample is of nano-regime.

INTRODUCTION

The field of nanotechnology is one of the most active research areas in modern materials science. Nanoparticles exhibit new or improved properties based on specific characteristics such as size, distribution and morphology. New applications of nanoparticles and nanomaterials are increasing rapidly. Fungi have been widely used for the biosynthesis of nanoparticles and the mechanistic aspects governing the nanoparticle formation have also

V. Rajendran, P. Prabu and K.E. Geckeler (eds.)
Nano Biomaterials, pp. 227–234 (2012).

been documented for a few of them. In addition to monodispersity, nanoparticles with well defined dimensions can be obtained using fungi. Compared to bacteria, fungi could be used as a source for the production of large amount of nanoparticles. This is due to the fact that fungi secrete more amounts of proteins which directly translate to higher productivity of nanoparticle formation.

Yeast, belonging to the class ascomycetes of fungi has shown to have good potential for the synthesis of nanoparticles. Gold nanoparticles have been synthesized intracellularly using the fungi *V.luteoalbum*. Mycelia free spent medium of the fungus, *Cladosporium cladosporioides* was used to synthesize silver nanoparticles by extracellular matrices. It was hypothesized that proteins, polysaccharides and organic acids released by the fungus were able to differentiate different crystal shapes and were able to direct their growth into extended spherical crystals. *Fusarium oxysporum* has been reported to synthesize silver nanoparticles extracellularly. Instead of fungi culture, isolated proteins from them have also been used successfully in nanoparticles production. Nanocrystalline zirconia was produced at room temperature by cationic proteins while were similar to silicatein secreted by *F. oxysporum.* The use of specific enzymes secreted by fungi in the synthesis of nanoparticles appears promising. Understanding the nature of the biogenic nanoparticle would be important. In the present study silver nano particles were synthesized using *Aspergillus Niger* which was isolated from contaminated soil samples. The silver nanoparticles were characterized by UV-visible spectrophotometer, XRD, FT-IR, SEM and TEM analysis.

EXPERIMENTAL

Collection of Soil Samples

Soil samples for the isolation of elite microorganisms were collected from drenched areas of dye house sludge at CLRI, Chennai. Soil samples were collected in polythene bags and sterilized containers and stored at 4°C in refrigerator until further analysis.

Isolation of Fungi

In order to ascertain microbiological bio system for nano particle synthesis by the *fungal sp.* and the same was isolated from the contaminated soil. The microorganisms were isolated by employing standard serial dilution technique. The isolated microorganisms were further purified by subsequent sub culturing and maintained in the slant culture at 4^0C.

Identification and Taxonomical Characterization of Isolated Fungi

The isolated fungal culture was grown on the potato dextrose agar medium for 5–7 days. The taxonomic position of the fungi was determined based on the conidial morphology, size and shape and was isolated and identified as *Aspergillus niger.* Black spores were seen on plates which is filamentous in nature and cause black mole disease.

Preparation of Extra Cellular Matrix from Isolated Fungal Sp.

Fungi was grown in yeast malt broth containing dextrose, 10 g/L; peptic digest of animal tissue, 5 g/L; yeast extract, 3 g/L; and malt extract, 3 g/L. The final pH was adjusted to 6.2 ± 0.2. The flasks were incubated in the environmental shaker at 200 rpm at 37°C. After 5 days of incubation, the mycelium mat was separated by filtration and washed thrice with Milli-Q deionized water. The washed mycelium mat was challenged with 100 mL of required concentration of silver nitrate solution and incubated in shaker at 200 rpm in dark condition at 37 °C. Simultaneously, a positive control of incubating the fungus mycelium with deionized water and a negative control containing only silver nitrate solution were maintained under same conditions.

RESULTS AND DISCUSSIONS

In the present study, we have investigated the extracellular biosynthesis of silver nanoparticles using *Aspergillus niger* which is isolated from contaminated soil samples. The synthesis process was quite effective and silver nanoparticles were formed within 96 hrs of silver ion coming in contact with the cell filtrate (After 24 h, test sample was turned to dark red colour from pale yellow colour). Silver nano particles are well defined nano crystalline structure. Detailed experimental results were discussed in detail below.

Effect of $AgNO_3$ Concentration in Biosynthesis of Nanoparticle

Effect of initial concentration silver nitrate on particle formation carried out in detail. The Colour change was observed and the particles were formed after 24 hours. The absorbance values were measured using UV-vis spectroscopy. The particles were then centrifuged at 10000 rpm and washed with ethanol to remove any water molecules and then was subjected to thermal treatment using muffle furnace at a temperature of 500°C for 3 hours. Finally the pure form of silver coated particles was obtained. Figures 1 and 2 illustrate the color change after adding $AgNO_3$ after 24 h and particle formation in different concentrations respectively.

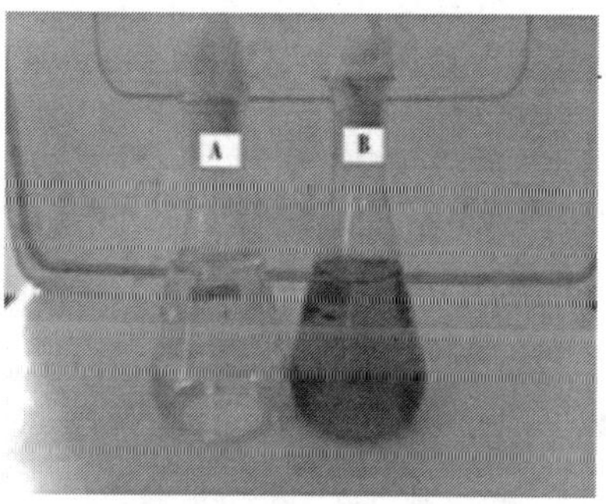

Fig. 1: A Digital Photograph Showing (A) Control (B) Change in Colour after Adding $AgNO_3$ after Reaction Time of 24 h

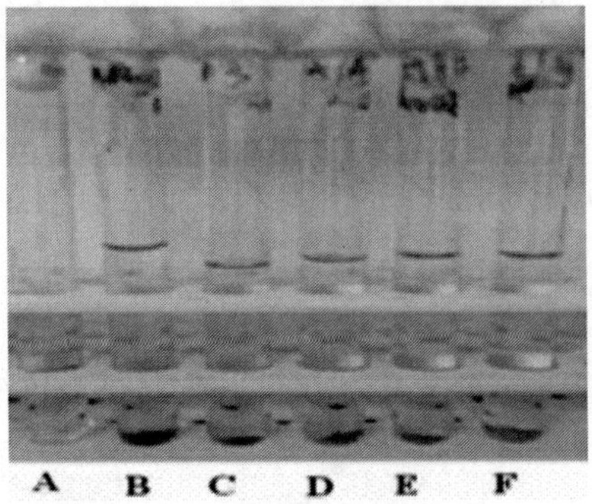

Fig. 2: Particle Formation in Different Concentrations

Characterization of Silver Nanoparticles by UV-Visible Spectroscopy

To observe the optical property of biosynthesized silver nanoparticles, samples were periodically analyzed for UV–visible spectroscopic studies at room temperature operated at a resolution of 1 nm between 200 and 800 nm ranges. It is well known that silver nanoparticles exhibit yellowish brown color in aqueous solution due to excitation of surface Plasmon vibrations in silver nanoparticles. The addition of fungal extracellular matrix to silver nitrate solution resulted in color change of the solution from transparent to dark yellow due to the production of silver nanoparticles. The surface plasmon resonance (SPR) of silver nanoparticles produced a peak centred near 413 nm. It was observed that absorbance peak was centred near 413 nm indicating reduction of $AgNO_3$ into silver nanoparticles. It was also observed that bio reduction of silver ions into nanoparticles started at the start of reaction and bio reduction was completed at almost 72 h indicating rapid biosynthesis of silver nanoparticles. Figure 3 shows the UV–Vis spectra recorded from the reaction medium after heating the solution at room temperature for 45 mins. Absorption spectra of silver nanoparticles formed in the reaction media has absorbance peak at 438 nm and broadening of peak indicated that the particles are poly dispersed and are to be favoured over the nucleation.

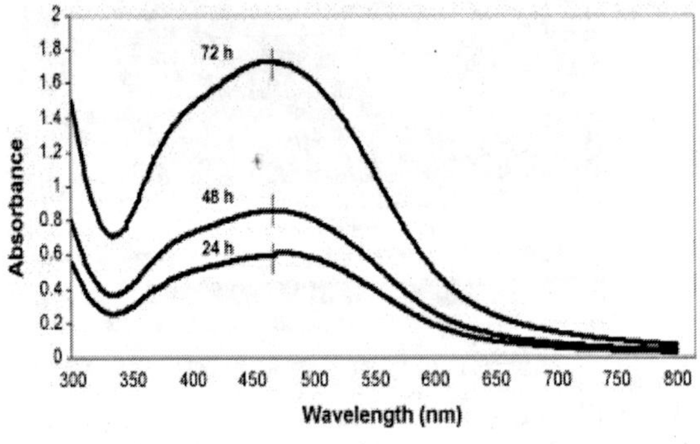

Fig. 3: UV-Vis Spectra of Aqueous Silver Nitrate with Asprgillus Niger at Different Time Intervals

Electrom Microscopy of the Prepared Silver Nanoparticles

The prepared silver nanoparticles were surface characterized by Scanning Electron Microscope (SEM) and Transmission Electron Microscope (TEM). The three-dimensional SEM images were shown in Figure 4 whereas the two-dimensional TEM image was presented in Figure 5. The Ag nano particles are spherical in shape with a smooth surface morphology. The diameter of the nano particles wass found to be 15–30 nm. TEM image also proved that the produced nano particles were more or less uniform in size and shape.

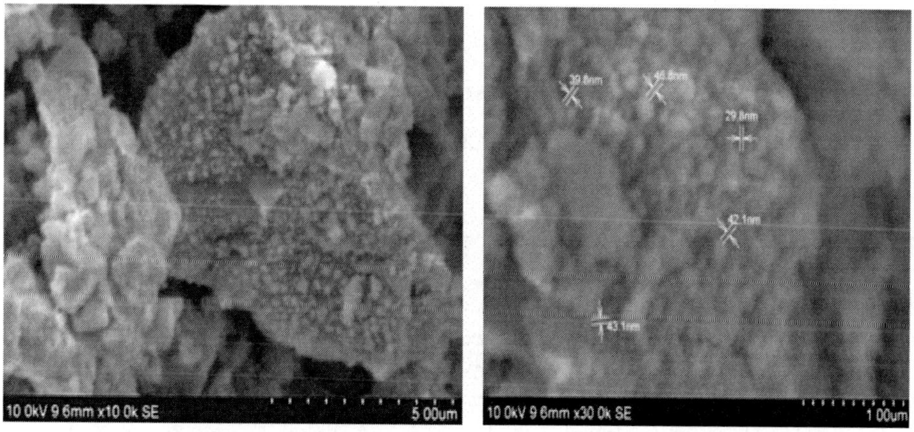

Fig. 4: SEM Images of Silver Nanoprarticles Prepared from Aspergillus Niger

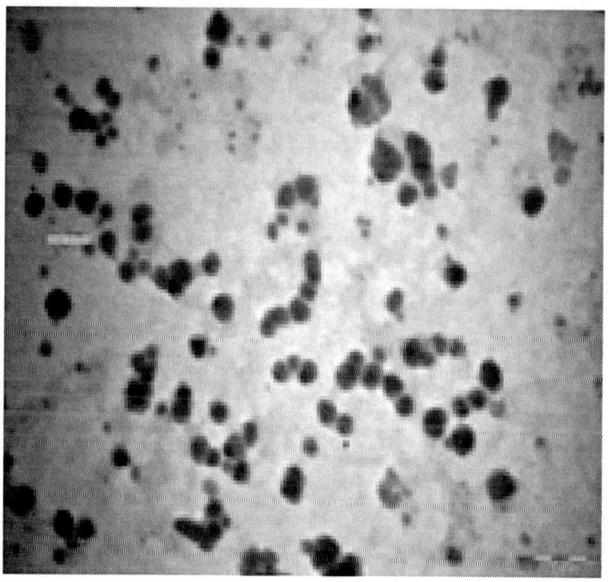

Fig. 5: TEM Images of Silver Nanoparticles Prepared from Aspergillus Niger

FTIR of the Prepared Silver Nanoparticles

FTIR analysis was used for the characterization of the the resulting nanoparticles. FTIR of silver nanoparticles prepared from *Aspergillus niger* was presented in Figure 6. Absorbance bands were observed at 1619, 1522, 1340, 1160 cm^{-1}. These absorbance bands are known to be associated with the stretching vibrations for –C–C– [(in-ring) aromatic], C–O–C (ethers) and C– O (–C–OH). The total disappearance of this band after the bio-reduction may be due to the fact that the polyols are mainly responsible for the reduction of Ag ions, whereby they themselves get oxidized to unsaturated carbonyl groups leading to a broad peak at 1660 cm^{-1} (for reduction of Ag).

XRD of the Prepared Silver Nanoparticles

The *fungi*-mediated synthesized Ag nanostructure was confirmed by the characteristic peaks observed in the XRD image which was shown in Figure 7. All diffraction peaks correspond to the characteristic face centered cubic (FCC) silver lines. These diffraction lines observed at 2θ angle 32.80, 38.20, 55.10° and 65.70 respectively, have been indexed as (111), (200), (220) and (311) respectively. XRD patterns were analyzed to determine peak intensity, position and width, full-width at half-maximum (FWHM) data was used with the Scherer formula explained in section materials and method. The typical XRD pattern revealed that the sample contains a mixed phase (cubic and hexagonal) structures of silver nanoparticles. The average estimated particle size of this sample was 35 nm derived from the FWHM of peak corresponding to 111 plane with cubic and hexagonal shape. From Figure 7, the particles were confirmed as elemental Ag (0) using XRD. A couple of Bragg's reflections are distinctly exhibited, which may be indexed on the basis of the face centered cubic structure of silver. It exhibits a sharp and intense peak at ~38° and ~32° corresponding to diffraction from the (111) and (101) planes of silver with FCC lattice (JCPDS no. 04–0783).

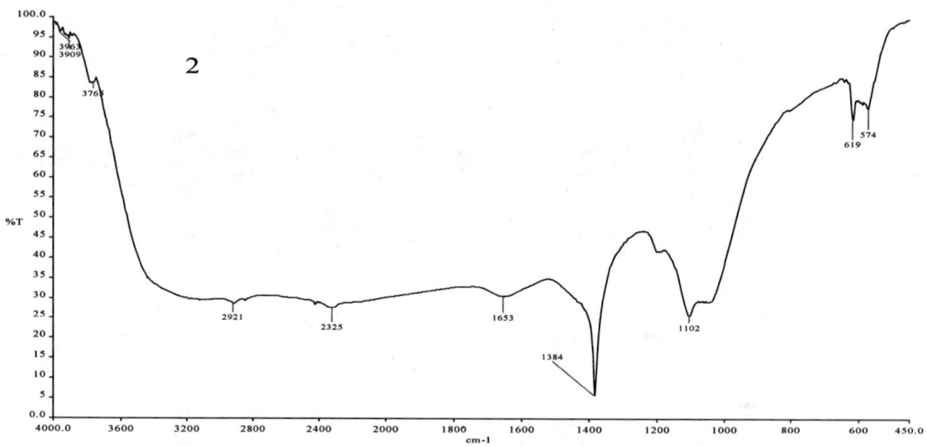

Fig. 6: FTIR of Silver Nanoparticles Prepared from Aspergillus Niger

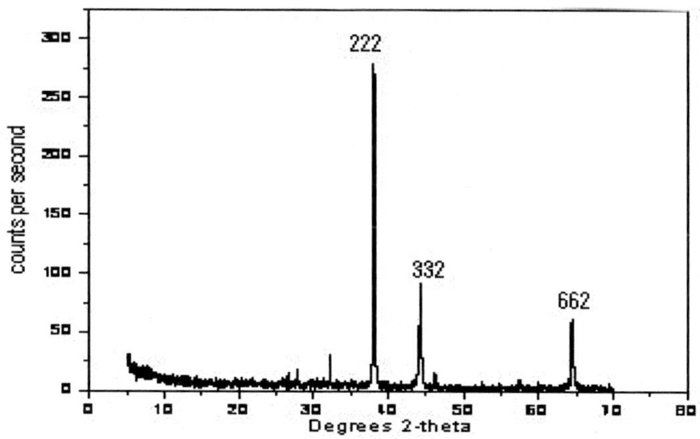

Fig. 7: XRD of Silver Nanoparticles Prepared from Aspergillus Niger

Mechanism of Formation of the Silver Nanoparticles

Since the bioreduction of silver ions and the formation of AgNPs are closely related with the biomolecular components of the extract, four main categories of the biomolecules (enzymes) before and after the reaction were quantitatively examined to explore the formation mechanism of the AgNPs. The contents of the reducing enzymes and extracellular components before the reaction were much higher than those of the plant based flavonoids and proteins. Therefore, the reducing enzymes and extracellular compounds were the two main components responsible for the bioreduction of the silver ions. It is worth noting that the difference in the contents of the reducing enzymes and extracellular components before and after the reaction were affected by initial $AgNO_3$ concentration.

CONCLUSIONS

The present investigation gives us a clear overview of synthesis and characterization of silver nanoparticles using *Aspergillus niger.* In brief the procedure explains all about sample collection from the contaminated soils and isolating the desired *fungal sp.* from the contaminated soil sample, subjecting to serial dilution, plating, streaking techniques etc which was followed by culturing, sub culturing. The pure culture was obtained from the sub cultured culture. The pure culture was inoculated in the fungal broth and left to see for its growth under the incubation of 30°C. Mycelial mat formation was observed after seven days and future it was subjected to 2 cycles of filtration with the help of Whattmann No.1 filter paper. The mat solution was set under ultrasonification to obtain pure form of extracellular matrix. To the matrix solution appropriate amount of substrate $AgNO_3$ is added and the particle formation was seen after 7 days. The or samples were centrifuged and set to sintering and given further characteristic analysis SEM, TEM, FTIR, UV-VIS SPECTROSCOPY and XRD which gave promising results of 30–40 nm sized silver nanoparticles.

REFERENCES

[1] Brady, M.J., Lisay, C.M., Yurkovetskiy, A.V. and Sawan, S.P., Persistent silver disinfectant for the environmental control of pathogenic bacteria, *Am. J. Infect. Cont.,* 2002, 31, 208–214.

[2] Nuorti, J., Butler, J., Crutcher, J., An outbreak of multidrag-resistant pneumonia and bacteremia among anvaccinated nursing home residents. *New Engl. J. Med.* 1998, 338, 1861–1868.

[3] Quake, S.R., Scherer, A., From Micro- to Nanofabrication with Soft Materials. *Science* 2000, 290, 1536–1540.

[4] Green, M. and Brien, P.O., Recent advances in the preparation of semiconductors as isolated nanometric particles: new routes to quantum dots. *Chem. Commun.,* 1999, 2235–2236.

[5] Joerger, R., Klaus, T. and Granquist, C.G., "Biologically Produced Silvercarbon Composite Materials for Optically Functional Thin-Film Coatings," *Advanced Materials,* Vol. 12, 2000, p. 407.

[6] Ankamwar, B., Damle, C., Ahmad, A. and Sastry, M., "Bio-synthesis of Gold and Silver Nanoparticles Using Emblica Officinalis Fruit Extract," *Journal of Nanoscience and Nanotechnology*, Vol. 5, 2005, p. 1665.

PREPARATION AND ANTIMICROBIAL OBSERVATIONS OF ZINC DOPED NANOHYDROXYAPATITE

C. DEEPA, A. NISHARA BEGUM[1] AND S. ARAVINDAN[2]

Department of Physics, Vellalar College for Women (Autonomous), Erode
[1]Department of Physics, Chikkaiah Naicker College, Erode
[2]Department of Physics, Government Arts College (Autonomous), Salem
E-mail: anishara@yahoo.com

ABSTRACT

Hydroxyapatite [$Ca_{10}(PO_4)_6(OH)_2$, HAp] is the major constituents of bones and hard tissues in mammals. In the present decade, HAp nanoparticles are used as ideal tool for the transformation of human hard tissues due to their high biocompatibility and easy biodegradability. Nowadays, metals are widely used in orthopedics to increase the bioactivity of hydroxyapatite. Among the metals, zinc is an essential trace element present in human bones and teeth. It plays important roles in increasing osteoblasts adhesion and alkaline phosphate activity of bone cells. In the present study, different amount of pure and ZnO (0%–43%) doped nano Hydroxyapatite powders were synthesized by sol-gel method. The properties of pure and Zn doped nHAp powders were characterized using X-ray diffraction (XRD), Scanning electron microscopy (SEM), and Energy dispersive X-ray analysis (EDAX). The results of X-ray diffraction studies reveal the progressive increase in the average crystallite size with increasing ZnO concentration found to be 49–100 nm. The In vitro Antimicrobial activity of the synthesized pure and Zn doped nHAp powders were investigated against gram-negative bacterial strains using disc diffusion method. The antimicrobial activity of pure and doped nHAp samples were observed irrespective of the content of ZnO.

INTRODUCTION

In the recent years, the development of high quality hydroxyapatite [$(Ca)_{10}(PO_4)_6(OH)_2$, HAp] with various additives was one of the primary aims in bone tissue engineering mainly

V. Rajendran, P. Prabu and K.E. Geckeler (eds.)
Nano Biomaterials, pp. 235–242 (2012).

to obtain suitable materials for artificial bone substitution. Biological apatite is not pure HAp, but a non-stoichiometric HAp with a variety of trace ions such as Ca^{2+}, Mg^{2+}, Sr^{2+}, Ba^{2+}, Pb^{2+}, Zn^{2+}, Cu^{2+}, Na^{+}, K^{+}, Fe^{3+}, etc. also can be prepared over the atomic ratio (trace ion/P) of 1.5–1.67.[1] The biological performance of HAp is found to be improved due to the incorporation of such a trace metal ions.[2] Among the metal ions, zinc is the one abundantly present as trace element in bone minerals,[3] in addition, zinc promotes the bone density and prevent the bone loss.[4]

Recently, the nano-sized particles gained importance in biomedical field since the size reduced and surface activated particles able to exhibit entirely new behaviour during dissolutions and precipitations. Earlier reports [5] reveal the successful incorporation of metal ions with HAp, but they failed to demonstrate antimicrobial effects. The aim of this research work is to synthesize zinc oxide doped nano-HAp powders and to examine their crystallinity, *in vitro* and antimicrobial activity against the content of zinc oxide.

MATERIALS AND METHODS

The pure nano-HAp powders were prepared employing sol-gel method with 0.55 M of calcium hydroxide $Ca(OH)_2$ (95% SIGMA ALDRICH) and 0.33M of di ammonium hydrogen phosphate $(NH_4)_2.HPO_4$ (99% SIGMA ALDRICH) respectively, used as sources for calcium and phosphorous. Ca $(OH)_2$ and $(NH_4)_2HPO_4$ were dissolved in 500ml of de-ionized water separately for obtaining stoichiometric mole ratio equal to 1.67. The pH of each solution was maintained at ~11 by the addition of ammonium hydroxide solution NH_4OH (99%, SIGMA ALDRICH) [6]. The gelatinous white precipitate was produced by the drop wise addition of $(NH_4)_2HPO_4$ solution to $Ca(OH)_2$ solution stirred vigorously at 333K for an hour. Then, the precipitate was aged for 24 h at room temperature followed by washing three times with de-ionized water and dried in hot air oven at 423 K for 10h. Then, the dried powder was milled using mortar and pestle and finally calcined in alumina crucible using muffle furnace at 673K for 5h.

The nano-HAp incorporated with varied content of zinc oxide was synthesized using the mixed aqueous solutions of prepared pure nano-HAp powder and zinc oxide. The dopants were used in the amount of 0, 10, 25, 50 and 75wt. % (here after named as (a) Z_0H, (b) $Z_{10}H$, (c) $Z_{25}H$, (d) $Z_{50}H$ and (e)$Z_{75}H$). The mixed solution was stirred for an hour. The obtained suspension was filtered and dried in hot air oven at 373K for 5h. The resultant pure and ZnO doped nano-HAp powders were characterized by XRD and SEM-EDAX.

X-ray Diffraction (XRD)

X-ray diffraction (XRD) studies were used to observe the structural properties of prepared nano-HAp powder. The analysis was carried out using PAN analytical X'PERT-PRO diffractometer with high intensity of CuKα (wavelength $\lambda = 1.54060$ Å) radiation. The diffraction spectra were recorded in the angle 2θ ranges from 10° to 90°. The average crystallite size of pure and ZnO doped HAp nano powders was calculated using Debye-Scherrer's formula,[7]

$$D = 0.9\,\lambda/\beta\,\cos\theta \qquad \ldots (1)$$

where D is the crystallite size in nanometer, λ is the wavelength of CuKα radiation (1.5406 Å), β full width of the peak at half of the maximum (FWHM) and θ the diffraction angle of the corresponding reflection. The unit cell volume (V) and lattice parameters a and c of pure and ZnO incorporated HAp nano powders are calculated using the equations (2 and 3).

$$V = 2.589\, a^2 c \quad \text{... (2)}$$

$$1/d^2_{hkl} = 4/3\, [h^2+hk+k^2]/a^2 + l^2/c^2 \quad \text{... (3)}$$

where d is the crystallite size, h, k and l are miller indices of the plane, a and c are lattice constants.

Morphology and Elemental Analysis (SEM/EDAX)

The Scanning Electron Microscopy (SEM) was performed using a JEOL JSM 6390, SEI model to determine microstructure and particle size on the surface of nano-HAp powders. The elemental analysis on the surface of pure and doped nano-HAp powders was performed by Energy dispersive X-ray analysis.

Antimicrobial Activity

The antibacterial activity of the prepared nano-HAp samples was tested against the bacterial strain *Klebsiella pneumonia* (ATCC 13883) by the modified disc-diffusion method.[8] The bacterial species were cultured for 24h at 310K and thereafter, the growth inhibition zones around the disc were measured.

RESULTS AND DISCUSSION

XRD Characterization

Figure 1 shows the XRD patterns of the pure and ZnO incorporated HAp samples. The observed diffraction peaks of all prepared samples ensured the emergence of calcium rich phosphate apatite phases by matching well with the peaks in the standard JCPDS file (09–0432) data. Intensity of the diffraction peak of plane (211) of HAp is found increases with the ZnO content. It is mainly because of the preferential orientation of ZnO in the apatite crystal. The measured crystallite size, lattice parameter values and unit cell volumes are shown in Table 1. It confirms the emergence of apatite mineral in nano meter scale. The measured average crystallite size and unit cell volume of the HAp increases with increasing ZnO content. Particularly, the cell parameter 'a' gradually increases with ZnO while another parameter 'c' fluctuates with the incorporation of ZnO above 25 wt%. According to Fuzeng Ren *et al.*,[7] the lattice parameters 'a' and 'c' of crystalline HAp increases with the incorporation of ZnO content. In addition to that, the substitution of H_2O lattice for OH sites in ZnO added HAp results in the increase of lattice parameter value.[9]

Table 1: Structural Parameters of the Zinc Free and Added HAp Crystalline Samples

Sample Code	*Average Crystallite Size "D" (nm)*	*Cell Parameters*		*Unit cell Volume V* $(\AA)^3$
		a $(\AA)$	*c* $(\AA)$	
Z_0H	49.789	9.4226	6.833	525.377
$Z_{10}H$	77.4167	9.4225	6.8666	527.964
$Z_{25}H$	82.2215	9.4395	6.8536	528.852
$Z_{50}H$	86.5311	9.4561	6.8692	531.931
$Z_{75}H$	99.7987	9.4901	6.864	535.352

SEM-EDAX Analysis

The morphology and elemental analysis of pure and ZnO added nano-HAp sample were examined employing scanning electron microscopy (SEM) and energy dispersive x-ray (EDAX) analysis. The SEM micrographs ensure the emergence of rod like crystalline species of ZnO added samples. The micrograph confirms the obtained particle size ranges from 49–67 nm with the incorporation of ZnO.[10] The result of elemental analysis (EDAX) of pure and ZnO added nano HAp confirms the existence of Ca, P, O and Zn elements. The observed elemental concentration confirm the Ca-P rich apatite crystalline phase. As expected, the elemental concentration of Zn is found increases while increasing the ZnO content in HAp. The resultant Ca/P ratio of pure HAp is observed as 1.45 while in ZnO incorporated samples it increases till the addition of 25 wt% and decreases with further additions of ZnO upto 75 wt%.

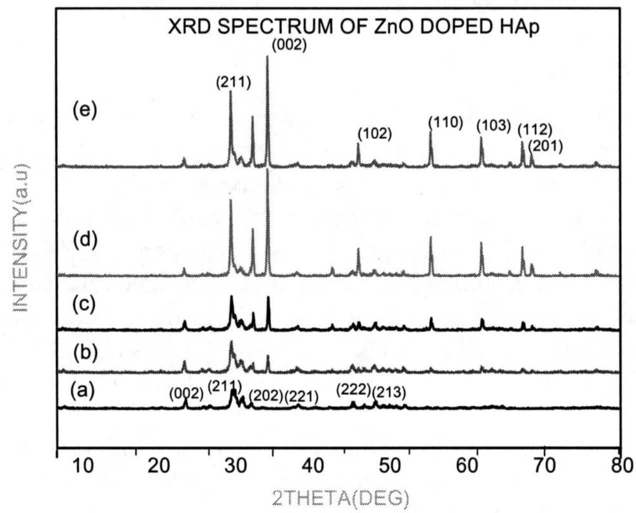

Fig. 1: XRD Pattern Of pure and Zn Added HAp Samples ($a-Z_0H$, $b-Z_{10}H$, $c-Z_{25}H$, $d-Z_{50}H$, $e-Z_{75}H$)

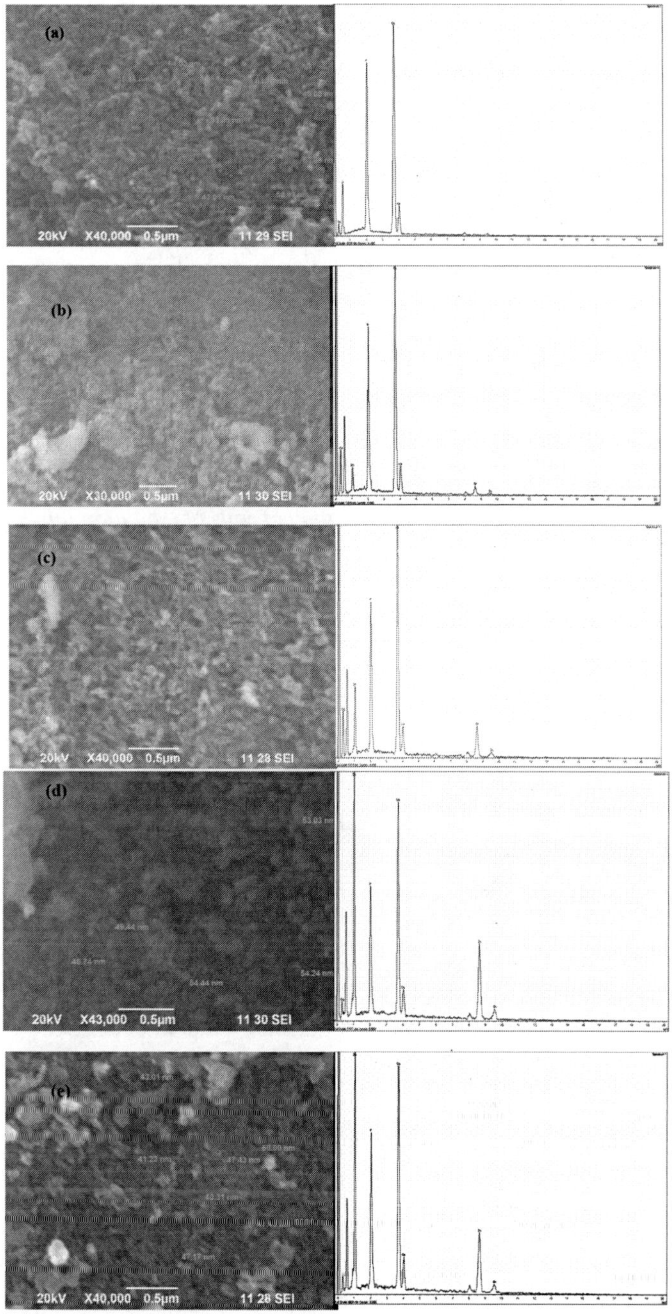

Fig. 2: SEM and EDAX Images of Zinc Free and Added HAp Samples ($a-Z_0H$, b- $Z_{10}H$, c- $Z_{25}H$, d- $Z_{50}H$, e- $Z_{75}H$)

Antimicrobial Activity

In vitro antibacterial activity of the synthesized pure and Zn incorporated nanoHAp samples have been examined against Gram-negative *Klebsiella Pneumoniae* bacteria. Irrespective of the ZnO content, all the prepared HAp samples exhibited the antimicrobial activity. The resultant data are shown in Table 2. It reveals the influence of ZnO in decreasing the antimicrobial activity of the HAp as shown in Figure 3. At the concentration of 200 μg/ml, the sample Z_0H recorded the highest level of antimicrobial activity against *Klepsilla pneumonia* by exhibiting highest inhibition zone diameter and the zone diameter decreases with increase in ZnO, as shown in Figure 4. It is obvious that the antibacterial property of apatite samples is dependent on the molecular structure of compound and also depends on the types of bacterial strains and solvent used.[11] The incorporation of ZnO leads to modify the structure of the nano HAp and thereby exhibit newer properties with zinc oxide incorporations.

Table 2: Antibacterial Activity of pure and ZnO Added Nano HAp

Organism	*Sample Name (representation of zone of inhibition (diameter in mm))*				
Klebsiella pneumonia	Z_0H	$Z_{10}H$	$Z_{25}H$	$Z_{50}H$	$Z_{75}H$
	200 μg/ml	200 μg/ml	200 μg/ml	200 μg/ml	200 μg/ml
	15	14	12	12	11

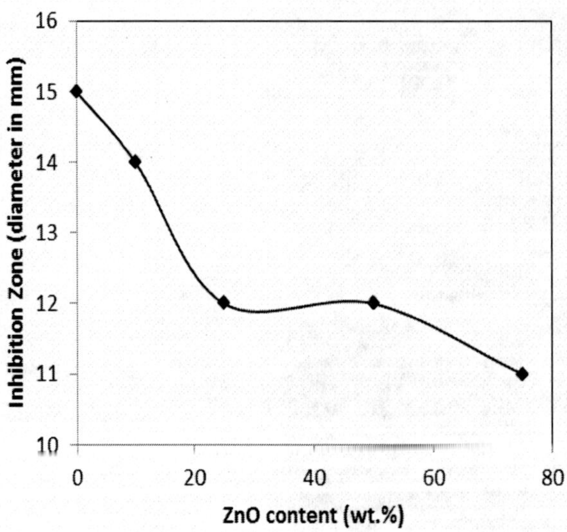

Fig. 3: Antimicrobial Activity of HAp against ZnO Additions

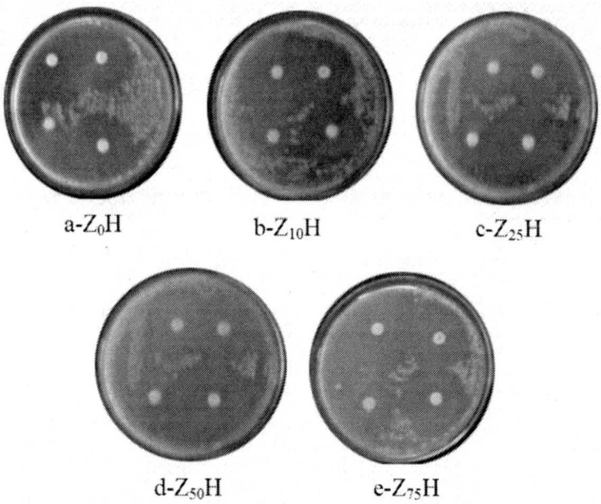

Fig. 4: Antibacterial Activity of Pure and ZnO-HAp against *Klepsilla Pneumonia*

CONCLUSIONS

A pure and ZnO incorporated nano HAp samples were prepared employing sol-gel method. The obtained XRD results confirm the crystallite size of the samples in nanoscale. The influence of ZnO is used to increase the average crystallite size and unit cell volume of the Ca-P apatite crystal. The ZnO incorporations modify the morphology of HAp in rod like structure. Elemental compositions confirm the emergence of calcium rich apatite crystalline phases. The in vitro antimicrobial activity of HAp reveals the role of zinc oxide in decreasing its antimicrobial activity.

REFERENCES

[1] Guo, D.G., Wang, A.H., Han, Y. and Xu, K.W., *Acta Biomaterialia*, 5 (2009), 3512–3523.

[2] Gemma Mestres, Clemence Le Van, Maria-Pau Ginebra, *Acta Biomaterialia*, 8 (2012), 1169–1179.

[3] Ito, A., Otsuka, M., Kawamura, H., Ikeuchi, M., Ohgushi, H. and Sogo, Y. *et al.*, *Curr Appl phys*, 5 (2005) 402–6.

[4] Legeros, R.Z., *US Patent,* 7, (2008), 419, 680, September 2.

[5] Kim, T.N., Feng, Q.L., Kim, J.O., Wu, J., Wang, H., Chen, G.C. and Cui, F.Z., *J. Mater. Sci. Mater. Med.*, 9 (1998), 129–134.

[6] Anee, T.K., Ashok, M., Palinichamy, M. and Kalkura, S. Narayana, *Materials Chemistry and Physics*, 80 (2003) 725–730.

[7] Ren, Fuzeng, Xin, Renlong, Ge, Xiang and Leng, Yang, *Acta Biomaterialia*, 5 (2009) 3141–3149.

[8] Baur, A.W., Kirby, W.M.M., Scherris, J.C. and Turck, M.,. *Amer. J. Cl. Pathol* (1966) 45: 493–496.

[9] Miyaji, F., Kono, Y. and Suyama, Y., *Mater. Res. Bull*, 40 (2005), 209–220.

[10] Maria Helena Santos, Luis Guilherme Dias Heneine, *Material Science and Engineering C*, 28 (2008), 563–571.

[11] Tripathi, Anjali, Saravanan, Sekaran and Pattnaik, Soumitri, *International Journal of Biological Macromolecules*, 50 (2012) 294–299.

INSITU PREPARATION AND CHARACTERIZATION OF HYDROXYAPATITE/PEG NANOCOMPOSITE AND ITS ELECTRO CHEMICAL STUDIES

C.P. Dhanalakshmi, R. Suresh, K. Giribabu, L. Vijayalakshmi and V. Narayanan

Department of Inorganic Chemistry, University of Madras, Guindy Maraimalai Campus, Chennai
E-mail: vnnara@yahoo.co.in

ABSTRACT

HAp/PEG nanocomposites of varying composition for biomaterial applications have been synthesized. The HAp/PEG nanocomposite materials were characterized by XRD, FTIR, and HRTEM. Hydroxyapatite (HAp) nano rod embedded composite was prepared using poly (ethylene glycol) (PEG) as a matrix with different weight percentages (wt%). The results indicated that the size and crystallinity of HAp nano particles decreases with increase in PEG concentration in the composite. HRTEM confirms the presence of HAp nano rod crystals in PEG matrix. Cyclic voltammograms (CVs) were recorded to evaluate the electrocatalytic behaviors of the modified electrode towards the oxidation of 4-NP in the potential range of 0 to +1.0 V. A glassy carbon electrode was modified with hydroxyapatite nanopowder (HAp-PEG-NP) and characterized in terms of electrochemical oxidation of 4-nitrophenol (4-NP) via cyclic voltammetry. The catalytic reaction facilitates electron transfer between 4-NP and the modified electrode, as a result the electrochemical oxidation of 4-NP becomes easier. The reason for this is that the HAp-PEG 40 can act as a promoter to increase the rate of electron transfer, lower the overpotential of 4-NP at the bare electrode, and the anodic peak shifts less positive potential. Thus, it is clear that HAp-PEG 40 modified GCE can be successfully used for the determination of organic pollutant.

INTRODUCTION

A critical challenge in the design of these hybrid inorganic-organic systems is a control of the mixing between the two dissimilar phases. A composite material is a macroscopic combination of

V. Rajendran, P. Prabu and K.E. Geckeler (eds.)
Nano Biomaterials, pp. 243–250 (2012).

two or more distinct materials, having a recognizable interface between them. Composites are used not only for their structural properties, but also for electrical, thermal, tribological, and environmental applications. Modern composite materials are usually optimized to achieve a particular balance of properties for a given range of applications.

4-NP is widely used as intermediates in the production of pharmaceuticals, pesticides and dyestuffs, such as parathion insecticide which can also reversely hydrolyze to form 4-NP. In addition, 4-NP can also be used as leather fungicide and acid-base indicator. Therefore, 4-NP will be inevitably released into environment to cause pollution in the process of fabrication and application. Based on the above description, it is important to develop simple and reliable method for determination of trace amounts of 4-NP in environments.[1,2] Up to now, various techniques have been investigated to determine 4-NP, such as spectrophotometry, fluorescence, high-performance liquid chromatography,[3] liquid chromatography with electrochemical detection, capillary zone electrophoresis,[4] fiber optode[5]and electrochemical method.[6]Among them, electrochemical methods have received considerable attention because of the advantage of fast response, cheap instrument, low cost, simple operation, timesaving, high sensitivity and selectivity, real-time detection in in-situ condition. However, until now, many electrochemical determination techniques mainly focus on utilizing the reduction of nitryl in 4-NP, not the oxidation of hydroxide radical. Thus, the determination would be interfered by the oxygen molecule dissolved in solvent. In order to eliminate it, purging nitrogen for a certain time was employed. But this will make the determination more complicated and time-consuming. Moreover, the reduction of nitryl is more complicated than the oxidation of hydroxide radical. Therefore, electrochemical determination of 4-NP using the oxidation signal should be an appropriate alternative. Nevertheless, the use of bare electrode has revealed the drawback of weak electrochemical response and low sensitivity. Consequently, chemically modified electrodes (CMEs) have been widely investigated with various modification materials, especially nanophase materials, such as MWCNT,[7] SWCNT,[8] gold nanoparticles,[9] TiO_2,[10] zeolite,[11] sodium montmorillonite [12] and HAp.[13,14]Among them, nano HAp has advantages of abundant in nature, cheap, readily available, and stable in water and nontoxicity. Due to the high surface area and strong adsorption ability, a film of silica/HRP-HAp has been deposited on GCE surface to investigate the direct electron transfer of immobilized HRP and H_2O_2 determination.[15] El Mhammedi *et al.* have evaluated the analytical performance of nano HAp-CPE for the detection of trace lead (II), paraquat and 4-NP. They found that HAp can effectively improve the electrochemical response due to its excellent adsorption property.[16,17] So, nano HAp should be a suitable electrode modified material for pollution determination.

The aim of the present work is to develop a simple, reliable and sensitive electrochemical method for the determination of 4-NP(4-nitrophenol) and 2-NP(2-nitrophenol) based on the unusual properties of polymer composites of nano HAp-PEG -NP modified electrode. The modified electrode showed good electrochemical oxidation to 4-NP with an increase of the oxidation peak current. The electrocatalytic determination of nitrophenols were carried out using the polymer nanocomposite modified electrodes and found to be good sensors for the detection of nitro organic compounds.

EXPERIMENTAL

Materials

The raw materials required to start the processing of the composite were: analytical grade calcium hydroxide ($Ca(OH)_2$) and ammonium dihydrogen phosphate ($(NH_4)H_2PO_4$) obtained from Merck (India) and PEG (MW 20000) purchased from Loba (India). Doubly distilled water was used as the solvent.

Methods

Synthesis of Nano HAp

The nano HAp was synthesized by following a modified wet chemical method. At 25°C, 7.48 g of $Ca(OH)_2$ was first dissolved in a 100 ml volume of an ethanol–water mixture (50:50%, v/v) and stirred for 3 h. A solution of 6.7 g $(NH_4)H_2PO_4$ was dissolved in 100 ml volume of water and then added to the $Ca(OH)_2$ solution over a period of 24 h. The amount of reagents in the solution was calculated to obtain a Ca/P molar ratio equals to a 1.67 value, corresponding to a stoichiometric HAp. The pH of the slurry was measured digitally during the precipitation reaction, reaching a final value of pH 11.

Synthesis of HAp-HAp Nanocomposites

The HAp-PEG nanocomposites were coded as nano HAp-PEG 20, nano HAp-PEG 40, nano HAp-PEG 60 and nano HAp-PEG 80 where number denotes the wt.% of PEG matrix (20,000) used in the processing. Water was used as the solvent to prepare the polymer solution. PEG was dissolved by using magnetic stirrer for 3h and the polymer solution was left overnight in room temperature to remove the air bubbles trapped in the viscous solution. Then suitable amount of HAp was dispersed in deionised water by 30 min ultrasonication (Metal Power Analytical (I) Pvt. Ltd. Maharashtra, India). Ultrasonication was necessary to avoid agglomeration of ceramic powder and to achieve proper dispersion. HAp in water was mixed with polymer solution under agitation. The homogeneously mixed solution is immediately taken to deep freeze (AC MAS Technology (p) Ltd, India) at –18°C. After 48 h freezing the samples were freeze dried.

Physical-Chemical Characterization

The prepared samples were studied by Fourier Transform Infrared Spectroscopy (FTIR) using a Schimadzu FT-IR 300 series instrument (Shimadzu Scientific Instruments, USA). The FTIR spectra were obtained over the region 450–4000 cm^{-1} in pellet form for 1 mg powder samples mixed with 200 mg spectroscopic grade KBr (). Spectra were recorded at 4 cm^{-1} resolution averaging 80 scans. The structure of the samples were analyzed by X-Ray Diffraction (XRD) using a Rich Siefert 3000 diffractometer (Seifert, Germany) with Cu-Kα_1 radiation (λ – 1.5418 Å). In this work a JEOL FX 2000 Transmission Electron Microscope and FEI Teccani

F30 High Resolution Transmission Electron Microscope were used to study the morphology of the nanoparticles with a resolution of 5 Å–10 Å. The powder samples were dispersed in Gatan G1 epoxy resin and dried at 353 K to get a hard and thick disc. Then thick discs were sliced to small pieces and mechanically polished to ~ 0.15 μm.

RESULTS AND DISCUSSION

Characterization of HAp-PEG Nanocomposites

FT-IR Analysis

The FTIR spectra of pure nano HAp and HAp-PEG nanocomposites are shown in Figure 1. The ν_2 phosphate stretching mode appearing at 471–472 cm^{-1} corresponds to PO_4^{3-} group in nano HAp . The bands located at 1032–1048 and 562–570 cm^{-1} are attributed respectively to the ν_3 and ν_4 P-O vibration modes of regular tetrahedral PO_4^{3-} groups.[18] The band at 602 cm^{-1} corresponds to O-P-O bending and ν_1 symmetric P-O stretching modes . The ν_1 symmetric stretching mode of phosphate group is observed at 962 cm^{-1}.[19] The observed bands at 1384 cm^{-1} is due to the stretching mode of carbonate, which may be due to the acquisition of air during mineral precipitation. Similarly, the observed bands at 1416 and 857–874 cm^{-1} are assigned to carbonate ions. The bands observed in the region between 2067 and 2069 cm^{-1} are related to their harmonic overtones and/or combination bands. The peak due to lattice H_2O exists in the range of 1603–1608 cm^{-1}, while the bands observed at 3400–3569 cm^{-1} overlap with the –OH group.[20] The band observed between 2942– 2944 cm^{-1} corresponds to C-H stretching band of PEG. A new peak of stretching band is observed at 2944 cm^{-1}, when PEG is added. This indicates the chemical interactions between nano HAp and PEG.

X-ray Diffraction Analysis

The XRD patterns of nano HAp and HAp-PEG nanocomposites were taken. The patterns indicate the presence of nano HAp. The broad peaks reveal that the particle sizes are very small in the range of 35 to 50 nm. The reflection planes corresponding to the characteristic XRD spectral peaks of pure nano HAp and HAp-PEG nanocomposites are as shown in Figure 2 The observed diffraction peaks were identified by standard Joint Committee on Powder Diffraction Standards (JCPDS) file (File no. 72–1243) and are assigned as crystalline nano HAp.[21,22] The XRD patterns show diffraction peaks with line broadening and high intensities, which confirms the nanosize with crystalline nature. Pure PEG has an obvious diffraction peak at 19° and 23°. After forming the composite, the crystallinity of the PEG phase decreased. The hydrogen bondings in PEG contribute to its crystallinity. During the formation of composite the interface bonding formed between nano HAp and PEG may result in a decrease in the number of hydrogen bonds, thus lessened PEG crystallinity. The crystallite size of the pure nano HAp using Scherrer's formula and found to be in range 30–50 nm with increase in PEG composition. Figure 2 reveals that the crystallite size decreases with increase in the composition of PEG.

The decrease in intensity of the nano HAp peaks are due to the presence of layers of PEG over HAp nanocomposites on in the use of HAp-PEG nanocomposites.

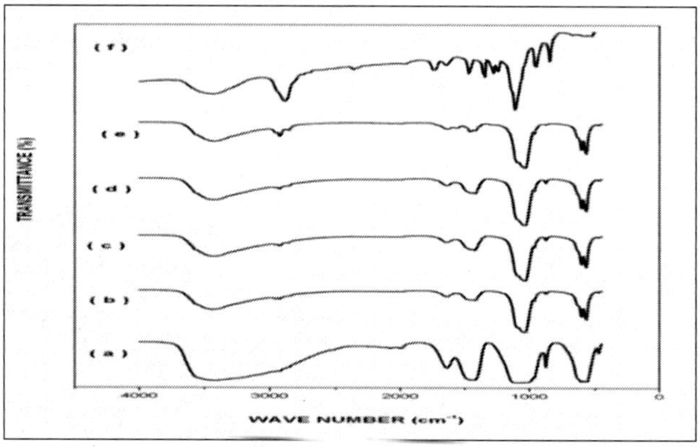

Fig. 1: FT-IR Spectra of (a) Nano HAp, (b) HAp-PEG 20, (c) HAp-PEG 40, (d) HAp-PEG 60, (e) HAp-PEG 80, and (f) PEG

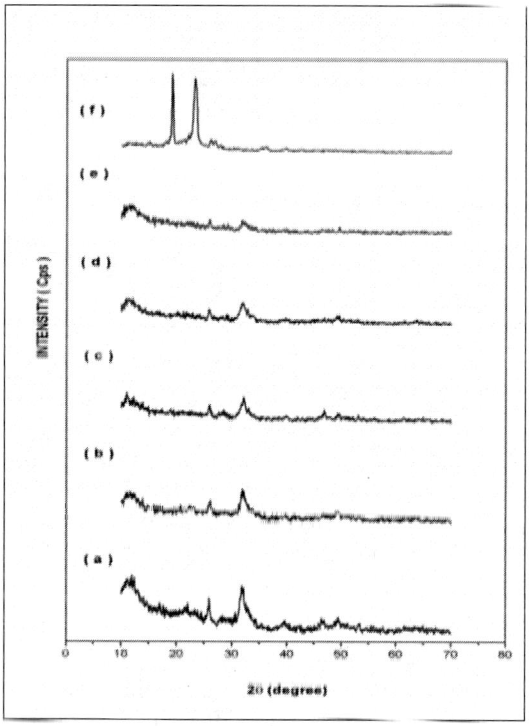

Fig. 2: XRD Pattern of (a) Nano HAp, (b) HAp-PEG 20, (c) HAp-PEG 40, (d) HAp-PEG 60, (e) HAp-PEG 80, and (f) PEG

HR TEM Analysis

Figure 3 shows the HRTEM photograph of HAp-PEG 20 nanocomposite powder. The CHAp-PEG 20 nanocomposite particles shown in Figure 3 are in nanometer scale. HAp-PEG 20 nanocomposite particles were a bit thicker (10–40 nm) and longer with clear contour. In addition the particles showed less agglomeration. In addition the selected area electron diffraction (SAED), in Figure 3 of the precipitates shows with diffraction ring of patterns, which implies that the precipitates are crystalline in nature, which is proved by XRD results.

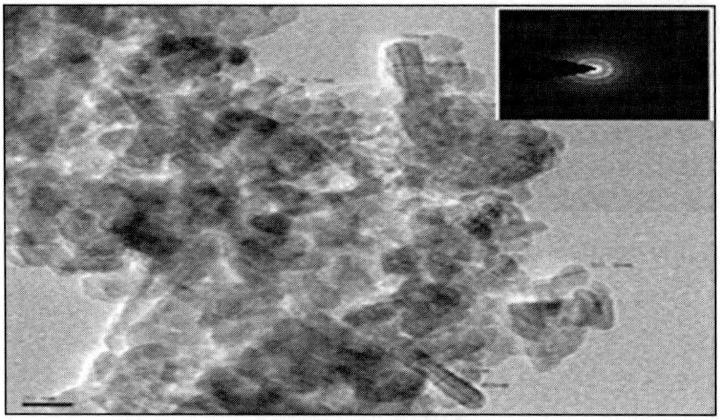

Fig. 3: HRTEM Image of CHAp-PEG20 Nanocomposite

Electrochemical Studies

Detection of 4-Nitrophenol

Cyclic voltammograms (CVs) were recorded to evaluate the electrocatalytic behaviors of the modified electrode towards the oxidation of 4-NP in the potential range of 0 to +1.0 V.[23] Figure 4 depicts the CV's of bare, nano HAp, nanocomposites of HAp-PEG 20, HAp-PEG 40, HAp-PEG 60 and HAp-PEG 80 modified GCE respectively in the presence of 0.5 mM 4-NP in 0.1 M PBS. The CV of 4-NP (Figure 4) at bare GCE shows a very broad peak at about +0.92 V. The 4-NP voltammogram obtained at nano HAp modified GCE showed an oxidation potential at +0.86 V (Figure 4). However, the 4-NP voltammogram obtained for nanocomposites of HAp-PEG 20, HAp-PEG 40, HAp-PEG 60 and HAp-PEG 80 modified GCE showed well defined oxidation wave of 4-NP with cathodic shift in potential and variation of the oxidation peak current compare to the bare and nano HAp modified GCE, indicating the electrocatalytic ability of the modified GCE.[24,25] From the Figure 4, it can be seen that the oxidation potential and current response of 4-NP varied with the amount of the added PEG. Especially HAp-PEG 40 nanocomposite modified GCE showed higher oxidation peak current with lower oxidation potential at +0.79 V. The overpotential had thus decreased by 0.13 V indicating the strong electrocatalytic effect of HAp-PEG 40 nanocomposite modified layer.

The above results indicate that catalytic reaction occurred between the HAp-PEG 40 nanocomposite with 4-NP. The catalytic reaction facilitates electron transfer between NPs and the modified electrode, as a result the electrochemical oxidation of NPs become easier. The reason for this is that the HAp-PEG 40 nanocomposite can act as a promoter to increase the rate of electron transfer, lower the overpotential of NPs at the bare electrode, and the anodic peak shifts less positive potential. Thus, it is clear that HAp-PEG 40 nanocomposite modified GCE can be successfully used for the determination of organic pollutant.

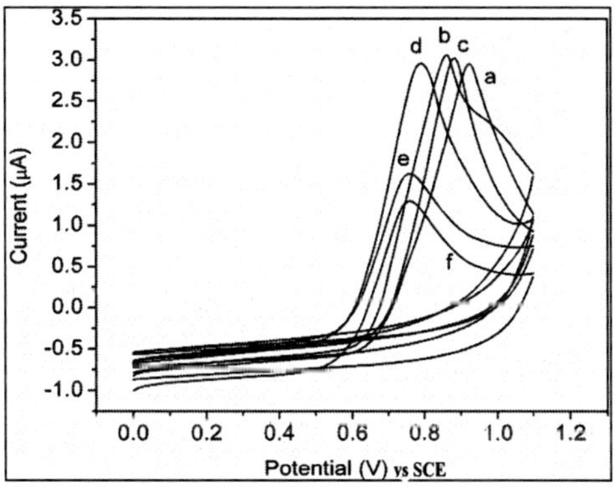

Fig. 4: Cyclic Voltammograms of 0.5 mM 4-NP at (a) bare, (b) nano HAp, (c) HAp-PEG 20, (d) HAp-PEG 40, (e) HAp-PEG 60, and (f) HAp-PEG 80 Modified GCE in the Scan Rate of 10 mV s^{-1}

CONCLUSIONS

The conclusions drawn from this work are the FTIR and XRD analysis of the HAp-PEG nano composites showed that both the components exist in the composite. The XRD pattern shows diffraction peaks with line broadening and high intensities, which confirms the nanosize with crystalline nature. HRTEM photograph of CHAp-PEG nanocomposite showed that the particles are in nanometer scale (10–40 nm). In addition, the particles showed less agglomeration. The selected area electron diffraction (SAED) studies showed that diffraction ring of patterns which implies that the ceramic particles are crystalline in nature, which is also proved by XRD results.The HAp-PEG nanocomposites modified GCE showed good electrocatalytic activity towards the detection of 4-NP. The oxidation peak potentials for the nitrophenols are shifted to lower anodic region for the modified electrodes compared to bare GCE. This study will be useful for the detection of environmental organic pollutants and also traces of explosives.

REFERENCES

[1] Niazi, A. and Yazdanipour, A., *J. Hazard. Mater.*, 146 (2007), 421.

[2] Nistor, C., Oubi, A., Marco, M.P. and Barceló, D., *J. Emnéus. Anal Chim Acta*. 426 (2001), 185.

[3] Thompson, M.J., Ballinger, L.N., Cross, S.E. and Roberts, M.S., *J. Chromatogr*. B. 677(1996), 117.

[4] Galeano-Diaz, T., Guiberteau-Cabanillas, A., Mora-Diez, N., Parrilla-Vazquez and P., Salinas-Lopez, F., *J. Agric. Food. Chem.*, 48 (2000), 4508.

[5] Guo, X., Wang, Z. and Zhou, S., *Talanta.,* 64 (2004), 135.

[6] Yang, X., Shen, G. and Yu, R., *Microchim. Acta.*, 136 (2001), 73.

[7] Huang, W., Yang, C. and S. Zhang. de. *Anal. Bioanal. Chem.* 375:703.

[8] Yang, C., *Microchim. Acta.*, 148 (2004), 87.

[9] Liu, Z., Du, J., Qiu, C., Huang, L., Ma, H., Shen, D. and Ding, Y., *Electrochem. Commun.*, 11 (2009), 1365.

[10] Kafi, A.K.M., *A.Chen. Talanta.,* 79 (2009), 97.

[11] M. Del Mar Cordero-Rando, Barea-Zamora, M., Barberá-Salvador, J.M., Naranjo-Rodríguez, I., Muñoz-Leyva, J.A., J.L. Hidalgo-Hidalgo de Cisneros. *Microchim Acta.*, 132, (1999), 7.

[12] Hu, S., Xu, C., Wang, G. and Cui, D., Talanta., 54 (2001), 115.

[13] El Mhammedi, M.A., Achak, M., Bakasse, M. and Chtaini, A., *J. Hazard. Mater.,* 163, (2009), 323.

[14] Lin, K., Pan, J., Chen, Y., Cheng, R. and Xu, X., *J. Hazard. Mater.*, 161 (2009), 231.

[15] Wang, B., Zhang, J.J., Pan, Z.Y., Tao, X.Q. and Wang, HS., *Biosens. Bioelectron.*, 24 (2009), 1141.

[16] El Mhammedi, M.A., Achak, M. and Chtaini, A., *J. Hazard Mater.*, 161 (2009), 55.

[18] El Mhammedi, M.A., Bakasse, M. and Chtaini, A., *Electroanalysis*, 19 (2007), 1727.

[19] Killoy, W.J., *Int. Dent. J.*, 48 (1998), 305.

[20] Raynaud, S., Champion, E. and Bernache-Assollant, D., *Biomaterials,* 2002 (2002), 1073.

[21] Fischer, J., Vanourkova, L., Danhel, A., Vyskocil, V., Cizek, K., Barek, J., Peckova, K., Yosypchuk, B. and Navratil, T., *Int. J. Electrochem. Sci.,* 2 (2007), 226.

[22] Choudhury, A., Bhowmick, A.K., C. Ong. *Polymer,* 50 (2009), 201.

[23] Yin, Huanshun, Zhou, Yunlei, Ai, Shiyun, Liu, Xianggang, Zhu, Lusheng and Lu., Linan *Microchim. Acta,* 169 (2010), 87.

[24] Dhanalakshmi, C.P., Vijayalakshmi, L. and Narayanan, V., *International Journal of Nano and Material Sciences,* 2012, 1(2): 81–96.

[25] Niaz, A., Fischer, J., Barek, J. and Yosypchuk, B., *Electroanalysis,* 21 (2009) 178.

STUDIES ON BIO-MEDICAL APPLICATION OF ZnO COATED ON PLASMA TREATED FABRIC

S. INBAKUMAR[1,2] AND A. ANUKALIANI[2]

[1]*Department of Physics, Kumaraguru College of Technology, Coimbatore, Tamil Nadu*
[2]*Department of Physics, Kongunadu Arts and Science College, Coimbatore, Tamil Nadu*
E-mail: anuplasma@gmail.com; s.inbakumar@gmail.com

ABSTRACT

Zno nanoparticles were synthesis by sol-gel method. Scanning electron microscopy (SEM), and X-ray powder diffractometry (XRD) were used to characterize the nanoparticles composition, their shape, size and crystallinity. Silver nanoparticles were then applied to plasma treated polyester to impart sunscreen activity to the treated textiles. The effectiveness of the treatment was assessed through UV–Vis spectrophotometry and ZnO coated fabrics were involved with the antibacterial analysis.

INTRODUCTION

Recently, much attention has been paid to new technologies, called "added-valuetechnologies", permitting the production of functionalised and so-called "intelligent textiles". The dynamic development of functionalised textile products is directly related to progress in chemistry and polymer processing, and in particular to nanotechnology. The functionalisation of textile products by the introduction of chemical nanomolecules into them or on the surface of their fibres has recently become the main direction of thedevelopment of textile product engineering, permitting the obtaining of products which show properties unachievable with the use of conventional technology.

Research work on the functionalisation of textile products has focused on the endowment of new properties or combinations of properties, e.g. antibacterial, self-cleaning, absorbing UV radiation or deodorising. Extremely promising agents for functionalisation are nanoparticles

V. Rajendran, P. Prabu and K.E. Geckeler (eds.)
Nano Biomaterials, pp. 251 258 (2012).

of metal oxides, in particular titanium dioxide and zinc oxide[1]. Zinc oxide, showing high absorption of UV radiation[2, 3] and antibacterial properties [4], is a very interesting modifier used in the textile industry[5] due to its prospective use for the production of safety garments (health service) and all kinds of fabrics for theconstruction industry[6]. Zinc oxide can occur in a variety of structures, including one-dimensional (e.g. nanowires, nanorods, nanorings or nanohelices) or two-dimensional ones (nanoplates, nanosheets), and can have a rich gamut of applications [7].

Literature gives many methods for obtaining zinc oxide with particles of different sizes and structures: the sol-gel process[8,9], precipitation[10–12], the emulsion route[13, 14],the hydrothermal process[15, 16], the organometallic method[17], the thermal decomposition of a precursor [18, 19] and many others. This paper is concerned the application of ZnO coated on plasma treated fabrics for ultra-violet protection and antibacterial materials.

EXPERIMENTAL METHODS

ZnO solid solutions were prepared by sol-gel method. The precursor sols were prepared from zinc acetate dihydrate ($Zn(CH_3COO)_2.2H_2O$, E-Merck, India) was dissolved in 1:1 mixture of ethanol and doubly distilled water. The 'sol' mixture was stirred for 6h at a temperature of 50°C. The resultant product was dried and calcined at 500°C for an hour to get a pure nano ZnO.

The crystallographic studies were performed using XRD (Shimadzu 6000 X-ray diffractometer) with CuKα wavelength and scanning in the 2θ range from 20 to 80°. Surface morphology analysis were taken by high resolution scanning electron microscope (HRSEM, FEI Quanta FEG 200) and the elemental compositions were obtained by energy dispersive X-ray spectrophotometer (EDS).

RESULTS AND DISCUSSION

Figure 1 shows the optical absorption spectrum of ZnO nanopowder prepared under aqueous conditions. The UV visible spectra for ZnO nanoparticles synthesized in aqueous media

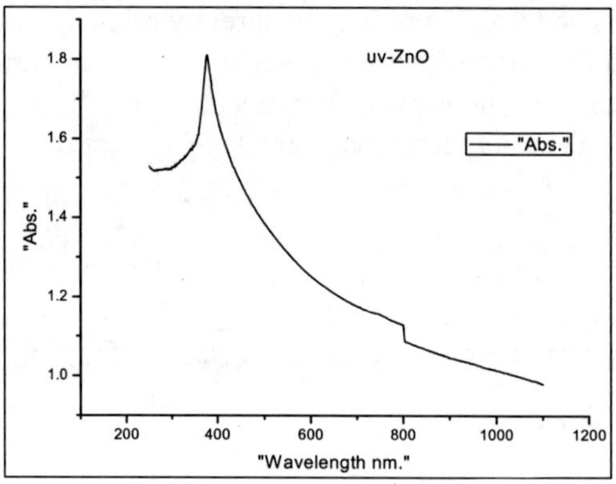

Fig. 1: UV Spectrum of ZnO

displayed excitonic absorption peak at 378 nm which implies the lower particle size ZnO. The bandgap calculated from the UV cut-off is found to be 3 · 41 eV ZnO nanoparticle. These band gap values blue shifted relative to the bulk zinc oxide of 3 · 37 eV (Huang *et al.*, 2001)[20].

Crystalline Analysis

The crystal structure and orientation of ZnO nanoparticles have been investigated by X-ray diffraction (XRD) method. The sharp and intense peaks from Figure 2 indicate that the samples are highly crystalline of ZnO and ZnO nanoparticles have polycrystalline structure[5]. The XRD peaks for (100), (002) and (101) planes indicates the formation of phase pure wurtzite structure of ZnO. The ZnO nanoparticles have a preferred growth orientation along 101 directions. No peaks were observed due to impurities, indicating high purity of the ZnO nanopowders obtained by sol-gel method. The values of a and c for ZnO calculated using the formula,

$$\frac{1}{d^2} = \frac{4}{3}\left\{\frac{h^2+hk+k^2}{a^2}\right\}+\frac{l^2}{c^2}$$

was found to be a = 3.232 Å and c = 5.177 Å. The lattice constants for hexagonal ZnO nanoparticles reported in Joint Committee on Powder Diffraction Standards (JCPDS) matches

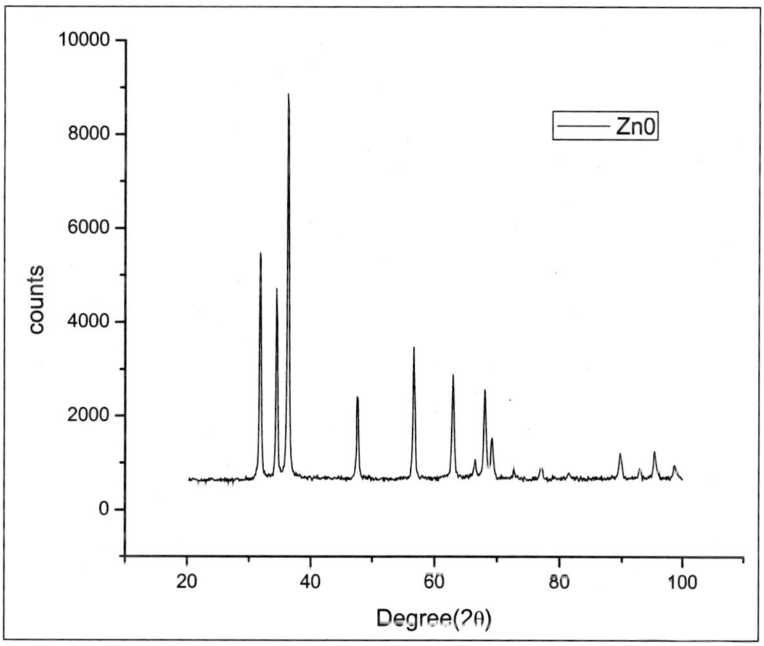

Fig. 2: XRD Pattern of ZnO

with the calculated one.[6] The crystalline size of ZnO can be calculated using Scherrar's formula,

$$D = \frac{K\lambda}{\beta \cos\theta}$$

where, the constant K is the shape factor = 0.94
'λ' the wavelength of X-rays
θ is the Bragg's angle
β is the full width at half maximum.

The average crystalline size is found to be 27 nm. Table 1 shows the XRD parameters of ZnO nanopowders. Figure 3 shows the EDX spectrum of ZnO nanopowders and Table 2 shows the ratio of ZnO elemental composition. EDX spectrum shows four peaks which are identified as zinc and oxygen. Hence, it can be seen that pure ZnO nano powders can be prepared using sol-gel method. Formation of a dominant rod like crystal the interplanar spacing values d were also calculated by using the formula,

$$D = n\,\lambda/2\sin\theta$$

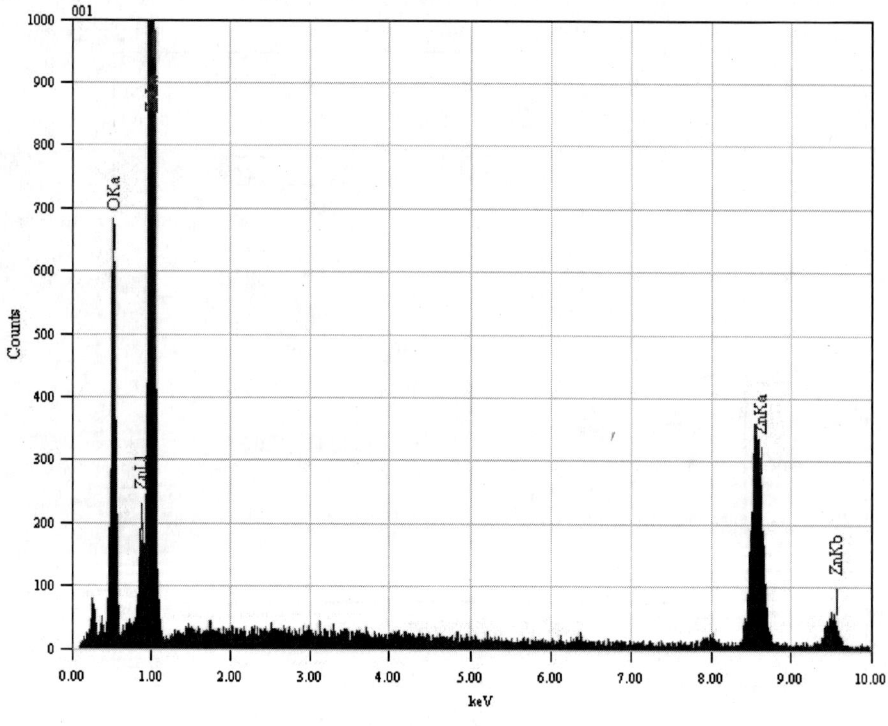

Fig. 3: EDX Spectrum of ZnO

Table 1: Particle Size Estimated from the Diffraction Spectrum

2(degree)			*Interplanar Distance(d) (A^0)*			*Crystalline (size) nm*
hkl	*Observed*	*JCPDS*	*Observed*	*JCPDS*	*FWHM*	
100	31.92	31.77	2.80	2.81	0.42	27.58
002	34.56	34.42	2.59	2.60	0.37	30.00
101	36.40	36.23	2.46	2.48	0.39	26.80
102	47.65	47.53	1.90	1.91	0.39	26.85
110	63.05	56.60	1.62	1.62	0.40	26.80

Table 2: Elements Composition of ZnO

Elements	*Experimental Results (Atomic%)*
Zn	23.14
O	76.86

Surface Morphology Analysis

SEM microphotographs of polyester fabric coated with ZnO with and without preliminary treatment with plasma are shown in Figure 4. The surface of unmodified polyester nonwoven fabric is smooth, sees Figure 4a, but has smallparticles in some places, probably impurities. In Figure 4, SEM micrographs show the nanoscaled ZnO particles on untreated polyester (b) and on plasma treated polyester (c) samples. The nanoparticles are well dispersed on the fiber surface in both cases, although some aggregated nanoparticles are still visible. The surface roughness plays a primary role in determining their adhesion to the fibers: it is reasonable to expect that the particle agglomerates will be easily removed from the smooth fiber surface, while the particles will penetrate deeper and adhere strongly into the rough surface of the fabric matrix

(a) (b) (c)

Fig. 4: SEM Image of (a) Untreated fabric, (b) ZnO + Fabric, (c) ZnO + Plasma Treated Fabric

Antibacterial Analysis

It is evident that the ZnO nanoparticles treatedfabric showed higher antibacterial activity when compared with plasma treated fabrics whereas the untreated fabrics showed no antibacterial activity. In general the antibacterial activity was higher against S.aureus than E.coli in both ZnO nanoparticles and plasma treated fabrics. The comparative antibacterial activity of ZnO nanoparticles, Zno plasma treated and untreated fabric (control) against S.aureus were shown in the Figure 5. The antibacterial activities of the ZnO particles were studied by Zhang *et al.*[21] It seems that active oxygen species generated by ZnO particles could be a mechanism although there is no direct evidence from the results of this study. The presence of active oxygen species has been detected by Yamamoto *et al.*[22]

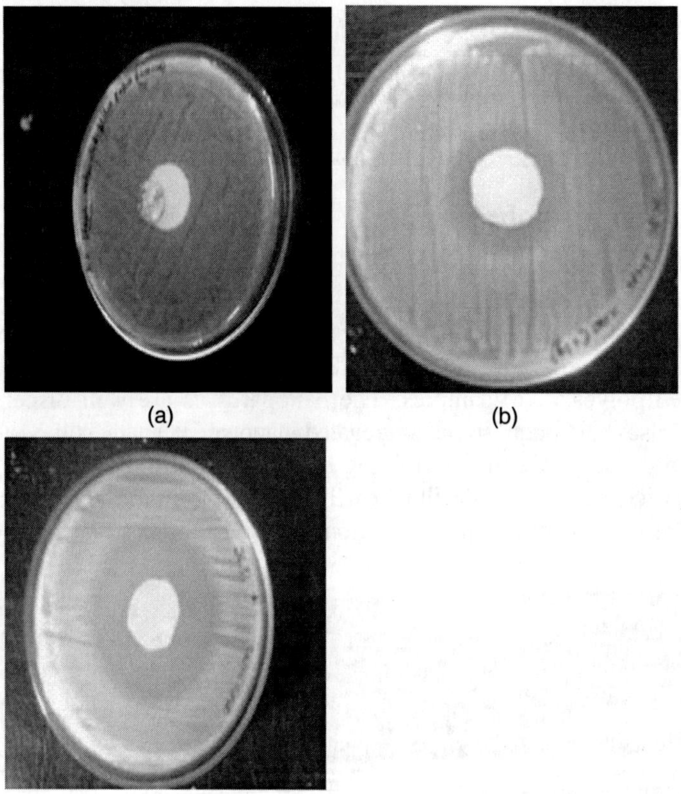

(a) (b) (c)

Fig. 5: (a): Control Sample (b): Plasma Treated Sample (c): Plasma +ZnO Sample

It has already been proved that both nano-sized and micron-sized ZnO suspensions are active in inhibiting the bacteria growth; the nano-sized ZnO suspension clearly has a much higher activity than the micron-sized ZnO suspension.[21] These results corresponds with the results of our study as ZnO nanoparticles treated fabrics and also the plasma treated fabrics showed antibacterial activity but the activity in the ZnO nanoparticles treated fabrics was much higher.

CONCLUSIONS

The proposed method of ZnO precipitation in an emulsion environment permits the obtainment of products of defined shape (particle spheres, wires). Such products are built of particle aggregates whose diameters vary in the range 160–955 nm. As follows from the results obtained, (SEM and EDS analyses) the preliminary treatment of nonwoven fabric with low-temperature plasma resulted in the attachment of a considerably greater amount of micronised zinc oxide particles to the fabric. Products from fabric preliminary treated with cold plasma and coated with paste containing ZnO can be classified as protecting against UV radiation.

Apart from the improvement of UV barrier properties, such products also show antibacterial activity. New textile products characterised by UV barrier and antibacterial properties can find a wide range of applications in the construction industry, furnishings and protective garments (medical personnel, catering establishments).

REFERENCES

[1] Wang, R., Xin, J.H., Tao, X.M. and Daoud, W.A., *Chem. Phys. Lett.,* 398, (2005), 250–255.

[2] Özgür, Ü., Alivov, Ya.I., Liu, C., Teke, A., Reshchikov, M.A., Doğan, S., Avrutin, V., Cho, S.-J. and Morkoç, H., *J. Appl. Phys.*, 98, (2005), 1–103.

[3] Yadav, A., Prasad, V., Kathe, A.A., Raj, S., Yadav, D., Sundaramoorthy, C. and Vigneshwaran, N., *Mater B.. Sci.*, 29, (2006), 641–645.

[4] Zhang, L., Jiang, Y., Ding, Y., Povey, M. and York, D., *J. Nanopart. Res.*, 9, (2007), 479–489.

[5] Sójka-Ledakowicz, J., Lewartowska, J., Kudzin, M., Jesionowski, T., Siwińska-Stefańska, K. and Krysztafkiewicz, A., *Fibres Text. East. Eur.,* 16, (2008), 116–116.

[6] Qian, L. and Hinestroza, J.P., *JTATM,* 4, (2004), 12–18.

[7] Kim, H.W., Kebede, M.A., Kim, H.S., Srinivasa, B,, Kim, D.Y., Park, J.Y. and Kim, S.S., *Curr. Appl. Phys.*, 10, (2010), 52–56.

[8] Kamalasanan, M.N. and Chandra S., *Thin Solid Film.*, 15, (1996),112–115.

[9] Suwanboon, S., *Sci. Asia,* 34, (2008), 31–34.

[10] Mayo, M.J., *Int. Mater. Rev.*, 41, (1998), 85–115.

[11] Jitianu, M. and Goia, D.V., *J. Colloid Interface.Sci.*, 309, (2007), 78–85.

[12] Sun, D., Wong, M., Sun, L., Li, Y., Miyatake N. and Sue H-J., *J. Sol-Gel Sci. Technol.*, 43, (2007), 237–243.

[13] Lopez-Quintela, M.A. and Rias, J., *J. Colloid.Interface Sci.,* 158, (1993), 446–451.

[14] Voroboyova, S.A., Lesnikovich, A.I. and Mushinskii, V.V., *Mater. Lett.*, 58(2004) 863–866.

[15] Chen, D., Jiao, X. and Cheng, G., *Solid State Commun.*, 113, (2000), 363–366.

[16] Lu, Ch.H. and Yeh, Ch.H; *Ceram. Int.*, 26, (2005), 363–366.

[17] Kahn, M.L., Monge, M., Colliere, V., Senocq, F., Maisonnat, A. and Chaudret, B., *Adv. Funct. Mater.*, 15, (2005), 458–468.

[18] Jing, L., Xu, Z., Shang, J., Sun, X., Cai, W. and Guo, H., *Mater. Sci. Eng.*, A332, (2002), 356–361.

[19] Jing, L., Xu, Z., Sun, X., Shang, J. and Cai, W., *Appl. Surf. Sci.*, 180, (2001), 308–314.

[20] Huang, M.H, Wu, Y, Feick, H and Tran, N., *Weber E and Yang Adv. Mater. B*, 13, (2001), 113.

[21] Zhang, L., Jiang, Y., Ding, Y., Daskalakis, N., Jeuken, L., Povey, M., Alex, J. O'Neill and YorkD, D.W., Mechanistic investigation into antibacterial behaviour of suspensions of ZnO nanoparticles against E.coli, *Journal of Nanoparticle Research,* (2009)DOI 10.1007/s11051-009-9711-1.

[22] Yamamoto, O., Sawai, J. and Sasamoto, T., Change in antibacterial characteristics with doping amount of ZnO in MgO–ZnO solid solution, *International Journal of Inorganic Materials*, 2, (2000), 451–454.

EVALUATION OF SURFACE ACETYLATED AND INTERNALLY QUATERNIZED POLY(PROPYLENE IMINE) DENDRIMER AS BIOCOMPATIBLE DRUG CARRIER USING NIMESULIDE AS A MODEL DRUG

E. MURUGAN AND D.P. GEETHA RANI

Department of Physical Chemistry, School of Chemical Sciences, University of Madras, Maraimalai Campus, Guindy, Chennai, Tamil Nadu
E-mail: dr.e.murugan68@gmail.com

ABSTRACT

This study demonstrated the application of new drug carrier with surface acetylated and internally quaternized generation 3 poly(propylene imine) dendrimer, viz., QPPI-NHAc (G3). The drug delivery potential of QPPI-NHAc (G3) dendrimer was evaluated using Nimesulide (NMD, an anti-inflammatory drug) as a model drug. Aqueous solubility, in vitro release of the drug and cytotoxicity study of the carrier were explored. The observed results showed that aqueous solubility of NMD has significantly increased in proportional to dendrimer concentration. It is noted that the complexation of NMD with QPPI-NHAc (G3) is responsible for increased solubility. The in vitro release of NMD in the presence of QPPI-NHAc (G3) dendrimer was remarkably slower compared to the release with the parent dendrimer. Cytotoxicity studies on Vero cell line (NCCS) through MTT assay showed that the QPPI-NHAc (G3) dendrimer showed appreciable increase in cell viability, which reduces the cytotoxicity thereby improving the biocompatibility of modified dendrimer carrier. The sustained drug release behavior along with their solubilicing potential and reduced cytotoxicity indicated that QPPI-NHAc (G3) dendrimer can be considered as potential biocompatible drug carrier of Nimesulide under suitable conditions.

V. Rajendran, P. Prabu and K.E. Geckeler (eds.)
Nano Biomaterials, pp. 259–268 (2012).

INTRODUCTION

Nonsteroidal anti-inflammatory drugs (NSAIDs) are the most widely prescribed medications in the world. As a therapeutic class, NSAIDs exhibit analgesic, anti-inflammatory, antipyretic, and platelet inhibitory properties.[1, 2] However, these drugs have serious side effects such as gastrointestinal (GI) toxicities, gastric mucosal ulcerations and haemorrhage due to inhibition of prostaglandin production.[3, 4] The mechanism of action of NSAIDs has been attributed to their ability to inhibit the cyclooxygenase enzyme (Cox). Out of the 2 isoforms of Cox, cox-1 is responsible for mediating the production of prostaglandin while cox-2 is primarily associated with inflammation, pain and fever.[5, 6] The traditional NSAIDs are nonselective Cox inhibitors. Cox-2 selective NSAIDs are, therefore, ideal anti-inflammatory drugs with minimum drug-related side effects, since they spare cox-1 activity. Nimesulide is one such COX-2 selective non-steroidal anti-inflammatory drug (NSAID) with analgesic and antipyretic properties in addition to low toxicity, a moderate incidence of gastric side effects and a high therapeutic index.[7] It is mainly prescribed for the treatment of acute pain, the symptomatic treatment of osteoarthritis. However, it's very poor aqueous solubility (0.010 mg/ml) [8] gives rise to difficulties in the design of pharmaceutical formulations and leads to variable oral bioavailability. This restricts their use in topical and parenteral applications. Further, poorly water-soluble drugs often require high doses in order to reach therapeutic plasma concentrations after oral administration. Hence its solubility has to be increased. In order to improve the aqueous solubility of Nimesulide various methods were employed such as cyclodextrin-Nimesulide complexation,[9] micellar solubilisation [10] and solid dispersion.[11]

Dendrimers are new class of artificial macromolecules with tree-like topological structure, nano-scale size, excellent monodispersity and dense peripheral groups.[12] Dendrimers has generated much interest in diverse areas more especially for drug delivery due to their unique structure and chemical versatility.[13, 14] Surface functionalities of dendrimers can be functionalized with drugs, targeting moieties and biologically active components.[15] Drugs or genes can be either encapsulated within the dendrimers through non-covalent strategies such as hydrophobic, ionic and hydrogen bond interactions or conjugated to the peripheral groups of dendrimers via covalent methods.[16] Among the various dendrimers used for drug delivery, poly(amidoamine) (PAMAM) and poly(propylene imine) (PPI) dendrimers are the two commercially available and most extensively investigated for drug delivery applications.[17] A plethora of reports are available which evidence the drug carrying potential of dendrimers through properties like increased solubility of the hydrophobic drugs, sustained drug release behavior and increased efficiency of the drugs.[18–20] Dendrimer based drug-delivery systems provide an attractive platform for loading and release of conventional drug molecules which improve the pharmacodynamic and pharamacokinetic behaviors of several families of drugs. These reveal promising future of dendrimer based drug delivery systems. Dendrimers such as PAMAM and PPI possess cationic primary amine groups at the surface showed excellent drug delivery efficacy, but the cytotoxicity issue of these dendrimer is a serious setback which limits the clinical applications of most dendrimer based drug formulations.[21] Generally, the cytotoxicity of dendrimer is dependent on dendrimer generation, surface functionality/charge and concentration.[22] To overcome this, there are several studies undertaken by various researchers to neutralize the cationic surface group such as acetylation,[23] PEGylation,[24] glycosylation [25] and hydroxylation.[26] Surface acetylated dendrimers reduced the cytotoxicity significantly but at the same time such dendrimers failed to load the hydrophobic drugs due to the absence of cationic binding sites.[27] In a subsequent study, the neutral dendrimer namely, PAMAM-OH was employed to form Polyplex with DNA, with the analogy that the surface neutral PAMAM-OH dendrimer may be advantageous in terms of cytotoxicity. However, PAMAM-OH dendrimer is unable to form DNA Polyplex because of low pK_α value of its interior tertiary amines.[28] With a view to alleviate this, internal quaternary ammonium salt was introduced to the tertiary amines of PAMAM-OH dendrimers by methylation thereby provide cationic binding sites for negatively charged plasmid DNA.[29] Further, the synthesis of surface modified and internally quaternized PAMAM dendrimer was carried out and employed as efficient siRNA delivery systems.[30] However to our knowledge there are no studies pertaining to the neutralization of surface amine groups of PPI dendrimers followed by internal quaternisation of terminal amine groups of the same which enabled the electrostatic interaction between the hydrophobic drug and the dendrimer. These characteristics make it a better choice as a carrier than cationic dendrimer in drug delivery systems.

EXPERIMENTAL

Materials

PPI dendrimer Diamino cored (DAB) with terminal amino groups of generation 3 was purchased from Symo Chem. (Netherlands) and used as received. All the reagents and solvents, Acetyl chloride, Potassium Carbonate, Methyl iodide, anhydrous DMF were of analytical grade and were procured from SRL. MTT (3-(4, 5-dimethyl-2-thiazolyl)-2, 5-diphenyl-2H-tetrazolium bromide) was purchased from Sigma-Aldrich, India. Amberlite resin IRA-402 Cl was from Sd fine Chem. Dialysis membranes (Float-A-lysers) were obtained from Spectra Pore. Membrane filter of pore size 0.45 μm was obtained from Himedia lab. Double distilled water was used for solubility and in vitro release studies. Nimesulide was obtained as gift sample from Micro labs, Bangalore, India. UV-Vis spectra were recorded on TECHCOMP 8500 instrument with kinetic software. Cytotoxicity study was done in King Institute of Preventive Medicine, Chennai, India.

Solubility Study of NMD Using QPPI-NHAC (G3) Dendrimer Carrier

This dendrimer carrier was prepared from generation 3 Poly(propylene imine) dendrimer PPI-NH_2(G3) by performing acetylation followed by quaternisation using the reported procedure[30] and named as QPPI-NHAc(G3) and it was characterized by FTIR, ^{1}H and ^{13}C NMR and MALDI-TOF spectral techniques. The observed spectral results strongly confirmed its structure. The newly prepared dendrimer based drug carrier, viz., QPPI-NHAc (G3) was tested for the solubility study of NMD under identical conditions. The degree of solubility of NMD was quantitatively estimated by adopting the previously described Phase solubility analysis method.[31] The experimental procedure adopted for this solubility study involves, using 7 different concentrations of drug carrier in the range from 0.05 mM to 0.35 mM. Then from each concentration, 5ml was taken in 20 ml vials and to which excess NMD (5 mg) was added. These vials were sonicated for 30 mins and subsequently shake mechanically for 24 h at 35°C. In Parallel, controlled experiment was also performed by taking 5mg of drug in 5 ml of double distilled water (without dendrimer). The vials were allowed to stand for 24 h to attain equilibrium. Then these solutions were centrifuged at 5000 rpm for 20 mins and filtered using Cellulose acetate membrane filter having pore size 0.45 μm. The respective filtrate was diluted appropriately with double distilled water and analyzed spectrophotometrically using UV spectrophotometer. The quantum of solubility of NMD was estimated by measuring the characteristic peak observed at 397 nm (λ_{max}). Though the increase in absorption of NMD is the indication for the extent of drug solubilised, but to get the exact amount of the drug solubilised in the respective dendrimer carrier solution, a calibration curve was constructed within the drug concentration range of 3.5 μM – 35 μM which is described by the equation, $y = (15602 \pm 0.006)x$, ($R^2 = 0.999$). The directional coefficient of the equation is equal to the absorption coefficient of NMD, ε_{max} = 15,602 M^{-1} cm^{-1}. Using this calibration curve, the amount of drug solubilised in each test solution was determined. The blanks were performed on the same concentrations of dendrimer delivery solutions so as to cancel any absorbance that may be exhibited by the dendrimer molecules. The results of phase solubility studies are shown in Table 1. The solubility experiments were repeated three times.

In-vitro Drug Release Studies

The extent of drug released from the dendrimer carrier, QPPI-NHAc (G3) dendrimer was studied by an equilibrium dialysis method under in vitro conditions as described in the references. The drug-dendrimer formulation for in vitro study was as follows. That is, 2 mg of NMD was dissolved in 5 ml of dendrimer carrier (0.02 mM) to get the formulation having drug concentration 1.3 mM. Using this formulation, in vitro release study was performed by taking 5ml of respective drug-dendrimer solution in dialysis bag (MW cutoff 1 kDa) and it was placed immediately in 150ml beaker, containing 100 ml of distilled water. The outer phase was stirred continuously to maintain sink conditions. At scheduled intervals, 3 ml of sample was withdrawn from outer phase and subsequently replenished with equal amount of fresh water. The withdrawn sample irrespective of type was analyzed spectrophotometrically with UV using the drug characteristic peak at 397 nm. Similarly, in vitro release of NMD from their parent dendrimer, viz., PPI-NH_2(G3) was also studied by adopting the same experimental condition. The quantum of drug released by the parent dendrimer was determined as before. The control experiment was also carried out by taking 2 mg of NMD in 5 ml of methanol to get the same drug concentration as in the drug-dendrimer formulations. This was also studied for in vitro release by usual UV analysis. Three repeats were conducted for each sample. Irrespective of the dendrimer carrier and the control, plots were drawn for the % of drug released vs time and the same is shown in Figure 1.

Cytotoxicity Assay

The cytotoxicity of QPPI-NHAc (G3) and its parent dendrimer as well as NMD-Dendrimer complexes was evaluated by the well established 3-(4,5-dimethylthiazol-2-yl)-2, 5-diphenyltrazolium bromide (MTT) colorimetric assay. Vero cell line (NCCS) was chosen for the MTT colorimetric assay. The assay procedure involves the incubation of cell line at 37°C in 5% CO_2 and minimum essential medium (MEM) with Earle's salts supplemented with 10% fetal calf serum. The cells were used when they reached 70–80% confluence. These cells were seeded in 96 well plates for 48 h before MTT assay to allow adherence of the cells to the plates. Then culture medium from the plates was replaced with fresh medium containing different concentrations of QPPI-NHAc (G3) and $PPI\text{-}NH_2$ (G3). The control study was performed by treating the cells only with the medium without the dendrimer samples. Following the standard protocol for MTT, the mediums were removed after 24 h from each of the well and the cells were incubated with MTT in PBS at a concentration of 1mg/ml (100 μl) for 4 h. After this, in each well, the MTT containing medium was replaced with DMSO to dissolve the resulted purple formazan crystals which was generated from MTT by the living cells. Absorbance of the solutions in each well irrespective of carrier were recorded at 570 nm using a micro plate reader (Thermo multiscan EX). The viability of cells was calculated with regard to the untreated cell control, which was considered as 100% viability. The viability of cells treated with the different dendrimer carrier is expressed as a percentage of the untreated cell control. The obtained cell viability was plotted against the concentration of the carrier and shown in Figure 2.

RESULTS AND DISCUSSION

Solubility Studies of NMD Using QPPI-NHAC (G3) Dendrimer

Drug solubility is one of the core factors, which influences the movement of a drug from the site of administration into the blood stream. It is widely acknowledged that insufficient drug solubility can lead to poor absorption.[32] The poor solubility of NMD can be addressed with newly developed QPPI-NHAc (G3) dendrimer by employing it as solubility enhancer.

Table 1: Solubility of NMD in QPPI-NHAc (G3) Dendrimer Solutions with Various Concentrations

S. No.	*[QPPI-NHAc(G3)] In M* $\times 10^{-4}$	*Absorbance at 397 nm* (λ_{max})	*[NMD] in M* $\times 10^{-4}$	*Sol in mg/ml (St)*	*Solubility Enhancement* St/S_0^*
1.	0.50	0.0826	5.378	0.1656	16.56
2.	1.00	0.1135	7.390	0.2276	22.76
3.	1.50	0.1468	9.558	0.2944	29.44
4.	2.00	0.1846	12.02	0.3702	37.02
5.	2.50	0.2412	15.71	0.4837	48.37
6.	3.00	0.2930	19.07	0.5876	58.76
7.	3.50	0.3417	22.25	0.6853	68.53

*S_0 the intrinsic solubility of NMD (0.01 mg/ml)

Solubility studies were performed by keeping the seven different concentrations in the range from 0.05 mM to 0.35 mM. The amount of solubilised drug was estimated in each concentration of dendrimer from the UV absorption and the calibration curve. The observed increase in drug solubility is tabulated in Table 1. To record precisely, at the maximum concentration (0.35 mM) of the dendrimer carrier, the solubility of NMD in aqueous medium has increased to the tune of 68 times as compared with its intrinsic solubility. This ensures that the water solubility of NMD has been significantly increased which is proportional to the concentration of dendrimer carrierThe effective solubility increase in the presence of QPPI-NHAc (G3) is due to (i) the availability of internal positive binding sites with neutral outer surface. (ii) The multi charged cations, viz., QPPI-NHAc(G3), attract the anion of NMD whose formation is facilitated by the formation of HCl, a stronger acid compared to NMD, (iii) Thus NMD binds electro statically with QPPI-NHAc(G3) in aqueous medium forming the complex, $(NMD\text{-}SO_2CH_3N^-)_n\ (QPPI\text{-}NHAc)^{n+}$ HCl, (iv) Thereby NMD gets internalized within the QPPI-NHAc(G3) dendrimer branches. This interaction substantially promotes the solubility of NMD. The increased solubility helps in increasing its bioavailability.

In-vitro Drug Release

The very purpose of drug carrier is to carry the drugs and deliver them at the site of action. One of the prerequisite of a good carrier is, it should deliver the drug in a sustained/delayed manner. Thereby the side effect of drug can be minimized by preventing the fluctuation of the therapeutic concentration of the drug in the body. It also eliminates the chance of over or under dosing. Normally, one of the methods to achieve this sustained/delayed release is via diffusion through a matrix. According to the previous reports, dendrimers are able to pack hydrophobic drugs and these formulations are suitable for different administration routes. These polymeric delivery systems with sustained release behaviors can improve the biocompatibility of drugs loaded in the carriers, simplify the dosing schedule and improve patient compliance.[33–36] To study the extent of drug release from drug-dendrimer formulations, we investigated the release profiles of NMD from QPPI-NHAc (G3) as well as from its parent dendrimer. The release pattern was plotted as percentage of drug released vs. time and the same is represented in Figure 1. It clearly shows that, after 1 h, 83% of drug gets released from the control formulations, whereas 40% of drug was released from $PPI\text{-}NH_2$ (G3)-NMD formulations. Further, in the case of QPPI-NHAc (G3)-NMD formulation the release gets sustained to 13% in the same span of time (1hr). As much as 90% drug gets released within 80mins from the control formulations, whereas it takes 180 mins for the same amount of drug to get released from the $PPI\text{-}NH_2$ (G3)-NMD formulations. However, the same amount (90%) of drug release from QPPI-NHAc (G3) NMD formulation gets extended to 390 mins. These results suggested that the QPPI-NHAc (G3) dendrimer significantly contributed for the sustained the release of NMD than their parent and control formulation.

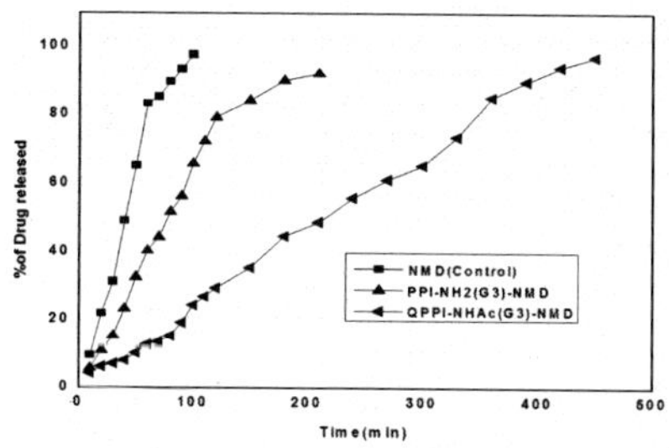

Fig. 1: In Vitro Release Profile of NMD from QPPI-NHAc (G3) and PPI-NH_2 (G3) Compared with the Pure NMD (control)

In general, the sustained/delayed release of drug by any delivery is an indication of effective internalization of drugs with delivery or formation of stable complex due to the interaction between deliveries and drugs. Since in our study also, the new dendrimer carrier, viz. QPPI-NHAc (G3) has delayed the release significantly, it is obviously due to the formation of stable complex $HCl(NMDNSO_2CH_3^-)_n$-$(QPPI\text{-}NHAc(G3)^{n+}$. Hence these observations has strongly proves that the QPPI-NHAc (G3) dendrimer is potential carrier for sustained release of the drug NMD.

Cytotoxicity Assay

An ideal polymeric drug carrier should be biocompatible by performing its expected functions without the concomitants of toxic and immunologic effects to normal cells.[37] The effect of dendrimers and their complexes with NMD on the viability of Vero cell lines were measured by the MTT assay. The variation of percentage cell viability incubated in Vero cell line with different concentrations of QPPI-NHAc (G3) and PPI-NH_2 (G3) are shown in Figure 2.

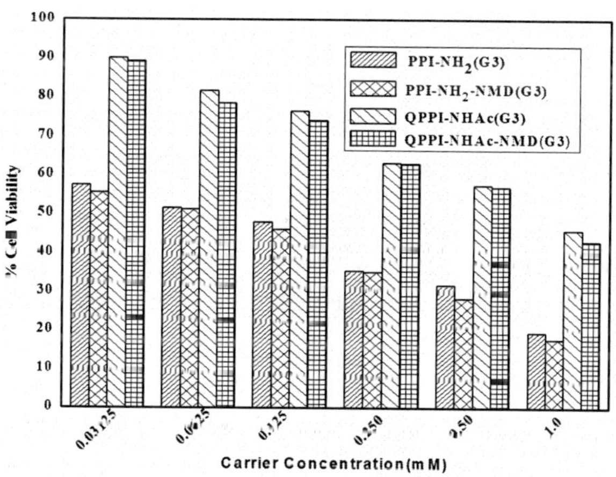

Fig. 2: Cytotoxicity of PPI-NH_2 (G3), PPI-NH_2 (G3)/NMD Complex, QPPI-NHAc (G3) and QPPI-NHAc (G3)/NMD Complex on Vero Cells at Different Concentrations

From the plot, it is observed that, the modified dendrimer carrier, viz., QPPI-NHAc (G3) dendrimer has proved to be significantly less cytotoxic than its parent dendrimer. The presence of NMD with the dendrimer carriers did not reduce cell viability on incubation over the tested concentration.

Photographs of Vero cells shown in Figure 3a-c, which reflect the cytotoxicity caused by 0.125 mM PPI-NH_2 (G3) dendrimers on Vero cells after 24h. The cells got distorted and a significant decrease in cell density can be found (Figure 3b). However, cells treated with QPPI-NHAc (G3) dendrimers are observed with retaining the cell structure and also considerable cell density (Figure 3c), suggesting that modified dendrimers are less toxic on Vero cells. Normally, under in vitro conditions, the degree of cytotoxicity usually been evaluated based on the IC_{50} values (concentrations where 50% of cells are viable) and this value (IC_{50}) increases significantly for QPPI-NHAc (G3) dendrimers. That is, IC_{50} value for PPI-NH_2 (G3) is less than 0.0625mM whereas for QPPI-NHAc (G3), it is 0.5 mM. These results show that QPPI-NHAc (G3) dendrimer can effectively increase the tolerance concentration during its administration. Previously reported studies reveal that electrostatic interaction between positively charged polymers and negatively charged cell membranes is responsible for cytotoxicity.[30, 38] Particularly, cationic dendrimers containing –NH_2 surface group, are prone to attract the negatively charged cell membrane that in turn damaged the cell lines and thus reflected more cytotoxic effect.

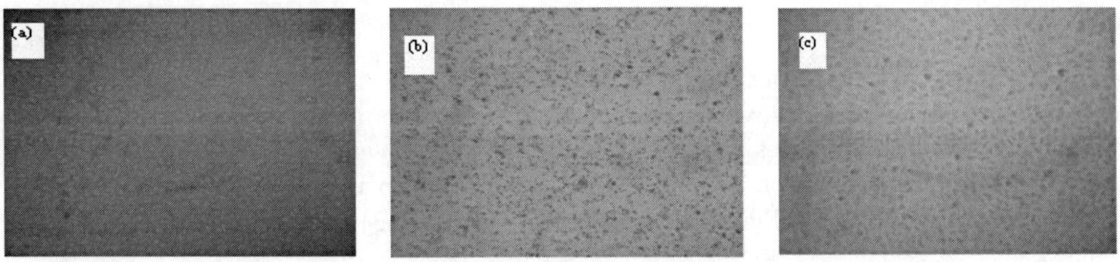

Fig. 3: Photographs of Vero Cells Incubated for 24h without Dendrimer, i.e. (a) Cell Control, (b) with 0.125 mM PPI-NH_2 (G3), (c)with 0.125 mM QPPI-NHAc (G3)

Therefore, it is necessary in any delivery, the cationic surface group of the polymer/ dendrimer must be neutralized so as to eliminate the electrostatic interaction of surface group with the negatively charged cell membranes and this in turn reduced cytotoxicity. On par with this, QPPI-NHAc (G3) is such a carrier containing neutral surface group, viz., -$COCH_3$ and CH_3 thereby their electrostatic interaction with Vero cell line was sizably reduced and hence lower cytotoxicity is noticed than its parent dendrimer. This proves that QPPI-NHAc (G3) dendrimer provide a potential platform to reduce the cytotoxicity which gives the pathway to increase its biocompatibility.

CONCLUSIONS

Dendrimer based drug delivery system provides an attractive platform to load and release conventional drug molecules which improves the pharmacodynamic and pharmacokinetic behavior of these drugs. In the present study, we investigated the potential of surface acetylated and interior quaternized PPI dendrimer, viz., QPPI-NHAc (G3) as drug carrier of the drug Nimesulide by aqueous solubility, in vitro release as well as cytotoxicity studies. Results showed that the significant drug solubilising potential which ensures promising NMD loading in this dendrimer. It is stated that complexation of NMD with QPPI-NHAc (G3) is responsible for drug loading and in turn the increased solubility. The drug release under in vitro conditions gets appreciably sustained in the presence of QPPI-NHAc (G3). The extent of sustained release reflected the magnitude of interaction between the dendrimer carrier and the drug molecule. QPPI-NHAc (G3) dendrimers showed increased cell viabilities on Vero cell lines which is correlated with its reduced cytotoxicity over the parent dendrimer and hence they are more biocompatible. The increase in IC_{50} value for QPPI-NHAc (G3) can effectively increase their tolerance concentration during the preparation of NMD-QPPI-NHAc (G3) dendrimer formulations. The results in this study suggest that the surface acetylated and internally quaternized dendrimer, QPPI-NHAc (G3) is promising candidate in the design of polymeric drug delivery systems with high performances and biocompatibility.

ACKNOWLEDGEMENTS

One of the authors Geetha Rani D.P. would like to thank University of Madras for providing financial support in the form of Junior Research Fellowship (TNJRF).

REFERENCES

[1] Lane, N.E. and Thompson, J.M., *Am J Med,* 103(6A), 25–35, 1997.
[2] Needleman, P. and Isakson, P.C., *J Rheumatol,* 24, 6–8, 1997.
[3] Bolten, W., *J Rheumatol,* 25, 2–7, 1998.
[4] Depre, M., Ehrich, E., Van Hecken, A., De Lepeleire, I., Dallob, A. *et al., Eur J Clin Pharmacol*, 56, 167–174, 2000.
[5] Hla, T. and Neilson, K., *Proc Natl Acad Sci USA*, 89, 7384–7388, 1992.
[6] Silas, S. and Clegg, D.O., *Bull Rheum Dis*, 40(2), 1–4, 1999.
[7] Davis, N. and Brogden, R.N., *Drugs*, 48, 431–454, 1994.
[8] Piel, G., Pirotte, B. and Delneuveille, I., *J Pharm Sci*, 86, 475–480, 1997.
[9] Vavia, P.R. and Adhage, N.A., *Drug Dev Ind Pharm*, 25, 543–545, 1999.
[10] Amidon, G.L., Jinno, J. and Crison, J.R., *J. Pharm. Sci*, 89, 268–274, 2000.
[11] Mutalik, S., Venkatesh and Udupa, N., *Indian J. Physiol. Pharmacol,* 46, 115–116, 2002.
[12] Gillies, E.R. and Fréchet, J., *Drug Discov Today,* 10, 35–43, 2005.
[13] Svenson, S. and Tomalia, D.A., *Adv Drug Deliv Rev.*, 57, 2106–2129, 2005.
[14] Roseita, E. and Tomalia, D.A., *Drug Discov Today*, 6, 427–436, 2001.
[15] Grainger, D. and Okano, T., *AdV. Drug DeliVery ReV*, 55, 311–313, 2003.
[16] D'Emanuele, A. and Attwood, D., *Adv Drug Deliv Rev*, 57, 2147–2162, 2005.
[17] Menjoge, A.R., Kannan, R.M. and Tomalia, D.A., *Drug Discov Today,* 15, 171–185, 2010.

[18] Lee, C.C., Mackay, J.A., Fréchet, J.M.J. and Szoka, F.C., *Nat Biotechnol,* 23, 1517–26, 2005.

[19] Boisselier, E., Liang, L., Dalko-Csiba, M., Ruiz, J. and Astruc, D., *Chem Eur J*, 16, 6056–68, 2010.

[20] Gupta, U., Agashe, H.B., Asthana, A. and Jain, N.K., *Biomacromolecules*, 7, 649–658, 2006A.

[21] Duncan, R. and Izzo, L., *Adv Drug Deliv Rev*, 57, 221–237, 2005.

[22] Jevprasesphant, R., Penny, J., Jalal, R., Attwood, D., McKeown, N. and D'Emanuele, A., *Int. J. Pharm*, 252, 263–266, 2003.

[23] Kolhatkar, R.B., Kitchens, K.M., Swaan, P.W. and Ghandehari, H., *Bioconjugate Chem,* 18, 2054–2060, 2007.

[24] Zhang, Y.Q., Sun, Y.H., Xu, X.P., Zhang, X.Z., Zhu, H. *et al.*, J. Med. Chem, 53, 3262–3272, 2010.

[25] Majoros, I.J., Thomas, T.P., Mehta, C.B. and Baker, J.R., *J. Med.Chem*, 48, 5892–5899, 2005.

[26] Roy, R. and Baek, M.G., *Rev, Mol. Biotechnol,* 90, 291–309, 2002.

[27] Yang, K., Weng, L., Cheng, Y.Y., Zhang, H.F., Zhang, J.H. *et al., J Phys Chem B*, 115, 2185–2195, 2011.

[28] Tomalia, D.A., Baker, H., Dewald, J., Hall, M., Kallos, G. *et al., Polym.J*, 17, 177– 182, 1985.

[29] Zhang, Y.Q., Sun, Y.H., Xu, X.P., Zhu, H., Huang, L.L. *et al., Bioorg.Med. Chem. Lett*, 20, 927–931, 2010.

[30] Patil, M.L., Zhang, M., Betigeri, S., Taratula, O. and Minko, T., *Bioconjug Chem*, 19, 1396–1403, 2008.

[31] Higuchi, T. and Connors, K.A., *Interscience*, Vol. 4, New York, 117–212, 1965.

[32] Zhao,Y., Abraham, M., Lee, J., Luscomble, C. and Beck, G. *et al.*, *Pharm. Res,* 9(10), 1446–1457, 2002.

[33] Cheng, Y.Y. and Xu, T.W., *Eur. J. Med. Chem*, 43, 2291–2297, 2008.

[34] Yiyun Cheng, Jinrong Wang, Tingling Rao, Xiaoxiao He, Tongwen Xu, *Front. Biosci,* 13, 1447–1471, 2008.

[35] Wu, Q.L., Cheng, Y.Y., Hu, J.J., Zhao, L.B. and Xu, T.W., *J. Phys. Chem. B*, 113, 12934–943, 2009

[36] Hu, J.J., Fang, M., Cheng, Y.Y., Zhang, J.H., Wu, Q.L. *et al., J. Phys. Chem. B,* 114, 7148–7157, 2010.

[37] Duncan, R. and Izzo, L., *Adv. Drug Delivery Rev*, 57, 2215–2237, 2005.

[38] Lee, J.H., Lim, Y.B., Choi, J.S., Lee, Y. and Kim, T. *et al., Bioconjugate Chem,* 14, 1214–1221, 2003.

SYNTHESIS OF QUATERNARY AMMONIUM FUNCTIONALIZED POLY(PROPYLENE IMINE) DENDRIMER AND ITS POSSIBLE APPLICATION AS DRUG CARRIER USING PIROXICAM AS A MODEL DRUG

E. MURUGAN, D.P. GEETHA RANI AND V. YOGARAJ

Department of Physical Chemistry, School of Chemical Sciences,
University of Madras, Maraimalai Campus, Guindy, Chennai, Tamil Nadu
E-mail: dr.e.murugan@gmail.com

ABSTRACT

Potential dendrimer drug delivery, viz., poly(quaternary ammonium) PPI dendrimer (PQPPI-$N(CH_3)_3$ (G3)) was prepared, characterized and used as a solubility enhancer for poorly soluble drug Piroxicam(PXM, an anti-inflammatory drug) in aqueous medium. The quantum of solubility of PXM in different concentrations of PQPPI-$N(CH_3)_3$ (G3) was studied by measuring the absorbance of the drug characteristic peak at 355 nm. The observed results reveal that the aqueous solubility of PXM has been gradually increased on increasing the [PQPPI-$N(CH_3)_3$ (G3)]. Particularly, on comparing the intrinsic solubility of PXM in aqueous (control) medium with 0.35 mM of PQPPI-$N(CH_3)_3$ (G3) dendrimer medium, it is observed that the latter one are able to solubilize PXM drug to the tune of 43 fold. It is further noticed that the complexation of PXM with PQPPI-$N(CH_3)_3$ (G3) is responsible for increased solubility. This complexation was evidenced through FTIR and UV-vis spectral techniques. Thermodynamic parameters such as ΔG, ΔH and ΔS were also determined for drug-dendrimer complexes. These were utilized to elucidate the mechanism of dendrimer mediated solubility enhancement. The observed thermodynamic results prove that the PQPPI-$N(CH_3)_3$ (G3) is one of the new promising candidate for effective delivery of PXM.

V. Rajendran, P. Prabu and K.E. Geckeler (eds.)
Nano Biomaterials, pp. 269–278 (2012).

INTRODUCTION

Dendrimers are outstanding candidates to act as drug carriers due to their well defined nanostructures and chemical versatility.[1, 2] As compared to the traditional polymeric drug vehicles, dendrimer shows excellent advantages more especially for drug delivery.[3] Their unique characteristics like stable monomolecular micellar structure, well defined number of surface functionalities and monodispersity have enabled them as devices for pharmaceutical and biomedical applications.[4–7] Poly(amidoamine) (PAMAM) and Poly(propylene imine) (PPI) dendrimers are the extensively used ones among the numerous dendrimers used for drug delivery. These dendrimers were modified to quaternary amine groups for a pH sensitive controlled release systems,[8] a dendritic anion conductor,[9] effective antimicrobials [10] and a photosensitizer for photodynamic therapy.[11] It is known from the literature that dendrimers possessing quaternary ammonium groups have not been employed as drug delivery system. An important feature of quaternized PPI-NH_2 (G3) dendrimers is that exchange of primary amines and tertiary amines with quaternary ammonium salts greatly reduces the cytotoxicity and also giving hydrophilic nature.[12–15] Moreover, their surface density and charge density can be increased by methylation of primary amine groups and thus provides potential applications by enhancing the solubility of low aqueous soluble drugs in aqueous medium and as delivery systems for bioactive materials.[16] Although, several dendrimer based drug delivery system are already studied, but dendrimers containing low molecular weight with higher generation are not studied so far. Particularly, quaternized PPI-NH_2 (G3) dendrimer with methyl groups are not studied as a drug delivery system. Further, it is worth to mention here that any dendrimer having numerous poly (quaternary ammonium) ions is expected to serve as a efficient drug delivery. Therefore, to the best of our knowledge, there are no studies devoted exclusively for the synthesis of low molecular weight dendrimer containing poly (quaternary ammonium) ions for effective solubility of poorly soluble drugs in aqueous medium. This may emerge as promising candidate for safer and more effective compared to traditional drug delivery system Piroxicam (PXM), an acidic nonsteriodal anti-inflammatory drug, is quite efficient in the short or long term treatment of rheumatoid arthritis, osteoarthritis and other painful inflammatory disorders.[17–19] But there are a number of aspects that limit its pharmaceutical application, the chief being its aqueous insolubility that immensely creates formulation tribulations.[20] Apart from this, oral administration shows low absorption and gastrointestinal tract associated adverse effects, which displays poor bioavailability.[21] Taking in to consideration for the pre requirement of efficient dendrimer drug delivery, viz., low molecular weight poly(quaternary ammonium) ions and poor solubility of PXM in aqueous medium, we reported a new poly (quaternary ammonium) poly(propylene imine) PQPPI-$N(CH_3)_3$ (G3) dendrimer delivery and investigated the same for effective solubility of poorly soluble PXM drug.

EXPERIMENTAL METHODS

Materials

PPI dendrimer Diamino cored (DAB) with terminal amino groups of generation 3 was purchased from Symo Chem (Netherlands) and used as received. Methyl iodide, anhydrous

N, N-Dimethylformamide (DMF) were of analytical grade and were procured from SRL. Amberlite resin IRA-402 Cl was received from SD fine Chem. Dialysis membranes (Float-a-lyzer) were obtained from Spectra Pore. Membrane filter of pore size 0.45 μm was obtained from Himedia lab. Double distilled water was used for solubility studies. Piroxicam (PXM) was obtained as gift sample from Mesha Pharma (Pvt) Limited Mumbai, India. UV-vis spectra were recorded on TECHCOMP 8500 instrument. Fourier Transform Infrared (FTIR) spectra were recorded on a Bruker Tensor-27 FTIR spectrometer.

Synthesis and Characterization of PQPPI-N(CH_3)$_3$ (G3) Dendrimer

Quaternary ammonium groups were introduced to poly (propylene imine) dendrimer viz., PPI-NH_2 (G3) by performing extensive methylation using the reported procedure [22] with slight modification and thus obtained poly(quaternary ammonium) PPI dendrimer with methyl as a surface group, PQPPI-N(CH_3)$_3$ (G3). This product was characterized by FTIR, ^{1}H and ^{13}C NMR and MALDI-TOF spectral techniques. The observed spectral results synergistically support the quaternization and thus confirmed its structure. The newly prepared dendrimer based drug carrier viz., PQPPI-N(CH_3)$_3$ (G3) was examined in enhancing solubility of PXM as a model drug.

Solubility Study of PXM Using PQPPI-N(CH_3)$_3$ (G3) Dendrimer

The solubility of PXM was quantitatively estimated by adopting the previously described phase solubility analysis method.[23] That is, seven different concentrations of drug carrier, viz., PQPPI-N(CH_3)$_3$ (G3) was prepared in the range from 0.05 mM to 0.35 mM. Then from each concentration, 5 ml was taken in 20 ml vials and to which excess PXM (5 mg) was added. These vials were sonicated for 30 mins and subsequently shaken mechanically for 24 h at 35°C. Parallely controlled experiment was also performed by taking 5 mg of drug in 5 ml of double distilled water (without dendrimer). The vials were allowed to stand for 24h to attain equilibrium. Then these solutions were centrifuged at 5000 rpm for 20 mins and filtered using cellulose acetate membrane filter having pore size 0.45 μm. The respective filtrate was diluted appropriately with double distilled water and analyzed spectrophotometrically using UV spectrophotometer. The amount of PXM solubilised in aqueous medium in the presence of drug carrier, viz., PQPPI-N(CH_3)$_3$ (G3) was estimated by measuring the absorbance of characteristic peak observed at 355 nm (λ_{max}). The quantitative estimation of PXM was done using the calibration curve constructed within the drug concentration range of 3.47 μM to 34.7 μM which is described by the equation, $y = (17545 \pm 0.003)x$ and ($R^2 = 0.999$). The directional coefficient of the equation is equal to the absorption coefficient of PXM, $\varepsilon_{max} = 17545\ M^{-1}cm^{-1}$. Using this calibration curve, the amount of PXM drug solubilised in each dendrimer solution was determined. The blank was performed on the same concentrations of dendrimer delivery solutions so as to cancel any absorbance that may be exhibited by the dendrimer molecules. The results of phase solubility studies are given in Table 1. The solubility experiments were conducted repeatedly three times so as to ensure the concordant values. From the results of phase solubility studies, phase solubility diagrams were constructed by plotting drug solubility versus molar concentration of each dendrimer solutions and shown in Figure 1. Using this plot, apparent stability constant Ks′ for the complexation of PXM with PQPPI-$N(CH_3)_3$ (G3) was calculated according to equation, i.e., Ks′ = slope/S_0 (2^{n+3}-2-slope).[24] Where, S_0 is the intrinsic solubility of drug; slope was obtained from phase solubility curve, n is the dendrimer generation number. Then from the obtained Ks′ value, the thermodynamic parameters such as ΔG, ΔH and ΔS were calculated using the equations mentioned below.[25, 26]

S. No.	Parameters	Equation
1.	ΔG (Free energy)	-2.303 RT log Ks^1
2.	ΔH (Enthalpy)	4.606 RT log Ks^1
3.	ΔS (Entropy)	ΔS = ΔH – ΔG/T

Characterization of PXM-PQPPI-$N(CH_3)_3$ (G3) Dendrimer Complex

The association of PXM to PQPPI-$N(CH_3)_3$ (G3) dendrimer delivery was confirmed by FTIR spectroscopy. The measurements were carried out with an accumulation of 16 scans over the range of 400–4000 cm^{-1}. The FTIR spectra of PXM, PQPPI- $N(CH_3)_3$ (G3) and PXM-PQPPI-$N(CH_3)_3$ (G3) dendrimer complex were shown in Figure 2. Similarly, the said three samples were also analyzed with UV-vis by measuring the absorbance in the range between 250–700 nm. The absorbance of PXM at its characteristic wavelength (355 nm) was related with the amount/solubility of PXM in the dendrimer solutions or distilled water.

RESULTS AND DISCUSSION

Synthesis and Characterization of PQPPI-$N(CH_3)_3$ (G3) Dendrimer

Synthesis of new drug delivery using dendrimer as a precursor is a fascinating area of current research. Earlier studies reveal that the commercial PAMAM (G3) has been modified with surface functionalization of methyl groups and investigated the same for different applications and found convincing results. It is in this background, we have modified the commercial PPI-NH_2 (G3) in to PQPPI-$N(CH_3)_3$ (G3) via procedure followed in the earlier study. That is, the surface functionalization of methyl on PPI-NH_2 (G3) was performed by said procedure and known through characterization of the resultant product with spectral techniques like FTIR, ^{1}H and ^{13}C NMR and MALDI-TOF. The FTIR spectrum of PQPPI-$N(CH_3)_3$ (G3) (Figure 2b) dendrimer has characteristic peaks at 2969 $cm^{-1}$and 2780 cm^{-1} for C-H (str), 1472 $cm^{-1}$for C-H (bend) and 1023 cm^{-1} for C-N (str). ^{1}H NMR spectrum shows a sharp singlet at δ 2.734 with respect to N-CH_3. ^{13}C NMR spectrum shows peaks at δ 46.839 (N-CH_3). MALDI-TOF spectrum shows characteristic peak corresponding to m/z value at 2358.All these results have confirmed the surface functionalization of methyl in PPI-NH_2 (G3) and thus proves that the PPI-$N(CH_3)_3$ (G3) containing poly (quaternary ammonium) ions. This new poly(quaternary ammonium) PPI dendrimer delivery in turn was examined for the solubility of PXM drug in aqueous medium.

Solubility Studies of PXM Using PQPPI-N(CH_3)$_3$ (G3) Dendrimer

Drug solubility is one of the core factors, which influences the movement of a drug from the site of administration into the blood stream. It is widely acknowledged that insufficient drug solubility can lead to poor absorption.[27] The poor solubility of PXM can be addressed with newly developed PQPPI-N(CH_3)$_3$ (G3) dendrimer by employing it as solubility enhancer. Solubility studies were performed by preparing the dendrimer solution with seven different concentrations in the range from 0.05 mM to 0.35 mM using double distilled water. The amount of drug solubilised in dendrimer solution was estimated quantitatively by measuring the characteristic absorbance of the drug noticed at 355 nm in UV-vis spectrum irrespective of the concentration of dendrimer. The absorbance measured in each concentration of dendrimer was compared with standard calibration curve and which in turn determined the extent of solubility of PXM. The observed drug solubility values are given in Table 1. From these results, it is observed that the amount of solubility of PXM has been increased gradually on par with [PQPPI-N(CH_3)$_3$ (G3)]. More particularly, it is worth to highlight here that in comparison with the intrinsic solubility of PXM in water (control), at maximum concentration of PQPPI-N(CH_3)$_3$ (G3) (0.35 mM), the PXM solubility has increased nearly 43 folds in aqueous medium.

Table 1: Solubility of PXM in Aqueous PQPPI-N(CH_3)$_3$ (G3) Dendrimer Solutions with Various Concentrations

S.No.	*[PQPPI-N(CH_3)$_3$ (G3)] In M × 10^{-4}*	*Absorbance (λ_{max})*	*[PXM] In M × 10^{-4}*	*Sol in mg/ml (S_t)*	*Solubility Increment S_t/S_0**
1	0.50	0.1225	7.068	0.2339	10.61
2	1.00	0.1743	10.005	0.333	14.46
3	1.50	0.2117	12.21	0.404	17.57
4	2.00	0.2783	16.04	0.531	23.09
5	2.50	0.3417	19.70	0.652	28.35
6	3.00	0.4246	24.47	0.810	35.23
7	3.50	0.5135	29.61	0.979	42.61

S_0* is the intrinsic solubility of PXM (0.023 mg/ml)

This is because, the PQPPI-N(CH_3)$_3$ (G3) has (i) more number of internal positive binding sites with neutral outer surface. (ii) Availability of the multi charged cations thus in turn attracted the anion of PXM. (iii) The PXM binds electro statically with PQPPI-N(CH_3)$_3$ (G3) in aqueous medium forming the complex viz., $(PXM\text{-}O^-)_n$ $(PQPPI\text{-}N(CH_3)_3)^{n+}$ HCl. (iv) thereby, PXM gets internalized within the PQPPI-N(CH_3)$_3$ (G3) dendrimer branches. Therefore the factors described from (i) to (iv) are collectively responsible for the promotion of solubility of PXM. Further it also necessary to mention here that the increased solubility will increase the bioavailability.

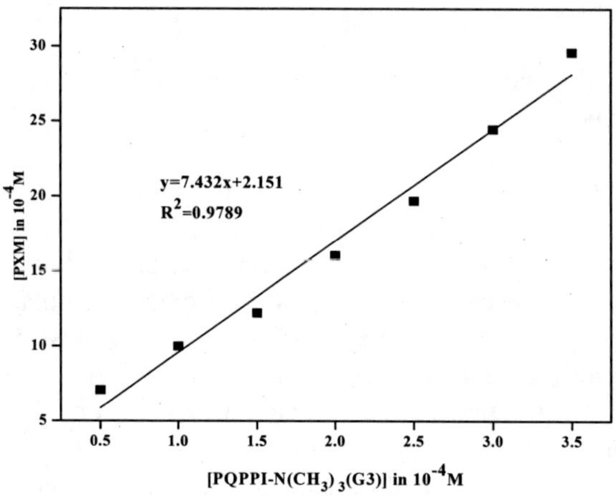

Fig. 1: Solubility of PXM vs. [PQPPI-N(CH_3)$_3$ (G3)]

From the phase solubility diagram, it is understand that the aqueous solubility of PXM has been significantly increased in proportional to the concentration of dendrimer carrier. Using this plot, apparent stability constant, viz., Ks' for the complexation of PXM with PQPPI-N(CH_3)$_3$ (G3) was evaluated. Further, using the Ks' values various thermodynamic parameters were also calculated by substituting in to relevant equations and the obtained values are given in Table 2.

Table 2: Thermodynamic Parameters of PXM-PQPPI-N(CH_3)$_3$ (G3) Dendrimer Complex

Drug	*Stability Constant Ks' in M^{-1}*	*Complexation Efficiency, CE (S_0 × Ks')*	*Free Energy Change ΔG (kJ/mole)*	*Enthalpy Change ΔH(kJ/mole)*	*Entropy Change TΔS (kJ/mole)*
Piroxicam	1715.38	0.408	–16.906	33.812	50.718

The observed thermodynamic parameter values reveal that the negative ΔG for PXM-PQPPI-N(CH_3)$_3$ (G3) dendrimer complex is an indication of its stability. The ΔH and TΔS values indicate that the solubilisation of PXM is entropy driven process. Therefore, the observation of ppositive ΔS value has provided an evidence that the electrostatic interaction and hydrophobic interaction between PXM and PQPPI-N(CH_3)$_3$ (G3) dendrimer is responsible for effective solubility.[25, 26] Increased ΔS value also indicates the more disorder or randomness in the system. Further, the electrostatic or hydrophobic interaction between PXM and dendrimer is possible, provided the ordered cluster sheets of water present around both drug and dendrimer break. This might be the origin for more disorderness or positive ΔS value.[28, 29] A positive value for ΔH indicated endothermic reaction and the absorbed heat have been used to produce randomness in ordered sheets of water.

Characterization of PXM-PQPPI-N(CH_3)$_3$ (G3) Dendrimer Complex

The use of dendrimer as a drug delivery carrier depends on its ability to form a complex with the drug. The ability of complex formation depends on the nature of functional groups of drug and dendrimer and the interaction between them.

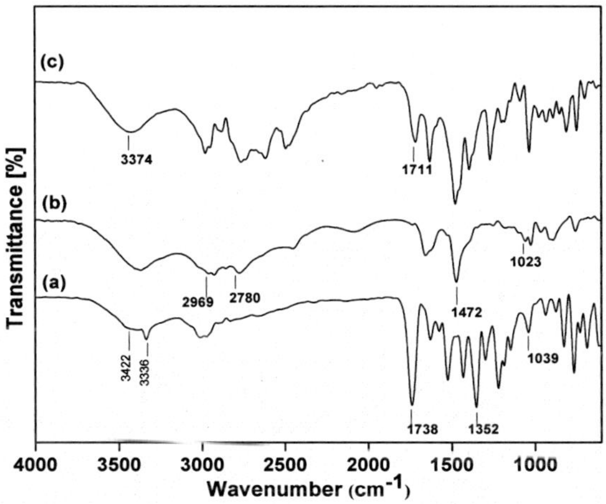

Fig. 2: FTIR Spectra of (a) PXM, (b) PQPPI-N (CH_3)$_3$ (G3), and (c) PXM-PQPPI- N (CH_3)$_3$ (G3)

The FTIR spectrum of pure PXM (Figure 2a) shows the characteristic peaks at 3422 cm^{-1}, 3336 cm^{-1}, 1738 cm^{-1}, 1352 cm^{-1} and 1039 cm^{-1} for O-H (str), N-H (str), C=O (str), O-H (bend), and C=O (str) respectively. In case of PQPPI-N(CH_3)$_3$ (G3) (Figure 2b), dendrimer shows the characteristic peaks at 2969 cm^{-1}and 2780 cm^{-1} for C-H (str), 1472 cm^{-1} for C-H (bend) and 1023 cm^{-1} for C-N (str). Further, in the case of FTIR spectrum of PXM- PQPPI-N(CH_3)$_3$ (G3)) (Figure 2c) dendrimer complex, a broad peak at 3374cm^{-1} was observed, and appearance of this peak confirms the electrostatic association of PXM with PQPPI-N(CH_2)$_3$ (G3). Also, an appreciable shift in C=O (str) from 1738 cm^{-1} to 1711 cm^{-1} and O-H from 3422 cm^{-1} to 3374 cm^{-1} as compared with pure PXM have been noticed. The shifting of these characteristic peaks justifies the formation of complex PXM-PQPPI-N(CH_3)$_3$ (G3). The observed results are in agreement with the previous reports.[30,31] The interactions between PXM with PQPPI N(CH_3)$_3$ (G3) dendrimer molecules were also determined by UV-vis spectroscopy. The maximum UV absorption (λ_{max}) for pure PXM was noticed at 355 nm. But, after the additions of PQPPI-N(CH_3)$_3$ (G3) dendrimers, these (λ_{max})s were shifted to 345–347 nm (blue shifts) (Figure 3).

In general, the occurrence of blue shift noticed in UV-vis (Figure 3) has strongly supported the formation of complex through electrostatic interaction.[32] In addition, hydrophobic encapsulation of PXM by the cavities of PQPPI-N(CH_3)$_3$ (G3) dendrimer is also contributed for blue shift.[32]

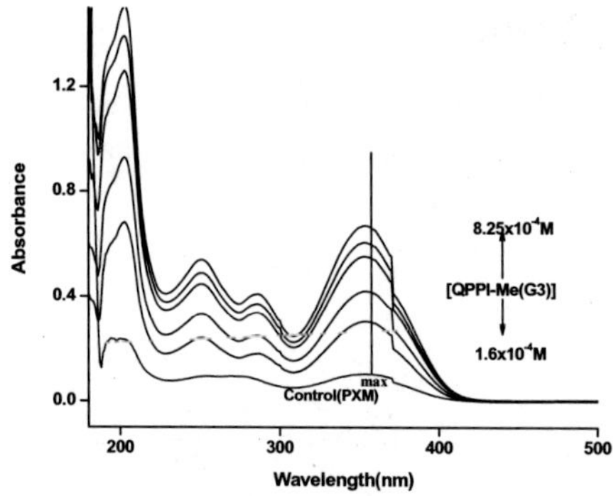

Fig. 3: UV-vis Spectrum of Pure PXM (dotted line) and PXM at Different PQPPI-N $(CH_3)_3$ (G3) Dendrimer Concentrations (bold lines)

CONCLUSIONS

Dendrimer based drug delivery system provides an attractive platform to enhance the drug solubility and act as drug carrier. In the present study, we prepared a poly (quaternary ammonium) PPI dendrimer, viz., PQPPI-$N(CH_3)_3$ (G3) as solubility enhancer of the drug Piroxicam. The structure of the said dendrimer delivery was established through various spectral techniques. This dendrimer was explored to investigate the degree of solubility of PXM drug in aqueous medium. The degree of solubility of PXM in dendrimer solution was quantitatively estimated through measuring the UV absorbance noticed at 355 nm and compared the same with intrinsic solubility of drug (control). The observed solubility results reveal that on increasing the [PQPPI-$N(CH_3)_3$ (G3)], the solubility of PXM is significantly increased. The FTIR and UV-vis spectrum proves that the complexation of PXM with PQPPI-$N(CH_3)_3$ (G3). The apparent stability constant Ks' value was used in different equations and calculated the ΔG, ΔH and ΔS values and it is observed as -16.906kJ/mole, 33.812 kJ/mole and 50.718 kJ/mole respectively. It was concluded that electrostatic and hydrophobic interactions played a major role in solubility enhancement. The results in this study suggest that the PQPPI- $N(CH_3)_3$ (G3) is a promising candidate in the design of polymeric drug delivery systems with high performances and to enhance the bioavailability of the conventional drugs.

ACKNOWLEDGEMENTS

One of the authors E. Murugan wants to record thanks to the UGC for providing financial support.

REFERENCES

[1] Svenson, S. and Tomalia, D.A., Dendrimers in biomedical applications-reflection on the field, *Adv Drug Deliv Rev*., 57, 2106–2129, 2005.

[2] Roseita, E. and Tomalia, D.A., Poly(amidoamine) (PAMAM) dendrimers from biomimicry to drug delivery and biomedical applications, *Drug Discovery Today,* 6, 427–436, 2001.

[3] Anil, K.P., Majoros, I.J. and Baker, J.R., Dendritic polymer macromolecular carriers for drug delivery, *Curr. Opin. Chem. Biol,* 6, 466–471, 2002.

[4] Newkome, G.R., Moorefield, C.N. and Vogtle, F., Dendrimers and Dendrons Concepts, Syntheses, Applications, Wiley-VCH, Weinheim, Germany, 2001.

[5] Fréchet, J.M.J. and Tomalia, D.A. (Eds.), Dendrimers and other Dendritic Polymers, Wiley, West Sussex, UK, 2001.

[6] Patri, A.K., Majoros, I. and Baker, J.R.J., Dendritic polymer macromolecular carriers for drug delivery, *Curr. Opin. Chem. Biol,* 6, 466– 471, 2002.

[7] Mintzer, MA. and Grinstaff, M.W., Biomedical applications of dendrimers a tutorial, *Chem Soc Rev.* 40, 173–190, 2011.

[8] Sideratou, Z., Tsiourvas, D. and Paleos, C.M., Quaternized Poly(propylene imine) Dendrimers as Novel pH- Sensitive Controlled-Release Systems, *Langmuir*, 16, 1766 –1769, 2000.

[9] Gong, A., Liu, C., Chen, Y., Zhang, X., Chen, C. *et al.,* A novel dendritic anion conductor: quaternary ammonium salt of poly(amidoamine), *Macromol. RapidCommun,* 20, 492–496, 1999.

[10] Chen, C.Z., Beck-Tan, N.C., Dhurjati, P., van Dyk, T.K. and LaRossa, R.A. *et al.,* Quaternary ammonium functionalized poly(propylene imine) dendrimers as effective antimicrobials: structure-activity studies, *Biomacromolecules,* 1, 473–480, 2000.

[11] Nishiyama, N., Stapert, H.R., Zhang, G.D., Takasu, D. and Jiang, D.L., Light harvesting Ionic Dendrimer Porphorins as New Photosensitizers for Photo dynamic Therapy, *Bioconjug Chem,* 14, 58–66, 2003.

[12] Dekie, L., Toncheva, V., Dubruel, P., Schacht, E.H. and Barrett, L. *et al.,* Poly-L-glutamic acid derivatives as vectors for gene delivery, *J. Contr. Release*, 65, 187–202, 2000.

[13] Fischer, D., Lib, Y., Ahlemeyerc, B., Krieglsteinc, J. and Kissel, T., In vitro cytotoxicity testing of polycations, *Biomaterials*, 24, 121–1131, 2003.

[14] Kim, S., Choi, J.S., Jang, H.S., Suh, H. and Park, J.S., Hydrophobic Modification of Polyethylene imine for Gene Transfectants, *Bull. Korean Chem. Soc*, 22, 1069–1075, 2001.

[15] Choi, J.H., Choi, J.S., Suh, H. and Park, J.S., Effect of Poly (ethylene glycol) Grafting on Polyethyleneimine as a gene Transfer Vector in Vitro, *Bull Korean Chem, Soc*, 22, 46–52, 2001.

[16] Newkome, G.R., Moorefield, C.N. and Baker, G.R., Alkane cascade polymers processing micellar topology: micellanoic acid derivatives, *Angewandte Chemie*, 30, 1178–1180, 1991.

[17] Twomey, T.M., Banolucci, S.R. and Hobbs, D.C., Analysis of piroxicam in plasma by high-performance liquid chromatography, *J. Chomatogr,* 183, 104–108, 1980.

[18] Chi, S.C. and Jun, H.W., Anti-inflammatory activity of ketoprofen gel on carrageenan-induced paw edema in rats, *J. Pharm. Sci,* 79, 974–977, 1990.

[19] Lin, S.Z., Wouessidijewe, D., Poelman, M.C. and Duchone, D., In vivo evaluation of indomethacin/cyclodextrin complex, *Int. J. Pharm*, 106, 63–67, 1994.

[20] Martinadale, The Complete Drug Reference, 34th ed, Pharmaceutical Press: London, 2005.

[21] Charumanee, S., Okonki, S. and Sirithunyalug, J., Improvement of the dissolution rate of piroxicam by surface dispersion. CMU J. 3, 77–84, 2004.

[22] Lee, J., Lim, Y., Sig Choi, J., Myung-un, C., Chul-hak, Y. *et al.*, Quaternised Polyamidoamine Dendrimers as Novel Gene Delivery System, *Bull. Korean Chem. Soc,* 24, 1637–1640, 2003.

[23] Higuchi, T. and Connors, K.A., Advances in Analytical Chemistry Instrumentation, Vol. 4, Interscience, New York, 117–212, 1965.

[24] Jingjing Hu, Yiyun Cheng, Yanrui Ma, Qinglin Wu and Tongwen Xu, Host-guest chemistry and physicochemical properties of the dendrimer-mycophenolic acid complex, *J. Phys. Chem.* B, 113, 64–74, 2009.

[25] Martin, A., Thermodynamics, Physical Pharmacy, 4th edition, New Delhi 1999.

[26] Tang, B., Ma, L., Wang, H.Y. and Zhang, G.Y., Study on the supramolecular interaction of curcumin and beta-cyclodextrin by spectrophotometry and its analytical application, J Agric Food Chem, 50, 1355–1356.

[27] Zhao,Y., Abraham, M., Lee, J., Luscomble, C. and Beck, G. *et al.,* Rate-limited steps of human oral absorption and QSAR studies, *Pharm. Res.,* 9(10), 1446–1457, 2002.

[28] Komiyama, M. and Bender, M.L., Importance of apolar binding in complex formation of cyclodextrins with adamantanecarboxylate, *J. Am Chem Soc.* 100, 2259–2260, 1978.

[29] Ventura, C.A., Giannone, I., Paolino, D., Pistora, V. and Corsaro, A. *et al.,* Preparation of celecoxib-dimethyl-β-cyclodextrin inclusion complex, *Eur J Med Chem,* 40 624–631 2005.

[30] Hitesh, K., Pooja, D. and Prajapati and S.K., Chauhan, Performance evaluation of PAMAM dendrimer based simvastatin formulations, *Int. j. pharm,* 405, 203–209, 2011.

[31] Asthana, A., Chauhan, A.S., Diwan, P.V. and Jain, N.K., Poly amidoamine Dendritic Nanostructutres for controlled Site-specific Delivery of Acidic Anti-inflammatory Active Ingredient, *AAPS Pharm. Sci. Technol,* 27, 536–542, 2005.

[32] Yiyun, C., Qu, H., Ma, M., Xu, Z. and Xu, P. *et al.,* Polyamidoamine dendrimers as bio-compatible carriers of quinolone antimicrobials: An in vitro study, *Eur J Med Chem*, 42, 1032–1038, 2007.

FACILE GREEN SYNTHESIS OF SILVER NANOPARTICLES AND THEIR ANTIVIRAL ACTIVITY

T. ASHOKKUMAR, S. TAMILSELVAN, R. GEETHA, K. GOVINDARAJU[1], A. MOHAMED SADIQ[2] AND G. SINGARAVELU

Nanoscience Division, Department of Zoology, Thiruvalluvar University, Vellore
[1] Centre for Ocean Research, Sathyabama University, Chennai
[2]Department of Biochemistry, Adhiparasakthi College of Arts and Science, Kalavai, Tamil Nadu
E-mail:gsvelu@gmail.com

ABSTRACT

Synthesis and functionalization of nanomaterials is one of the key issues in nanoscience research. There is a growing need to generate biocompatible nanoparticles to avoid adverse effects on environment. In the present investigation, an eco-friendly technology for the synthesis of silver nanoparticles has been achieved using a marine alga Sargassum ilicifolium. Newly synthesized silver nanoparticles have been characterized using UV spectral analysis, FT-IR, XRD, SEM and TEM. The newly genre silver nanoparticles exhibit in vivo antiviral activity against Bombyx mori Nuclear Polyhedrosis Virus (BmNPV).

INTRODUCTION

Silver nanoparticles is a blooming field of research and has been attracting interest for a wide range of biomedical applications. Nanosilver comprises nano sized structures formed from silver atoms that are metallically bonded together. At the nanoscale, particles exhibit different physical, optical and chemical properties owing to the dominance of quantum mechanics. Studies have shown that the size morphology stability and physical-chemical properties of the metal nanoparticles are strongly influenced by the experimental conditions, the kinetics of the interaction of metal ions with reducing agents and the adsorption processes of stability agents

V. Rajendran, P. Prabu and K.E. Geckeler (eds.)
Nano Biomaterials, pp. 279–288 (2012).

with metal nanoparticles.[1,2] Hence, the design of the synthesis method in which the size morphology, stability and biological properties are controlled has become a major field of interest.[3] It has been reported that nanoparticles synthesis using chemical approach suffer from disadvantages, including high energy requirement, low yield, expensive, difficult purifications and the toxic nature of reducing agents.[4] Since, nanomaterials find their potential application in medical sector it is badly in need of biocompatible nanomaterials, particularly in concern with silver nanoparticles. Thus it is imperative to develop technologies for the synthesis of biocompatible nanomaterials which find their application in life sciences. In recent years great efforts are made to fabricate biocompatible nanomaterials of gold and silver.[5–7]

Nanosilver biosynthesis has been achieved and many publications have demonstrated the use of various species of bacteria and fungi.[8,9] These methods share the common methodology whereby silver nitrate solution is added to a microbial supernatant. Reducing agents, such as hydroquinones present in the microbial supernatant, reduce Ag^+ to nanosilver under constant conditions (e.g. Temperature). The obvious drawback of this method is the need to purify the sample and extract the nanosilver because pathogenic bacteria might contaminate nanosilver used in medical applications. Further bio-entities in nanoparticles synthesis is that the genetic manipulation of eukaryotic organisms as a means of over expressing specific enzymes (eg, the ones identified in synthesis of metallic nanoparticles) is relatively much more difficult than that in prokaryotes.[10] Foregoing facts discloses the drawbacks of microbes in metal nanoparticles synthesis. Using plants for nanoparticles synthesis can be advantageous over the other biological process because it eliminates the elaborate process of maintaining cell cultures and can also be suitably scale up for large-scale nanoparticles synthesis.[11] Recently, integration of green chemistry principles to nanoscience have become an important branch of nanotechnology. Phytochemicals may provide a better alternative to nanoparticles generation. It has been reported that biological synthesis nanoparticles using *Aloe vera*,[12] lemongrass,[13] neem,[14] tamarind [15] and seaweed.[6] Indeed, the fascinating potential application of metal nanoparticles are determined by their chemical compositions, size, shape and controlled dispersities. Based on this the idea of functionalizing silver nanoparticles as antiviral nanomaterial against BmNPV by synthesized with pharmacologically important seaweed has been conceived. In this respect, nature has provided many exciting possibilities for utilizing biological systems for this purpose. The results reported herein encompass the biological synthesis of silver nanoparticles using seaweed and their antiviral activity against BmNPV.

MATERIALS AND METHOD

Extraction and Synthesis of Silver Nanoparticles

Silver nitrate was used for synthesis of Ag nanoparticles. Seaweed *(Sargassum ilicifolium)* was collected from Mandapam camp East Coast of Tamil Nadu, India. Collected seaweed was washed with deionized water and then shade dried. Dried seaweed was ground to powder. 100 gm of seaweed powder was extracted with solvents system (Methanol, chloroform, ethylacetate, hexane and petroleum ether). Each extract was tested for bioreduction property, among them methanol, extract has potent candidate for strong bioreduction property rapidly. Different concentration of methanol extract of *S. ilicifolium* and 10^{-3} M aqueous $AgNO_3$ solutions was analyzed (0.5:4.5, 1.0:4.0, 1.5:3.5, 2.0:3.0 and 2.5:2.5).

Characterization of Silver Nanoparticles

Aliquots of the reaction solution absorptions were measured using a UV-2300 TECHCOMP spectrophotometer operated at a resolution of 1 nm. X ray diffraction (XRD) of bioreduced Ag nanoparticles was carriedout by Rich seifert p 300 instrument operated at a voltage of 40 kV and a current of 30 mA with Cu Kα radiation. The FTIR spectroscopy measurements were carriedout to identify the biomolecules for synthesis of silver nanoparticles. Dry powder of the biomass and silver nanoparticles solutions were centrifuged at 5000 rpm for 15 min and resulting suspensions was redisperesed in sterile distilled water. The purified pellets were dried and analyzed by Perkin-Elmer FT-IR spectrophotometer in the diffuse reflectance mode at a resolution of 4 cm^{-1} in KBr pellets. TEM samples of the silver nanoparticles were recorded by placing a drop of the suspension on carbon-coated copper grids and allowing water to evaporate. The observations of TEM were performed on JEOL 3010 operated at an accelerating voltage of 120 kV.

In vivo Antiviral Studies

Isolation and Purification of BmNPV Occlusion Bodies

Hemolymph was collected from the Nuclear Polyhedrosis (NPV) infected worms exhibiting typical symptoms, shortly before death by just cutting the front pair prolegs. The turbid milky hemolymph was carefully collected and purified according to reported protocol.[16] The polyhedral concentration was determined by counting polyhedral inclusion bodies (PIB) on Neubauers chamber slide under the microscope. The stock solution of virus inoculum was diluted to desired concentrations by adding sterile distilled water. The concentration was expressed as polyhedra/mL.[17]

Silkworm Rearing

Silkworm hybrid of PxNB4 D2 of uniform size and age were used as a test material. The egg cards were obtained from the Sericulture Department, Govt. of Tamilnadu, Vaniyambadi, Vellore District. The host plant used as food for the silkworm larvae was the V1 variety of mulberry *Morus indica*. Silkworm larvae were reared according to standard rearing conditions at 26 ± 1 °C with 75±% RH.[18]

Bioassay

Approximately, 2.22×10^6 polyhedra/mL was mixed with different concentration (1 μg, 3 μg, 5 μg and 10 μg) of *S. ilicifolium* synthesized silver nanoparticles and incubated at different time intervals (1, 5 and 10 min) at room temperature with occasional shaking. 700 silkworm larvae were divided into fourteen groups:

Group 1	Normal control
Group 2	BmNPV inoculated (2.22 × 10^6 polyhedra/mL)
Group 3	Silkworm larvae fed with BmNPV polyhedral (2.22 × 10^6 polyhedra/mL)-silver nanoparticles (1 μg) complex incubated at 1 min
Group 4	Silkworm larvae fed with BmNPV polyhedral (2.22 × 10^6 polyhedra/mL)-silver nanoparticles (1 μg) complex incubated at 5 min
Group 5	Silkworm larvae fed with BmNPV polyhedral (2.22 × 10^6 polyhedra/mL)-silver nanoparticles (1 μg) complex incubated at 10 min
Group 6	Silkworm larvae fed with BmNPV polyhedral (2.22 × 10^6 polyhedra/mL)-silver nanoparticles (3 μg) complex incubated at 1 min
Group 7	Silkworm larvae fed with BmNPV polyhedral (2.22 × 10^6 polyhedra/mL)-silver nanoparticles (3 μg) complex incubated at 5 min
Group 8	Silkworm larvae fed with BmNPV polyhedral (2.22 × 10^6 polyhedra/mL)-silver nanoparticles (3 μg) complex incubated at 10 min
Group 9	Silkworm larvae fed with BmNPV polyhedral (2.22 × 10^6 polyhedra/mL)-silver nanoparticles (5 μg) complex incubated at 1 min
Group 10	Silkworm larvae fed with BmNPV polyhedral (2.22 × 10^6 polyhedra/mL)-silver nanoparticles (5 μg) complex incubated at 5 min
Group 11	Silkworm larvae fed with BmNPV polyhedral (2.22 × 10^6 polyhedra/mL)-silver nanoparticles (5 μg) complex incubated at 10 min
Group 12	Silkworm larvae fed with BmNPV polyhedral (2.22 × 10^6 polyhedra/mL)-silver nanoparticles (10 μg) complex incubated at 1 min
Group 13	Silkworm larvae fed with BmNPV polyhedral (2.22 × 10^6 polyhedra/mL)-silver nanoparticles (10 μg) complex incubated at 5 min
Group 14	Silkworm larvae fed with BmNPV polyhedral (2.22 × 10^6 polyhedra/mL)-silver nanoparticles (10 μg) complex incubated at 10 min.

Polyhedra-silver nanoparticles complex was smeared over host plant leaves mulberry and fed to the first day of 5th instar silkworm *Bombyx mori* larvae one time only. Mortalities were recorded for each day, and the experiment was carried out up to end of the fifth instar larvae.

RESULTS AND DISCUSSION

The reduction of silver nitrate into silver nanoparticles during exposure to the aqueous extract of marine algae *S. ilicifolium* is followed by a gradual increase in colour development from clear to yellowish brown as a result of the surface plasmon resonance (SPR) phenomenon. Figure 1, shows the UV-vis spectra of the newly formed silver nanoparticles and the characteristic SPR band occurs at ca 422 nm and at 1 h that attained the maximum intensity. Metal nanoparticles have free electrons, which provide surface plasmon resonance absorption band due to the combined vibration of electrons of metal nanoparticles in resonance with light wave.[19] FTIR analysis of the newly formed silver nanoparticles were accomplished to findout the possible biomolecules responsible for the reduction of the Ag+ ions and capping the biologically synthesized silver nanoparticles. Dry powder of, *S. ilicifolium,* where the strong bands were observed at 3432 cm^{-1} (H–bonded hydroxyl groups), 2925,2854 cm^{-1} (–OH stretching), 1630 cm^{-1} (C O stretching band of the carboxylic acid group), 1399, 1127 cm^{-1} (C–O stretching of alcoholic groups) and 600 cm^{-1} (C–OH vibration of the carbonyl group) (Figure curve a).The *S. ilicifolium* synthesized silver nanoparticles of showed intense peaks at 3412, 2923, 2853, 1613, 1402, 1384, 1165, 1114 and 604 cm^{-1}(Figure curve b). The peaks at 1127, 1399 (C–O stretching of alcoholic groups) are suppressed and the peaks at 1630 cm^{-1}, 3432 $cm^{-1}$are shifted to lower frequency [1613 cm^{-1}(C O stretching band of the carboxylic acid group), 3412 cm^{-1} (H–bonded hydroxyl groups)] which indicates carboxyl group reduce silver ions into silver nanoparticles and hydroxyl group to stabilize nanoparticles surface (Figure 2).

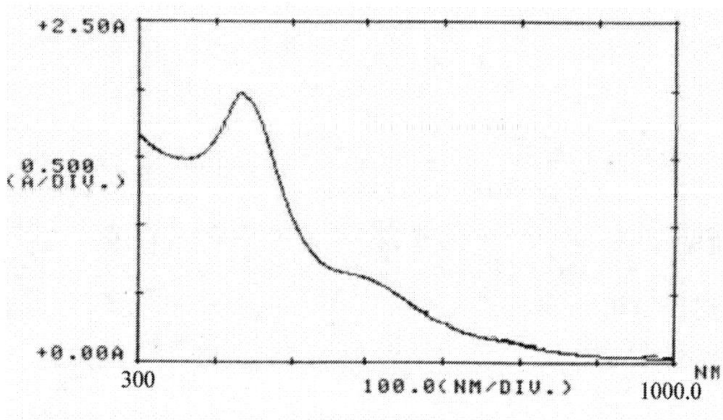

Fig. 1: UV-vis Spectra of *S. ilicifolium* Synthesized

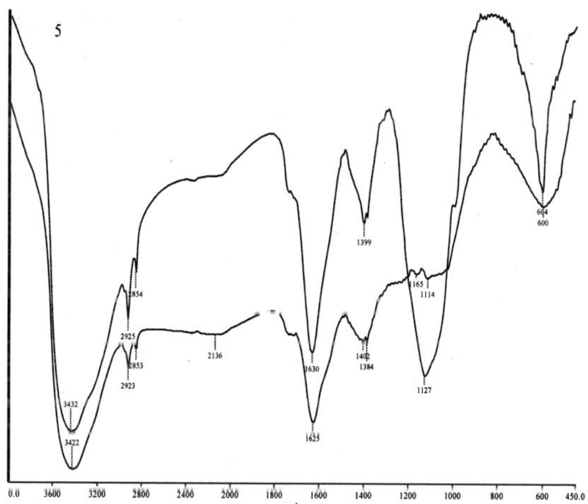

Fig. 2: FTIR Absorption Spectrum Obtained From (a) Aqueous Extract of *S. ilicifolium.* (b). Silver Nanoparticles Synthesized Using *S. ilicifolium*

Silver Nanoparticles

The crystalline nature of silver nanoparticles was confirmed from X-ray diffraction analysis. Figure 3. shows the XRD pattern of the dried nanoparticles obtained from colloid solution. Three peaks were observed at 38.2°, 44.3°, and 64.7° in the 2θ range 10°–80° which can be indexed to the (111), (200) and (220) which is confirmed fcc structure of silver nanoparticles. The results reveals that the Ag^+ reduced to Ag^0 by *S. ilicifolium* are face centered cubic crystalline nature. TEM images obtained by the reaction of *S. ilicifolium* aqueous extract and 1 mM $AgNO_3$ solution. The TEM microscopic images reveal the formation of monodisperse spherical silver nanoparticles with dispersion. The obtained nanoparticles are quite uniform in size and in the range of 2 to 30 nm (Figure 4). Previously we have reported that the use of seaweed *S. wightii* extract in the extracellular synthesis of silver nanoparticles. FTIR spectra reveals that alcoholic group and carboxylate ions reduce and stabilized silver nanoparticles formed.[5] In the present attempt, carboxyl group and hydroxyl ions of seaweed *S. ilicifolium* found to reduce and stabilized the newly formed silver nanoparticles. Marine plants represent a potential source of pharmaceutically active agent.[20,21] There has been a great deal of research on marine algal antiviral compounds [22–24] accordingly it is proposed to investigate the antiviral activity of seaweed *S. ilicifolium* synthesized silver nanoparticles against a double stranded DNA virus of silkworm *B. mori*. Silkworm larvae often suffer from viral infections causing heavy loses to the economy of the silk industry. The silkworm *Bombyx mori* is exploited both as a powerful biological model system and also as a tool to convert leaf protein into silk. Insects exhibit both humoral and cellular immune responses in addition to metabolic alterations that are effective against various pathogens like bacteria, fungi, protozoa, etc., but no insect immune response is effective against physiological adjustments during viral infections. Recent success on transgenesis of the silkworm has opened new prospects for this insect species.[25] The BmNPV is the most harmful virus in the sericulture industry, often causing severe economic losses.[26] Several researchers have try to control BmNPV infection using various methods and technologies. On the basis in the present study, nanotechnology intervention against BmNPV has been addressed.

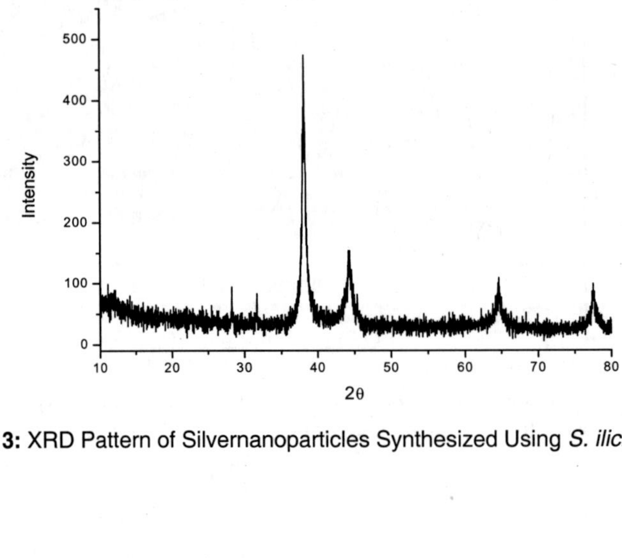

Fig. 3: XRD Pattern of Silvernanoparticles Synthesized Using *S. ilicifolium*

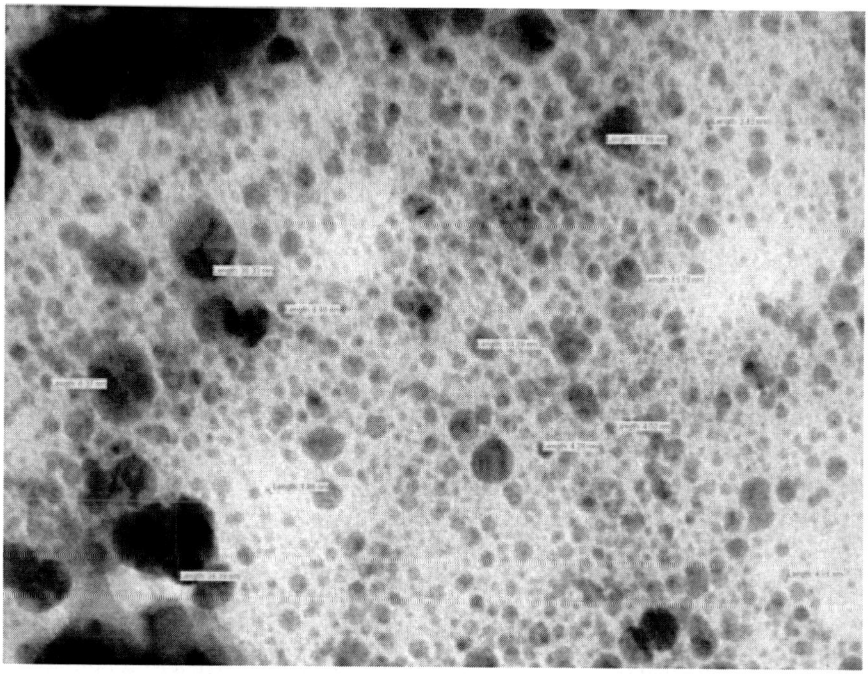

Fig. 4: TEM Image of Silver Nanoparticles Synthesized *S. ilicifolium*

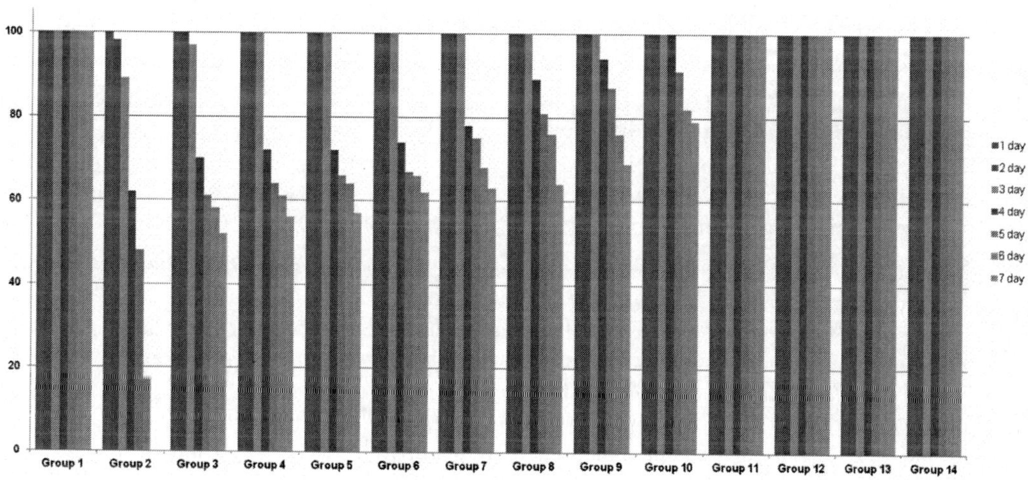

Fig. 5. *In vivo* Anti Viral Studies Using *S. ilicifolium* Synthesized Silver Nanoparticles

The physicochemical properties of nanoparticles are strongly dependent upon their interactions with capping agent molecules. Indeed, the surface chemistry of the nanoparticles can modify their interactions with external systems. For this reason we tested silver nanoparticles with biomolecules of seaweeds. The present results suggest that three possible mechanisms may explain the antiviral activity of these newly synthesized silver nanoparticles. (i) viral inactivation might occur by the reaction between the silvernanoparticles and the envelope proteins of the virus, which prevent the entry of the virus into the host. (ii) influence of the silver nanoparticles which prevents the multiplication of the virus in the host cell and (iii) silver nanoparticles might be the immunostimulant which enhances the innate immunity of silkworm against BmNPV. Further works on isolation and identification of bioactive compound of the seaweed *S. ilicifolium* which is responsible for reduction and stabilization of silver nanoparticles is in progress. The interactions of silver nanoparticles with biosystems are just beginning to be understood and potential biological applications are being discovered at an increasing rate. However, in order to realize the future promise of nanoscience, it is imperative that the toxicity and long-term health effects of exposure to nanomaterials be fully explored. The present work will enlighten the interaction of metal nanoparticles with various viruses.

ACKNOWLEDGEMENT

G.S. thank the Indian Council of Medical Research (ICMR), New Delhi for financial assistance.

REFERENCES

[1] Knoll, B. and Keilmann, F., Near-field probing of vibrational absorption for chemical microscopy, *Nature,* 399, 134–137, 1999.

[2] Sengupta, S., Eavarone, D., Capila, I., Zhao, G.L., Watson, N. and Kiziltepe, T., Temporal targeting of tumour cells and neovasculature with a nanoscale delivery system, *Nature,*436, 568–572, 2005.

[3] Wiley, B. J., Sun, Y. and Xia, Y., Synthesis of silver nanostructures with controlled shapes and properties, *Acc. Chem. Res,* 40, 1067–1076, 2007.

[4] Sadhasivam, S., Shanmugam, P. and Yun, K., Biosynthesis of silvernanoparticles by streptomyces hygroscopicus and antimicrobial activity against medically important pathogenic microorganisims., *Colloids Surf B: Biointerf.,* 81, 358–362, 2010.

[5] Shankar S, Ahmad A and Sastry, M., Geranium leaf assisted biosynthesis of silver nanoparticles. *Biotechnol. Prog.*, 19, 6:1627–31, 2003.

[6] Singaravelu G, Arockiamary J, Ganesh K, Govindaraju K., A novel extracellular synthesis of monodisperse gold nanoparticles using marine alga, *Sargassum wightii* Greville. *Colloids Surf. B Biointerf.,* 57:97–101, 2007.

[7] Khaleel Basha, S., Govindaraju, K., Manikandan, R., Ahn, J.S., Bae, E.Y and G.Singaravelu., Phytochemical mediated gold nanoparticles and their PTP 1B inhibitory activity. *Colloids and Surfaces B: Biointerfaces.*75:405–409, 2010.

[8] Shahverdi, A.R, Fakhimi, A, Shahverdi, H.R and Minaian S., Synthesis and effect of silver nanoparticles on the antibacterial activity of different antibiotics against *Staphylococcus aureus* and *Escherichia coli. Nanomedicine*, 3:168–71, 2007.

[9] Shaligram, S.N., Bule, M., Bhambure, R., Singhal, S.R., Singh, K.S., Szakacs, G., Pandey, A., Biosynthesis of silver nanoparticles using aqueous extract from the compactin producing fungal, *Process Biochemistry,* 44, 939–943, 2009.

[10] Kaushik, N., Thakkar, M.S., Snehit, S., Mhatre, M.S., Rasesh, Y. and Parikh, M.S., Biological synthesis of metallic nanoparticles, *Nanomadicine,* 2009.

[11] Jae, Y.S. and Beom, S.K., Bioprocess and Biosystems Engineering, 37, 79, 2009.

[12] Chandran, S.P, Chaudhary, M., Pasricha, R., Ahmad, A. and Sastry, M., Synthesis of gold nanotriangles and silver nanoparticles using *Aloe vera* plant extract. *Biotechnol Prog,* 22:577–83, 2006

[13] Shankar, S., Rai, A., Ankamwar, B., Singh, A., Ahmad, A. and Sastry, M., Biological synthesis of triangular gold nanoprisms. *Nat Mater*; 3: 482–8, 2004.

[14] Shankar, S.S., Rai, A., Ahmad, A. and Sastry, M., Rapid synthesis of Au, Ag and bimetallic Au core–Ag shell nanoparticles using Neem *(Azadirachta indica)* leaf broth. *J. Colloid Interface Sci,* 275:496–502, 2004.

[15] Ankamwar, B., Chaudhary, M. and Murali, S., Gold nanotriangles biologically synthesized using tamarind leaf extract and potential application in vapor sensing. *Synth React Inorg Met-Org Nanometal Chem,* 35:19–26, 2005.

[16] Govindan, R., Narayanaswamy, T.K and Devaiah, M.C., Principles of silkworm pathology. *Seri Scientific Publication*, Bangalore, pp 64–65,1998.

[17] Ratnasen Ashwath, S.K. and R.K. Datta., Inheritance of resistance to nuclear polyhedrosis virus in silkworm, *Bombyx mori. Int J Indust Entomol.,* 3:187–190, 2001.

[18] Krishnaswami, S., New technology of silk rearing. Bulletin no. 2., C.S.R. & T.I, Mysore, India, 1978.

[19] Haes, A. J., Zou, S. L., Schatz, G. C, Van Duyne, R. P., Nanoscale Optical Biosensor: Short Range Distance Dependence of the Localized Surface Plasmon Resonance of Noble Metal Nanoparticles., *J. Phys. Chem. B,* 108, 22, 6961–6968, 2004.

[20] Baslow, M.H., Marine pharmacology, The Williams & wilkins Co.ltd., Baltimore,257 pp. 1969.

[21] Hoppe, H.A., Levring, T. and Tanaka, Y., Marine alage in pharmaceutical science., Walter de Gryter, Berlin, New York, 1979.

[22] Nakashima, H., Kido, Y., Kobuyashi, N., Motoki, Y., Neushul, M., Yamamoto, N., Antiretroviral activity in a marine red alga: reverse trancriptase inhibition by an aqueous extract of schizymenia pacifica., *J. Cancer. Res. Clin. Oncol.*, 113, 413–416, 1987.

[23] Venkateswaran, P.S., Millman, I. and Blumberg, B.S., Interaction of fucoidan from pelvetia fastigiata with surfaces antigens of hepatits B and Woodchuck hepatitis virus, *Planta Medica,*55, 265–270, 1989.

[24] Premnathan, M., Nakashima, H., Kathiresan, K., Rajendran, N., Yamamoto, N., *In vitro* anti human immunodeficiency virus activity of mangrove plants., *Indian Journal of medical Research*, 103, 278–281, 1996.

[25] Tamura, T., Thibert, C., Royer, C., Kanda, T., Abraham, E., Kamba, M., Komoto, N., Thomas, J.L., Mauchamp, B., Chavancy, G., Germline transformation of the silkworm using a piggyback transposon-derived vector. *Nat. Biotechnol.*, 18, 1, 81–84, 2000.

[26] Ponnuvel, K.M., Nakazawa, H., Furukawa, S., Asaoka, A., Ishibashi, J., Tanaka, H., Yamakawa, M., A Lipase isolated from the silkworm shows antiviral activity against NPV., *J. Virol.*, 77, 19, 10725–10729, 2003.

GREEN SYNTHESIS OF GOLD NANOPARTICLES AND THEIR ANTICANCER ACTIVITY

R. Geetha, T. Ashokkumar, S. Tamilselvan, K. Govindaraju[1], A. Mohamed Sadiq[2] and G. Singaravelu

Nanoscience Division, Department of Zoology, Thiruvalluvar University, Vellore
[1]Centre for Ocean Research, Sathyabama University, Chennai

[2]Department of Biochemistry, Adhiparasakthi College of Arts and Science, Kalavai, Tamil Nadu

E-mail: gsvelu@gmail.com

ABSTRACT

As the nano revolution unfolds it is imperative to integrate nanoscience and medicine. The secret gleaned from nature have led to the generation of biogenic technologies for the fabrication of advanced nanomaterials. Present investigation discloses that pharmacologically important flower of the tree Couroupita guianensis biosynthesizing capability of gold nanoparticles. Rapid, cost-effective, one-step process of synthesis has been achieved. Newly genre gold nanoparticles was characterized by involving UV-vis spectroscopy, FT-IR, XRD, SEM and TEM analysis. Interestingly, as a result of extensive screening on the application of newly synthesized gold nanoparticles their anticancer potential has been discovered using MTT assay, DNA fragmentation, apoptosis by DAPI staining and Comet assay for DNA damage.

INTRODUCTION

Metal nanoparticles are of great importance due to their high surface area and a high fraction of surface atoms. The scientific and technological significance of metal nanoparticles has made them the subject of intensive research, given their special chemical and physical properties. In particular, gold nanoparticles are employed in many fields: biosensing,[1] catalysis,[2] optics, mechanics, electronics and biomedicine[3] among others. Presently, biological nanoscience has drawn increasing attention due to its avant-garde nature and the efficacy of produced

V. Rajendran, P. Prabu and K.E. Geckeler (eds.)
Nano Biomaterials, pp. 289–298 (2012).

nanoparticles in industrial, biomedical and electronic applications such as catalyst,[4] cancer detection [5] and bioimaging.[6] Among the loads of metal nanoparticles, gold nanoparticles have an especially wide application in DNA recognition, heredity medicine and nano-catalysis. It has been pointed out that, the divergent capabilities of these nanoparticles are controlled by their size.[7–9] Numerous methodologies are developed to synthesize noble metal nanoparticles of particular shape and size depending on specific requirements. Biomolecules as reductants are found to have significant advantage over chemical reductants due to their non biocompatible nature.[10] In the present cram, a simple biosynthetic method using flower extract of *C. guianensis* is addressed herein for synthesizing gold nanoparticles. The newly synthesized gold nanoparticles were characterized by employing standard technologies. The plant *C. guianensis* commonly known as cannon ball tree and it has innumerable medicinal applications including antibiotic, antiseptic, anti-inflammatory activity.[11] Indeed, it is expected that gold nanoparticles synthesized using the plant's flower *C. guianensis* might have medicinal properties. Accordingly, efforts are made to identify the newly generated gold nanoparticles medicinal value. As a result their antileukemic cancer activity has been conceived. Cancer is the leading cause of death worldwide after cardiovascular diseases.[12] Leukemia is a major type of cancer affecting a significant segment of the population and especially children. In fact, leukemia is the most frequent childhood cancer, with 26% of all cases, and 30% mortality.[13] Clinical trials in identifying new agents and treatment modalities has been studied and achieved. However, current treatments have many limitations related to their side effects and the development of acquired drug resistance.[14] The new therapeutic agents thus needed should be more active and produce fewer side effects and they also should act through a mechanism different from that of cytotoxic agents already used.

MATERIALS AND METHODS

Preparation of Flower Extract

Fresh flowers of *C. guianensis* were collected, washed thoroughly with double distilled water, air dried and powdered. About 1g of powdered flower was transferred into a 100 mL beaker containing 50 mL double distilled water and mixed well. The extract obtained was filtered through whatmann No.1 filter paper and the filterate was collected in a separate flask.

Synthesis of Gold Nanoparticles Using Plant Extract

1 mM solution of 100 mL chloroauric acid at concentration of 10^{-3} M prepared by dissolving double distilled water. Different reaction concentration of *C. guianensis* extract and $HAuCl_4$ solution (1:9, 2:8, 3:7, 4:6, and 5:5) was subjected respectively. The reduction of gold ions to gold nanoparticles was completed within 5mins. The pale yellow colour solution turns to ruby red colour and thus indicating the formation of gold nanoparticles. The reduction of metal ions was continuously monitored by visual inspection as well as by measuring with UV-visible spectrometer in the absorbance range 300–800 nm.

Characterization of Gold Nanoparticles

UV–visible spectrum was recorded as a function of the reaction time on a Techomp (UV 2300) spectrophotometer containing double beam in identical compartments each for reference and test solution fitted with 1 cm path length quartz cuvettes. For FTIR measurements, the newly synthesized gold nanoparticles were centrifuged at 8000 rpm for 10 mins and the resulting suspension was redispersed in distilled water. The purified pellets were dried and smirks with KBr pellets and analyzed by Thermo Nicolet Quator instrument in the diffuse reflectance mode operating at a resolution of 4 cm^{-1}. XRD measurements of the *C. guianensis* synthesized gold nanoparticles solution drop coated on to glass substrates was done on a Siefert X-ray diffractometer, operated at a voltage of 40Kv and tube current of 30 mA with Cu-Kα, radiation. Surface morphology of newly formed gold nanoparticles was determined using Scanning Electron Microscope (FEI Quanta 200 SEM). For TEM studies newly synthesized gold nanoparticles were prepared by placing a drop of nanoparticle solution on carbon coated copper grids and allowing water to evaporate. TEM measurements were performed on JEOL 3010 UHR TEM equipped with a Gatan Imaging Filter.

Assessment of Cell Morphology

HL-60 cells were cultured in 24-well plates and incubated at 37°C for 24 h before each well were individually treated with 0, 60, 90, 120, 150 and 180 µg/ml gold nanoparticles for 48 h. Determination of cell morphology was made using a phase contrast microscope (Lu *et al.,* 2010). [15]

Comet Assay for DNA Damage

Cells in 96 well plates were treated with or without 0, 60, 90, 120, 150 and 180 µg/ml gold nanoparticles for 48 h. Cells were then harvested and DNA damage was determined with the Comet assay as previously described (Chen *et al.,* 2009).[16]

DAPI Staining for Apoptosis

Cells in 12-well plates were treated with or without 0, 80, 120, 180 and 200 µg/ml gold nanoparticles for 48 h. Cells were then stained by using 4, 6-diamidino-2-phenylindole dihydrochloride (DAPI) as previously described (Chen *et al.,* 2006).[17]

DNA Gel Electrophoresis for DNA Fragmentation

Cells in 6 well plates were incubated with 0 and 150 µg/ml gold nanoparticles 48 h. Cells were harvested by centrifugation and DNA was isolated before DNA fragmentation was determined by DNA gel electrophoresis as previously described (Lu *et al.,* 2010).[15]

Effect of Gold Nanoparticles on Cytotoxicity

HL-60 cells were obtained from the NCCS, Pune, India and grown in RPMI1640 medium and supplemented with a 2mg mL^{-1} sodium bicarbonate, 4.5 mL^{-1} glucose, 100 µg ml^{-1} streptomycin sulphate, 40 µg ml^{-1} gentamycin, 100 UmL^{-1} penicillin as well as 10% heat inactivated fetal calf serum. An environment of humidified air containing 5% CO_2 was maintained at 37°C. Peripheral mononuclear cells (PBMC) were obtained from healthy donors and cultured as described.[17] Effects of gold nanoparticles on the proliferation of cells was determined by MTT [3-(4, 5-dimethyl thiazoyl-2-yl)-(2, 5-diphenyltetrazolium bromide] assay. Briefly, the cells were suspended at $3x10^5$ cell mL^{-1}. The cells were placed in 96 well microtitre plates (200µl $well^{-1}$) and incubated at 37°C in a CO_2 incubator in the presence of the gold nanoparticles. After 5 days, cell viability was measured by MTT assay, from which 50% cytotoxic concentration (CC_{50}) was calculated.[19]

RESULT AND DISCUSSION

Design and development of nanomaterials with unusual physico-chemical and optoelectronic features is a corner stone in nanoscience. Present report discloses ethnopharmacologically important

C. guianensis flower's extract bioreduction property of gold ions into gold nanoparticles. Synthesis of gold nanoparticles has been achieved in 5mins of reaction without any experimental protocols like stirring, heating, change in pH. Reduction and capping of newly formed gold nanoparticles was achieved with the flower extract of *C. guianensis*. As a result of extensive screening with different concentration of auric chloride and flower extract, synthesis has been succeeded with the concentration of 2:8 (2 ml of flower extract and 8 ml of auric chloride). Figure, 1, shows the UV-vis spectra of the newly formed gold nanoparticles and the characteristic SPR band occurs at ca. 534 and at 5 mins that attained the maximum intensity. Fourier Transform Infrared (FTIR) was made in order to identify the possible biomolecules responsible for the reduction of gold ions and capping of the bioreduced gold nanoparticles (Figure 2 a&b). FTIR spectrum of the aqueous extract of *C. guianensis* (Curve a) show peaks at 3414 (O–H stretch, H–bonded of alcohols/phenols), 2133 (–C (triple bond) C– stretch of alkynes), 1648 (–C=C– stretch of alkynes), 1383 (C–N stretching vibration of aliphatic amines), 1288 (C-O stretch of alcohols) and 1073 cm^{-1} (C–N stretching vibration of aliphatic amines or to alcohols/ phenols). The intense broad absorbance at 3414 cm^{-1} is the characteristic of the hydroxyl functional group in alcohols and phenolic compounds. The FTIR spectrum of the gold nanoparticles (curve b) showed peaks at 3615 and 3547 (O–H stretch, H–bonded of alcohols/phenols), 2924 and 2855 (C–H stretch), 2402 and 2129 (–C (triple bond) C– stretch of alkynes), 1744 (C=O stretch of carboxylic acids), 1378 (C–N stretching vibration of aliphatic amines) and 1071 cm^{-1} (C–N stretching vibration of aliphatic amines or to alcohols/phenols). Comparing both FT-IR spectra it can be identified that the changes in the –COOH group for –OH, i.e., hydroxyl group the peak appeared at 3414 cm^{-1} in raw material, but after encapsulation of nanoparticles, the peak is narrower and shifted to 3615

cm^{-1} and also for –C– of carboxylic group the peak intensity reduced after encapsulation of nanoparticles. The peaks at 1648, 1383, 1288 and 1073 cm^{-1} are reduced and the new peak appeared at 1744 cm^{-1} which indicates the alcoholic group is converted into aldehyde to reduce Au^{3+} to Au^{0}.

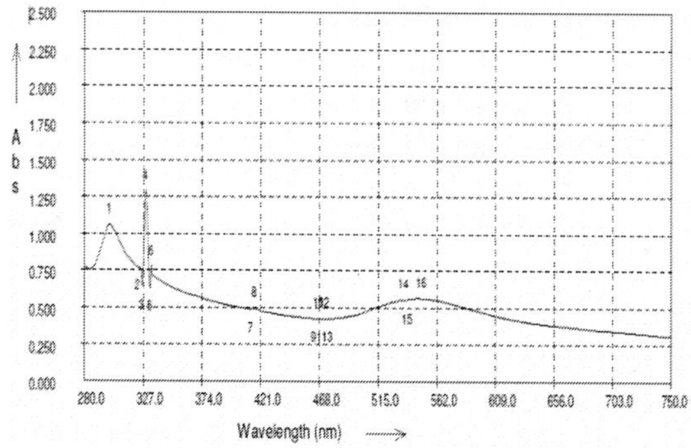

Fig. 1: UV-Visible Absorption Spectrum of Biosynthesized Gold Nanoparticles by *C. guinensis* Flower Extract

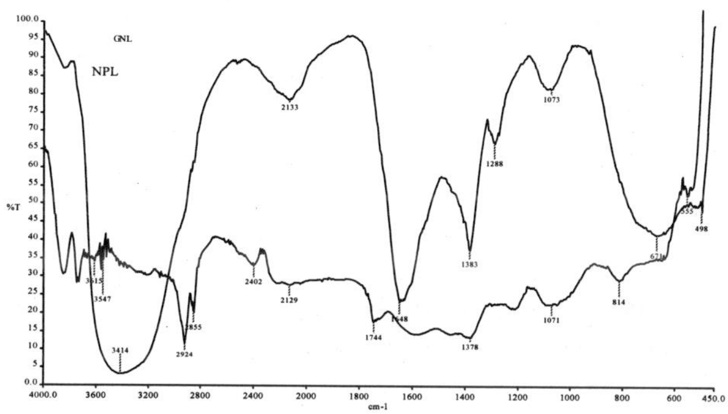

Fig. 2: FT-IR Spectra of (A) *Couroupita guinensis* Flower Extract, (B) Gold Nanoparticles Synthesized by *Couroupita guinensis* flower extract

A structural analysis of gold nanoparticles prepared from the sample was performed by XRD. Taking into account the angular positions of the bragg peaks (Figure 3.), a face centered cubic (fcc) structures was assigned to the gold nanoparticles. The XRD pattern thus clearly illustrates that the gold nanoparticles synthesized by *C. guianensis* are crystalline in nature. Figure, 4, shows the SEM images of the newly synthesized gold nanoparticles. It was observed that the surface morphology of gold nanoparticles synthesized from flower extract of *C. guianensis* solution clearly indicates that they are in different shapes and well dispersed. The TEM images confirm the formation of gold nanoparticles and it has been observed that the newly formed nanoparticles are polydispersed consisting of spherical, triangular, tetragonal and pentagonal with irregular contours. The size distribution of gold nanoparticles diameter is located between 7 to 48 nm (Figure 5). Generation of metal nanoparticles mainly rely on the use of lethal reducing agents. For the development of biocompatible nanoparticles various biomimetic approaches are being explored. Indeed, a discovery on the application of nanoparticles in medical sciences is exciting which is continually strengthened by the ever widening spectrum. This necessitates the need of biologically benign nanoparticles. Biological methods may provide an alternate to the conventional techniques in concerned with the production of environmentally benign nanoparticles. However, the limitations such as change in pH, incubation are accountable in the large scale production. Greener technology on the synthesis of nanoparticles deserves merit to overcome the drawbacks of microbial synthesis of nanoparticles. The result addressed herein is on the biological synthesis of gold nanoparticles using medicinally important plant *C. guianensis* flower extract. There are reports on the plant mediated synthesis of gold nanoparticles. Ghoreshi and Co-workers reported on the synthesis of gold nanoparticles using the flower extract of *Rosa damascene*. Studies also demonstrate flavanoids and polyphenols of the flower responsible for the formation of quasi-spherical nanoparticles.[20] Philip *et al.,* demonstrated, *Hibiscus* mediated synthesis of gold nanoparticles and it has been reported that carboxylic group of malic acid may be the active principle for the reduction and capping of gold nanoparticles.[21] *Cinnamomum zeylanicum* mediated synthesis of gold nanoparticles have been reported and possibly the amines of proteins bound the newly formed nanoparticles.[22] It has been illustrated that flavanoids of the plant *Syzygium aromaticum* found to have the property of synthesizing irregular shaped nanoparticles.[23] Sastry and Co-workers have made on the synthesis of gold nanoparticles using geranium leaves, neem, *Aloe vera* and lemon grass as green chemistry approach.[24–27]

HL-60 cells were treated with various concentrations of gold nanoparticles for different time-periods. The results shown in Figure 7 a&b indicates the nuclear morphology of dying HL-60 cells due to gold nanoparticles treatment and these effects were dose-dependent. Figure 8 a&b indicates that gold nanoparticles induced apoptosis in HL-60 cells, and the results showed less cell number compared to control.

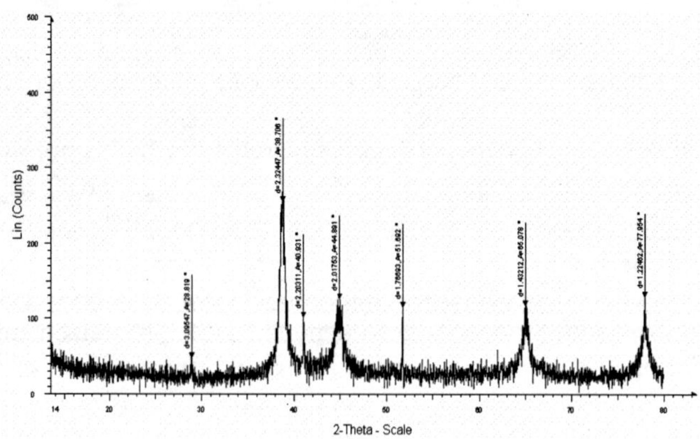

Fig. 3: XRD Pattern of Gold Nanoparticles Synthesized Using *C. guinensis*

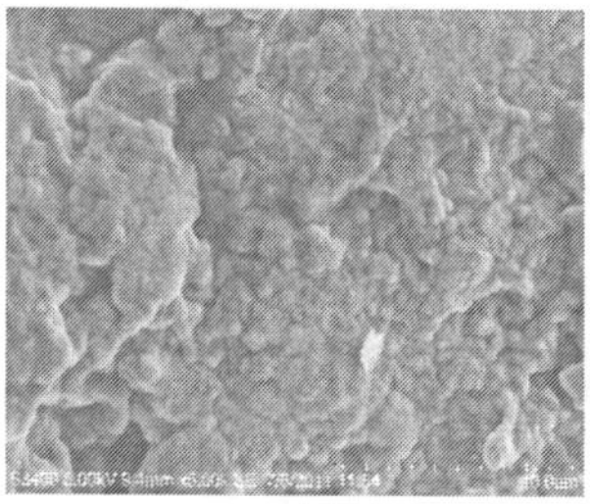

Fig. 4: SEM Photograph of *C. guinensis* Synthesized Gold Nanoparticles

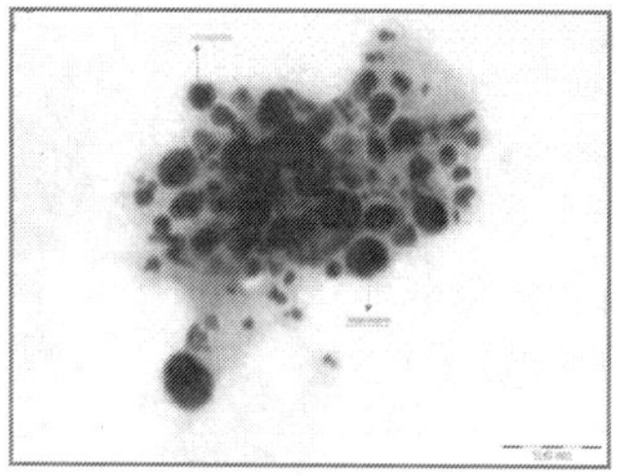

Fig. 5: TEM Micrograph of *C. guinensis* Synthesized Gold Nanoparticles

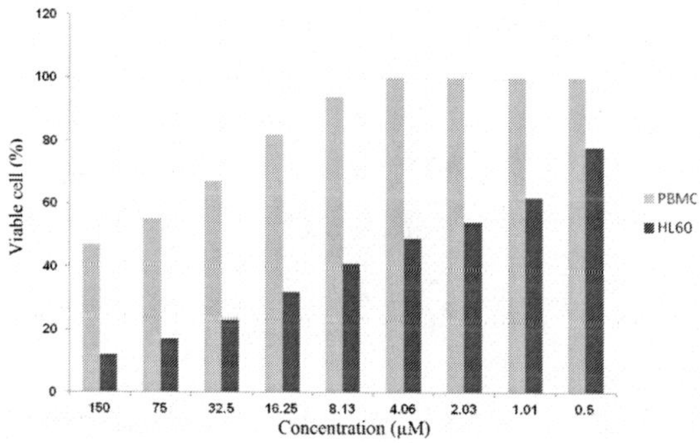

Fig. 6: Differential Cytotoxic Effects of *C. guinensis* Synthesized Gold Nanoparticles on Cancerous HL60 Cells and PBMC from a Representative Experiment is Presented, n = 3

(a) (b)

Fig. 7: Morphological Analysis of HL-60 Cells (a) Control and (b) *C. guinensis* Synthesized Gold Nanoparticles Treated with HL-60 Cells

(a) (b)

Fig. 8: (a) Control and (b) Gold Nanoparticles Induced Apoptosis

Fig. 9: (a) Control and (b) Assessment of DNA Damage by Comet Assay

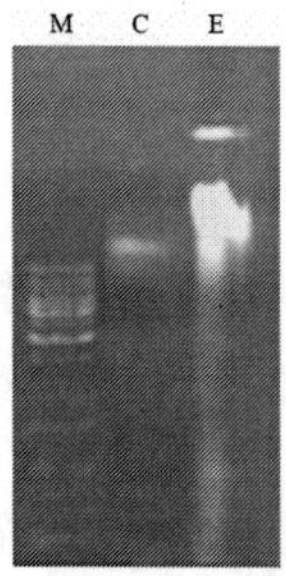

Fig. 10: DNA Fragmentation Analysis Using Agarose Gel Electrophoresis Lane 1: Marker; Lane 2: Control (HL-60); Lane 3: *Couroupita guinensis* Synthesized Gold Nanoparticles Treated with HL-60 Cells

Gold nanoparticles induced apoptosis which was consistent with data showing that DNA fragmentation occurred as seen in Figure 9 a&b Concentrations of gold nanoparticles led to a longer tail (DNA damage). Further supporting the fact that gold nanoparticles induced apoptosis in HL-60 cells and data in Figure 10 shown DNA gel electrophoresis of DNA fragmentation which was enhanced with increasing exposure to gold nanoparticles. As a result of our extensive screening on identifying the newly formed gold nanoparticles application, anticancer activity has been explored. The effect of newly synthesized gold nanoparticles using flower extract of *C. guianensis* on acute myeloblastic leukemic cells (HL-60) and peripheral blood mononuclear cells (PBMC) were analyzed. Cell viability determination was made using MTT assay which was conducted for 120h *C. guianensis* flower extract synthesized gold nanoparticles caused a significant cytotoxicity in HL-60 cells in a concentration dependent manner with the CC_{50} value of 5.14 μM as where for PBMC it is 113.25μM (Figure 6). Despite the widespread use of gold nanoparticles, there are only a few studies to determine the cytotoxic effects of biologically synthesized gold nanoparticles, particularly in the context of apoptosis. MTT assay was used to appraise the effect of gold nanoparticles on proliferation of HL-60 cells. This is the first study to evaluate *C. guianensis* based gold nanoparticles cytotoxicity against cancer cell lines. Thus we hypothesize that cell killing could be the possible mechanism induced by the cytotoxic effect of biosynthesized gold nanoparticles. Nanotechnology has attracted a great interest in recent years due to its expected impact on many areas such as energy, medicine, and electronics. The development of new materials with nanometer size including nanoparticles, nanotubes, nanowires, *etc.,* is the major activity. Among all, nanoparticles with the unique properties in chemistry, optics, electronics, and magnetics have led to an increasing interest in their synthesis. Much of the existing research in the development of anticancer therapeutics from the plants has been focused on investigating the molecular mechanism by which an agent induces cytotoxicity and apoptosis in cancer cells. This cram provides valuable insight into the mechanism of action of the gold nanoparticles and that are able to trigger apoptosis in cancer cells.

ACKNOWLEDGEMENT

The authors, G.S and R.G thank the DST/INSPIRE Fellowship, New Delhi, Government of India for financial assistance.

REFERENCES

[1] Mirkin, C.A., Letsinger, R.L, Mucic, R.C. and Storhoff, J.J.A, DNA-based Method for Rationally Assembling Nanoparticles into Macroscopic Materials, *Nature*, 382,607–609, 1996.

[2] Sen, I.K., Maity, K. and Islam, S.S., Green synthesis of gold nanoparticles using a glucan of an edible mushroom and study of catalytic activity, *Carbohydrate polymer.* DOI: doi:10.1016/j.carbpol.08.058, 2012.

[3] Klaus, T., Joerger, R., Olsson, E. and Granquist, C.G., Bacteria as workers in the living factory: metal accumulating bacteria and their potential for materials science. *Trends Biotechnol.*, 19:15–20, 2001.

[4] Hashmi, A.S.K. and Hutchings, G.J., Gold catalysis, *Angew. Chem. Int. Ed.*, 45, 7896–7936, 2006.

[5] Huang, X.H., Jain, P.K., EI-Sayed, I.H., M.A., Gold nanoparticles: interesting optical properties and recent applications in cancer diagnostic and therapy, *Nanomedicine,* 2, 681–693, 2007.

[6] Wu, Y.L., Y.N. Li, P. Liu, S. Gardner and Ong., B.S., Studies of gold nanoparticles as precursors to printed conductive features for thin-film transistors, *Chem. Mater.*,18, 4627–4632. 2006.

[7] Alivisatos, A.P., Perspectives on the physical chemistry of semiconductor nanocrystals, J. *Phy. sChem.* 100, 13226–13239, 1996.

[8] Jin, R.C., Cao, Y.W., Mirkin, C.A., Kelly, K.L., Schatz, G.C. and Zheng, J.G., Photoinduced conversion of silver nanospheres to nanoprisms, *Science* 294,1901–1903., 2001

[9] Aizpurua, J., Hanarp, P., Sutherland, D.S., Kall, M., Bryant, G.W. and García de Abajo, F.J., Optical properties of gold nanorings, *Phys. Rev. Lett.* 90, 057401 1 057401-4, 2003.

[10] Huang, J., Li, Q., Sun, D., Lu, Y., Su, Y., Yang, X., Wang, H., Shao, W., He, N., Hong, J. and Chen. C., Biosynthesis of silver and gold nanoparticles by using novel sun-dried *Cinnamomum camphora* leaves. *Nanotechnol*, 18105104, 2007.

[11] Geetha, M., Saluja, Shankar, A.K. and Mehta, R.S., Analgesic and antiinflammatory activity of *Couroupita guianensis Aubl*, *J. of Natural Remedies,* 4(1): 52–55, 2004.

[12] Garrett, M.D. and Workman, P., Discovering novel chemotherapeutic drugs for the third millennium. *Eur J. Cancer,* Vol. 35(14), pp. 2010–2030, 1999.

[13] Canadian Cancer Society—National Cancer Institute of Canada. *Canadian Cancer Statistics 2005,* Toronto, Canada, April, ISSN: 0835–2976, 2005.

[14] Robert, J. and Jarry, C., Multidrug resistance reversal agents. *J Med Chem,* 46(23):4805–4817, 2003.

[15] Lu, H.F., Lai, T.Y., Hsia, T.C., Tang, Y.J., Yang, J.S., Chiang, J.H., Lu, C.C., Liu, C.M., Wang, H.L. and Chung, J.G., Danthron induces DNA damage and inhibits DNA repair gene expressions in GBM 8401 human brain glioblastoma multiforms cells. *Neurochem. Res.*, 35, 1105–1110, 2010b.

[16] Chen, J.C., Lu, K.W., Lee, J.H., Yeh, C.C. and Chung, J.G., Gypenosides induced apoptosis in human colon cancer cells through the mitochondria-dependent pathways and activation of caspase-3. *Anticancer Res.* 26, 4313–4326, 2006.

[17] Chen, J.C., Lu, K.W., Tsai, M.L., Hsu, S.C., Kuo, C.L., Yang, J.S., Hsia, T.C., Yu, C.S., Chou, S.T., Kao, M.C., Chung, J.G., Wood, W.G., Gypenosides induced G0/G1 arrest via CHk2 and apoptosis through endoplasmic reticulum stress and mitochondria-dependent pathways in human tongue cancer SCC-4 cells, Oral Oncol. 45, 273–283, 2009a.

[18] Premnathan, M., Arakaki, R., Izumi, H., Kathiresan, K., Nakano, M., Yamamoto, N. and Nakashima, H., *Antiviral Res,*44,113–122, 1999.

[19] Premnathan, M., Nakashima, H., Kathiresan, K., Rajendran, N., Yamamoto, N., in vitro anti human immunodeficiency virus activity of mangrove plants., Indian journal of medical Research, 103, 278–281, 1996.

[20] Ghoreishi, S.M., Behpour, M. and Khayatkashni, M., Green synthesis of silver and gold nanoparticles using *Rosa damascene* and its primary application in electrochemistry, *Physica E.*44:97–104, 2011.

[21] Philip, D., Green synthesis of gold and silver nanoparticles using *Hibiscus rosa sinensis. Physica E* 42: 1417–1424, 2010.

[22] Smitha, S.L., Philip, D., and Gopchandran, K.G., Green synthesis of gold nanoparticles using Cinnamom zeylanicum leaf broth. *Spectrochim. Acta, Part A: mol and biomol spec* 74: 735–739, 2009.

[23] Deshpande, R., Bedre, M.D, Basavaraja, S., Sawle, B., Manjunath, S.Y. and Venkataraman, A., Rapid biosynthesis of irregular shaped gold nanoparticles from maccrated aqueous extracellular dried clove buds (*Sygyium aromaticum*) solution. *Colloids and surfaces B: Biointerf.,* 79:235–240, 2010.

[24] Ankamwar, B., Chaudhary, M and Sastry, M., Gold nanotriangles biologically synthesized using tamarind leaf extract and potential application in vapor sensing. *Synth React. Inorg. Metal-org Nano-metal Chem.*, 35:19–26, 2005.

[25] Shankar, S., Rai, A., Ahmed, A. and Sastry, M., Rapid synthesis of Au, Ag, and bimetallic Au core-Ag shell nanoparticles using Neem (Azadirachta indica) leaf broth. *Journal of colloid and Interf Sci.*, 275: 496–502, 2004.

[26] Chandran, S.P., Chaudhary, M., Pasricha, R., Ahmad, A. and Sastry, M., Synthesis of gold nanotriangles and silver nanoparticles using *Aloe vera* plant extract. *Biotechnol Prog.,* 22:577–583, 2006.

[27] Shankar, S.S., Rai, A., Ahmad, A. and Sastry, M., Controlling the optical properties of lemon grass extract synthesized gold nanotriangles and potential application in infrared-absorbing optical coatings. *Chem.Mater.*, 17:566–572, 2005.

GREEN SYNTHESIS OF SILVER NANOPARTICLES FROM *DODONAEA VISCOSA* LINN. AQUEOUS EXTRACT AND ITS ANTIBACTERIAL ACTIVITY

PANNERSELVAM BALASHANMUGAM AND
PUTHUPALAYAM THANGAVELU KALAICHELVAN

Centre for Advanced Studies in Botany, University of Madras, Chennai, Tamil Nadu
E-mail: biobala17@gmail.com

ABSTRACT

A green rapid biogenic synthesis of silver nanoparticles (Ag NPs) using Dodonaea viscosa (D. viscosa) aqueous extract was demonstrated in this present investigation. Cubic shape was observed from biosynthesized SNPs in the range of 30 to 50 nm. The formation of Ag NPs was confirmed by Surface plasmon resonance (SPR) at 435 nm using UV-VIS spectrophotometer. The reduction of silver ions to Ag NPs by D. viscosa extract was completed within 24 hrs which was evidenced potentiometrically. Synthesised Ag NPs was characterised using UV–vis spectroscopy, Fourier transformed infrared spectroscopy (FT-IR), Scanning Electron Microscopy (SEM), Energy dispersive X-ray Analysis (EDX), Powder X-ray diffraction (XRD) and also antibacterial activity using clinically important pathogens. Plant derived proteins or terpenoids are used as reductants and stabilizers as well. In addition, it showed significant antibacterial activity towards two Gram-positive bacteria (Bacillus subtilis, Enterococcus faecalis) and two Gram-negative bacteria (Proteus vulgaris, Klebsiella pneumonia). The highest zone of inhibition was observed in Klebsiella pneumonia (19 mm in diameter) and followed by Enterococcus faecalis (18 mm in diameter).

INTRODUCTION

The field of nanotechnology is one of the most active areas of research in modern materials science. Nanotechnology refers broadly to a field of applied science and technology whose

V. Rajendran, P. Prabu and K.E. Geckeler (eds.)
Nano Biomaterials, pp. 299–308 (2012).

unifying theme is the control of matter on the atomic and molecular scale.[1] Nanotechnology is highly interdisciplinary by nature and requires close collaboration between biologists, physical scientists and engineers.[15] Nanoparticles, generally considered as particles with a size of up to 100 nm, exhibit completely new or improved properties based on specific characteristics such as size, shape, surface area and distribution. As specific surface area of nanoparticles is increased, their biological activities of the nanoparticles are also increased.[3] Metal nanoparticles are used prominently because of thier antimicrobial activity. New applications of nanoparticles are emerging rapidly. Nanoscale materials possess unique electrical, optical as well as biological properties.[7] Nature has devised various processes for the synthesis of nano- and micro- length scaled inorganic materials which have contributed to the development of relatively new and largely unexplored area of research based on the biosynthesis of nanomaterials.[14] A number of approaches is available for the synthesis of silver nanoparticles viz., chemical and photochemical reactions in reverse micelles, thermal decomposition of silver compounds, radiation assisted, electrochemical, sonochemical and microwave assisted process and via green chemistry route.[11,21] Chemical synthesis methods lead to the presence of some toxic chemical species adsorbed on the surface that may have adverse effects in medical applications. This is not an issue when it comes to biosynthesized nanoparticles as they are eco friendly and biocompatible for pharmaceutical applications. Thus, there is a need for green chemistry that includes a clean, nontoxic, and environment friendly method for nanoparticle synthesis.[6] Plants synthesize nanoparticles with controlled size and shape due to its slower kinetics, better manipulation etc. *Dodonaea viscosa* is a family of *Sapindaceae,* distribution in tropical, subtropical and warm temperate regions of Africa, the Americas, southern Asia and Australia. In Traditional system of medicine

MATERIALS AND METHODS

Preparation of *Dodonaea Viscosa* Extract

All the chemicals and reagents used in the study were of analytical grade. Silver nitrate obtained from Himedia, E. Merck, and Sigma chemicals (U.S.A). In order to prepare *D. viscosa* extract, 5 gm of powder sample boiled with 100 ml of sterile distilled water for 15 min at 50°C. The crude extract was passed through Whatman No.1 filter paper and the filtrates were stored at 4°C for further use.

Synthesis of AG NPS

5 ml of *D. viscosa* leaf extract was added into 45 ml of aqueous solution of 1 mM silver nitrate for reduction into Ag+ ions and kept for incubation in darkroom 24 hrs at room temperature suitable controls were maintained throughout the experiments. The color change from yellowish to dark brown color indicated the formation of Ag NPs.

Characterization

UV-Visible Spectroscopy

The initial characterization of synthesized Ag nanocolloids carried using UV-Vis spectroscopy. The reduction of silver ions was monitored from 200 and 800 nm by (Hitachi-u-2900 spectrometer, Japan) using deionized water as a blank.

Fourier Transformed Infrared Spectroscopy (FTIR)

Purified Ag NPs in the form of powder were analysed using FT-IR spectral measurements. The measurements were carried out on a Jasco Ft-Ir 4100 instrument.

SEM and EDX Analysis

Purified Ag NPs in suspension were characterized for their size using Scanning Electron Microscope (Hitachi S-3400n, Japan). EDX (Energy Dispersive X-ray) analysis of purified Ag NPs was carried out using the same instrument for confirming the elemental composition of the sample.

X-Ray Diffraction

After mixing of extract and Ag–salt solution the mixture was allowed to complete conversion and Ag NPs were collected after centrifugation followed by filtration. The purified Ag NPs were subjected to X-ray diffraction analysis by (Seifert Jso- Debyeflex 2002).

Antibacterial Activity of AG NPs

The silver nanoparticles synthesized using *D. viscosa* leaf extract was tested for antibacterial activity by agar well- diffusion method against *Proteus vulgaris*, *Klebsiella pneumonia* (Gram-negative bacteria), *Bacillus subtilis* and *Enterococcus faecalis* (Gram-positive bacteria).[5] Four wells were made on 6 nm in diameter in Muller Hinton agar plates. The holes were filled with 40μl of synthesized silver nanoparticle solution'S', Plant extract 'C', Antibiotic 'A' (Streptomycin) and Silver Nitrate 'D', Plates were incubated at 37°C for 24 hrs to observe formation of zone of inhibition.

RESULT AND DISCUSSION

Synthesis of Nanoparticles from *D. Viscosa* Leaf Extract

The study on extracellular green synthesis of silver nanoparticles through plant extracts were carried out. It is well known that silver nanoparticles exhibit yellowish to dark brown color (Figure 1) in aqueous solution due to excitation of surface plasmon vibrations in silver nanoparticles.[20] This important observation indicates the reduction of the Ag^{+} ions takes place extracellularly. The appearance of a dark brown color in solution containing the

D.viscosa aqueous extract is a clear indication of the formation of silver nanoparticles in the reaction mixture.

Fig. 1: Biosynthesis of Silver Nanoparticles
A. without silver nitrate (Blank), B. Silver Nitrate (1 mM) Reduction after 24 hrs

UV–Vis Spectroscopic Analysis

The formation and stability of the reduced Ag NPs in the *D. viscosa* aqueous extract was monitored by using UV–Vis spectral analysis. This technique has proved to be very useful for the analysis of nanoparticles [16] Silver Nanoparticles suspension exhibits an intense dark brown color due to the surface Plasmon resonance (SPR). Absorption spectra of silver nanoparticles formed in the reaction solution has absorbance peak at 430 nm (Fig 2) broadening of peak indicated that the particles are mono dispersed and spherical silver particles.

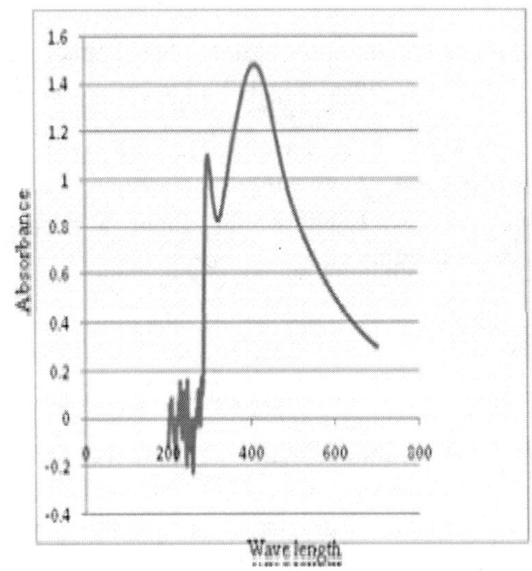

Fig. 2: UV-Vis Spectrum of Ag NPs

The FTIR spectrum shows several absorption bands (Figure 3) indicates the presence of active functional groups in the synthesized silver Nanoparticles. The intensity peaks are slightly increased for the period of silver nanoparticle synthesis like 3853.2, 3436.0, 2933.7, 1384.4, cm^{-1} as well as some intensity peaks decreased like 1384.4, 1075.9, and 668.8 cm^{-1}.The band at 3436 cm^{-1} corresponds to O-H stretching H-bonded alcohols and phenols. The peak at 2933.7 cm^{-1} corresponds to O-H stretch carboxylic acids. The assignment at 1636.0 cm^{-1} corresponds to N-H bend primary amines. The peak at 1384.4 cm^{-1} corresponds to C-N stretching of aromatic amine group and the bands observed at 1384.4, 1075.9, 668.8 cm^{-1} corresponds to C-N stretching alcohols, carboxylic acids, ethers and esters.[17] Therefore the synthesized nanoparticles were surrounded by proteins and metabolites such as terpenoids having functional groups of alcohols, ketones, aldehydes and carboxylic acids.

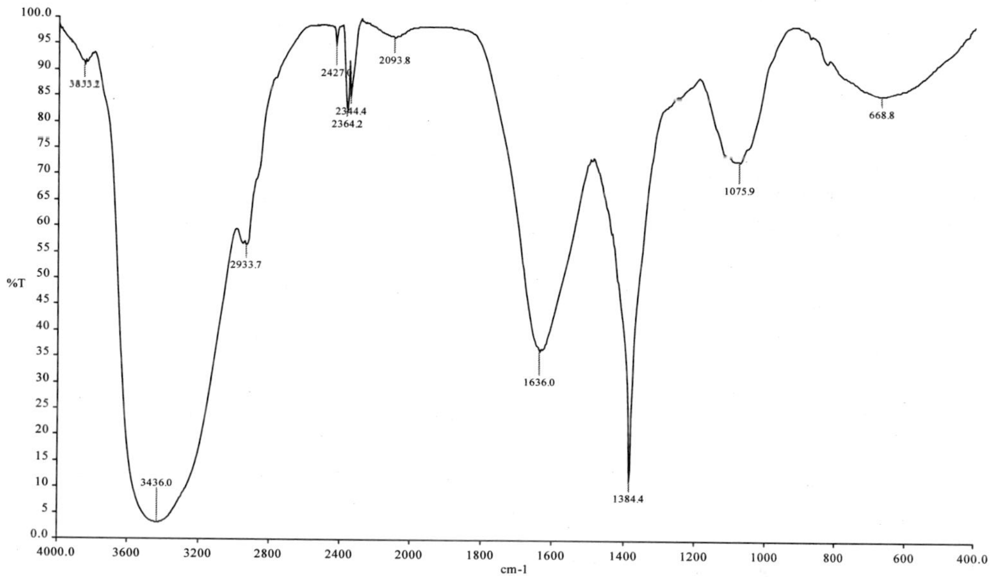

Fig. 3: Fourier Transform Infrared Spectroscopy (FTIR)

SEM and EDX Studies

Figure 4a represents the Scanning Electron Microscopy (SEM) image recording from the drop coated films of silver nanoparticles synthesized with *D. viscosa* leaf extract. The SEM image showed cubical and relatively uniform shape of nanoparticle formation with diameter range 30–50 nm and analysis through Energy dispersive X-ray Analysis (EDX) spectrometers confirmed the presence of elemental silver signals of the silver nanoparticles (Figure 4b). The vertical axis displays the number of X-ray counts whilst the horizontal axis displays energy in KeV [10] confirms the metallic nanoparticles due to surface plasmon resonance.

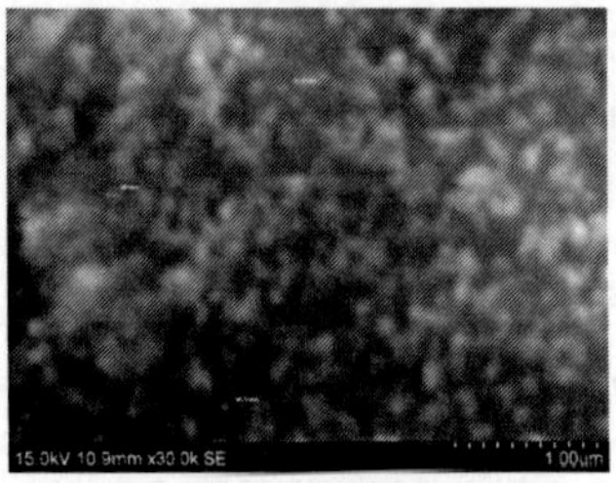

Fig. 4a: SEM

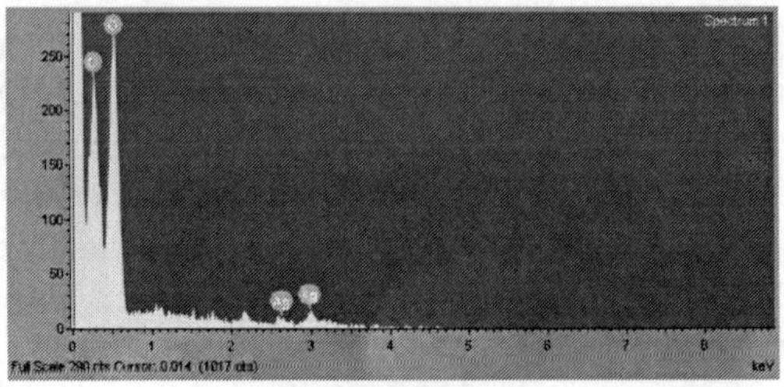

Fig. 4b: EDX

X-Ray Diffraction Analysis

XRD patterns were analyzed to determine peak intensity, position and full width at half maximum (FWHM) data was used with the Scherrer's formula to determine mean particle size. Scherrer's equation is given by $d = 0.9\lambda/\beta \mathrm{Cos}\theta$ where d is the mean diameter of the nanoparticles; λ is the wavelength of X-ray radiation source, β is the angular FWHM of the XRD peak at the diffraction angle θ. The distinct Bragg reflections corresponding to (111), (200) (220), (311), and sets of lattice planes were manifested in the X-ray diffraction patterns are shown in (Figure 5). They may be indexed on the basis of face-centered cubic structure of silver.[22] Average size of the particles synthesized with size range 30 to 50 nm.

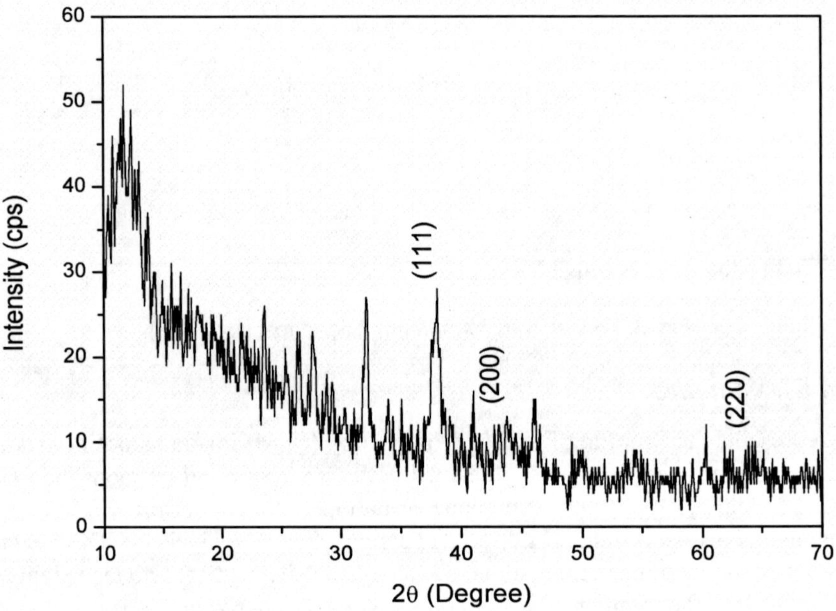

Fig. 5: XRD

Antibacterial Activity

In different types of nanomaterials like copper, zinc, titanium, magnesium, gold and silver have arising as antimicrobial materials, especially silver nanoparticles that are more efficient. These materials exhibit antimicrobial activity against bacteria, viruses and other eukaryotic microorganisms.[19] One such aspect proven for silver nanoparticles is that for antibacterial activity, size, morphology and concentrations are all important. For example, small particles have larger surface areas to be in contact with the bacterial cells, showing a larger activity. Biologically synthesized silver nanoparticles was tested and compared with positive control for antibacterial activity against *Enterococccus faecalis, Klebsiella pneumonia, Proteus vulgaris* and *Bcillus subtillis* by well diffusion method (Figure 6). The highest zone of inhibition was observed in *Klebsiella pneumonia* (19 mm in diameter) and followed by *Enterococcus faecalis* (18 mm in diameter), *Proteus vulgaris* and *Bacillus subtillis.* There is no zone of inhibition was observed in *D. viscosa* extract and silver nitrate alone treated well for throughout the experiment (Table 1) in all the four organism, except *K. pneumonia* showed activity in silver well.

PLATE 1

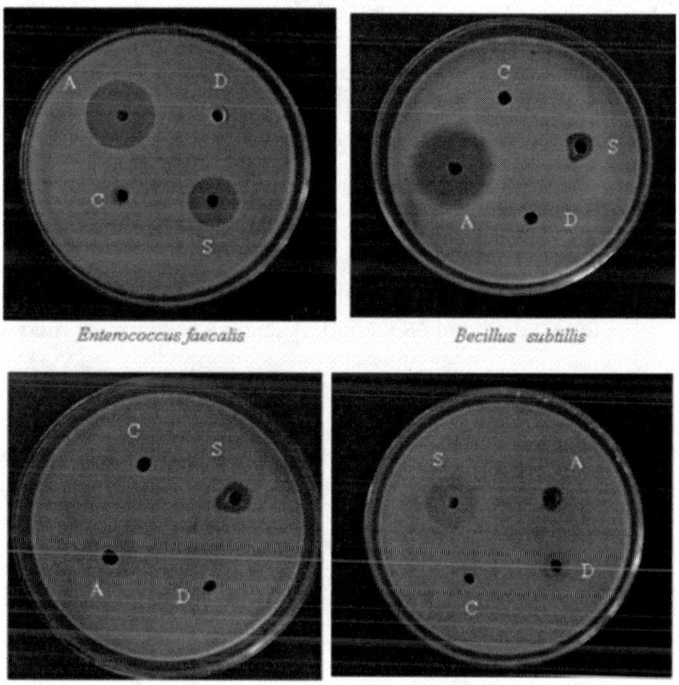

Proteus vulgaris *Klebsiella pneumoniae*

S-Biologically synthesized silver nanoparticles; C- plant extract
A-Streptomycin; D- silver Nitrate

Fig. 6: The Antibacterial Activity

Table 1: The Antibacterial Activity

Name of the Organism	*Zone Inhibition (mm in diameter)*			
	Ag NPs	*D. Viscosa Extract*	*Positive Control*	*Silver Nitrate*
Enterococccus faecalis	18	–	24	–
Klesiella pneumonia	19	–	30	7
Proteus vulgaris	10	–	22	–
Bacillus subtilis	9	–	18	–

CONCLUSIONS

A superficial rapid green eco-friendly synthesize of AgNPs using *D. viscosa* aqueous extract has been developed. The *D. viscosa* aqueous extract showed potential for extracellular synthesis of fairly monodispersed, AgNPs in the range of 30–50 nm. Protein and terpenoids serve both as reducing and capping agents. This method is a promising alternative to the traditional reduction routes to avoid usage of toxic chemicals. AgNPs obtained by this method exhibited antimicrobial activity towards test bacterial cultures. Structural analysis of AgNPs using XRD pattern strongly infers the transformation of elemental AgNPs instead of their oxides nanoparticles. This pristine method is rapid, convenient, less time consuming, environmentally safe and can be applied in a variety of existing applications.

ACKNOWLEDGEMENT

The authors wish to thank The Director, CAS in Botany, University of Madras, Chennai, for providing the laboratory facilities .We also thanks to NCNSNT, University of Madras, for providing characterization studies and UGC herbal science scheme for financial support.

REFERENCES

[1] Anima, Nanda and Saravanan, M. Biosynthesis of silver nanoparticles from *Staphylococcus aureus* and its antimicrobial activity against MRSA and MRSE, *Nanomedicine: Nanotechnology, Biology and Medicine*, Doi:10.1016/7nano-.01.012. 2009.

[2] Jain, D., Daima, H.K., Kachhwaha, S. and Kothari, S.L., Synthesis of silver nanoparticles using fruit papaya fruit extract and evaluation their antimicrobial activities, *Digest journal of Nanomaterials and Bio structures,* 4, 557– 563, 2009.

[3] Jhan, W., Chemical aspects of the use of gold clusters in structural biolog,. *J. Struct. Biol,* 127, 106–112, 1999.

[4] Jain, D., Daima, H.K., Kachhwaha, S. and Kothari, S.L., Synthesis of plant-mediated silver Nanoparticle using papaya fruit extract and evaluation of their anti microbial activities, *Digest journal of nanomaterials and biostructures,* 4, 557–563, 2009.

[5] Mudasir, A. Dar, Avinash Ingle and Mahendra, Rai, Enhanced antimicrobial activity of silver nanoparticles synthesized by Cryphonectria sp. evaluated singly and in combination with antibiotics, *Nanomedicine: Nanotechnology, Biology, and Medicine,* 2012.

[6] Mukherjee, P., Roy, M., Mandal, B., Dey, G., Mukherjee, P. and Ghatak, J., Green synthesis of highly stabilized nanocrystalline silver particles by anon-pathogenic and agriculturally important fungus T.asperellum, *Nanotechnology*, 19,75, 103–110, 2008.

[7] Mohanpuria, P., Rana, K.N. and Yadav, S.K., Biosynthesis of nanoparticles: Technological concepts and future applications, *J. Nanoparticle Res*, 10, 507–517, 2008.

[8] Mukherjee, P., Ahama, P., Mandal, D., Senapati, S., Sainkar, S.R., Khan, M.I., Parishcha, R., Ajaykumar, P.V., Alam, M. and Kumar, Sastry, R., *Nano Lett,* 1, 515, 2001.

[9] Manish Dubey, Seema, Bhaduria and B.S. Kushwah., Biosynthesis, characterization and antimicrobial studies of AgNPs extracts from *Bacopa monniera* whole plant, *Digest Journal of Nanomaterials and Biostructures,* 4, 537–543, 2009.

[10] Marimuthu Vivek, Palanisamy Senthil Kumar, Sesurajan Stusffi and Sellappa Sudha, Biogenic Silver Nanoparticles by Gelidiella accrosa Extract and their Antifungal Effects, *Avicenna Journal of Medical Biotechnology*, 3, 2011.

[11] Parashar, V., Parashar, R., Sharma, B. and Pandey Avinash, C., *Parthenium leaf* extract mediated synthesis of silver nanoparticles: A novel approach towards weed utilization, *Digest journal of Nanomaterials and biostrctures,* 4, 45–50, 2009.

[12] Prabhu, N., Divya, T. Raj., Yamuna Gowri, K., Ayisha Siddqua, S., Synthesis of silver phytonanoparticles and their antibacterial efficacy, *Digest journal of Nanomaterials an biostrctures,* 5, 185–189, 2010.

[13] Prasad, Elumalai, E.K. and Khateeja, S., Evaluation of the antimicrobial efficacy of phytogenic silver nanoparticles, *Asian Pacific Journal of Tropical Biomedicine*, 82–85, 2011.

[14] Raveendran, P., Fu, J. and Wallen, S.L., Completely green synthesis and stabilization of metal nanoparticles, *J.Am.Chem.Soc*, 125, 13940–13941, 2003.

[15] Singh, A., Jain, D., Upadhyay, M.K., Khandelwal and Verma, H.N., *Green synthesisof silver nanoparticles using Argemone Mexicana leaf extracts and evaluation of their antimicrobial activities. Digest Journal of Nanomaterials and Biostructures,* 5, 483–489, 2010.

[16] Sastry, M., Ahmad, A., Khan, I. and Kumar, R., Biosynthesis of metal nanoparticles using fungi and actinomycete, *Curr Sci*, 85,162–170, 2003.

[17] Sathyavathi, R., Krishna, M.B., Rao, S.V., Saritha, R. and Rao, D.N., Biosynthesis of Silver Nanoparticle Using *Coriandrum Sativum* Leaf Extract and their application in Nonlinear Optics, *Adv. Sci. Lett*, 31–6, 2010.

[18] Singh, A.K., Talat, M., Singh, D.P. and Srivastava, O.N., Biosynthesis of gold and silver Nanoparticle by natural precursor clove and their functionalization with amine group, *J. Nanopart Res*, 121667– 1675, 2010.

[19] Shankar, S.S., Rai, A., Ahmad, A. and Sastry, A., Rapid synthesis of Au, Ag, and bimetallic Au core–Ag shell Nanoparticle using Neem (*Azadirachta indica*) leaf broth, *J Colloid Interface Sci*, 2, 275– 496, 2004.

[20] Thirumurugan, A., Jiflin, G.A., Rajagomathi, G., Tomy, N.A., Ramachandran, S. and Jaiganesh., Biotechnological synthesis of gold Nanoparticle of *Azadirachta indica* leaf extract, *International Journal of Biological Technology*, 1, 75–77, 2010.

[21] Vorobyova, S.A., Lesnikovich, A.I. and Sobal, N.S., Preparations of silver nanoparticles by interphase reductions, *colloids and surfaces A: Physiochemical and Engineering Aspects*, 152, 375–379, 1999.

[22] Yamini, Sudhalakshmi, G., Green Synthesis of Silver Nanoparticles from *Cleome Viscosa*: Synthesis and Antimicrobial Activity, *International Conference on Bioscience, Biochemistry and Bioinformatics, IPCBEE,* Vol. 5, 2011.

BIOCOMPATIBLE TITANIA–CHITOSAN NANOCOMPOSITES FOR TISSUE ENGINEERING AND ORTHOPEDIC APPLICATIONS

K. KAVITHA, M. PRABHU, R. SURIYAPRABHA, P. PRABU, V. RAJENDRAN AND T. JAYAKUMAR[1]

Centre for Nanoscience and Technology, K.S. Rangasamy College of Technology, Tiruchengode, Tamil Nadu
[1]*Metallurgy and Materials Group, Indira Gandhi Centre for Atomic Research, Kalpakkam, Tamil Nadu*
E-mail: veerajendran@gmail.com

ABSTRACT

Nanostructured amorphous titania (TiO_2) has attracted extensive attention in biomedical research because of its numerous advantages. To produce good biocompatible biomaterial, we synthesised a nano-TiO_2–chitosan (CH) composite by in situ sol-gel method and comprehensively characterised that using conventional techniques. The resultant composites showed anatase phase with an irregular morphology and 4.5 nm particle size. Nanocomposites with 265 m^2/g surface area and high purity were obtained. The samples were further analysed for in vitro bioactivity (1.5 mM simulated body fluid (1.5 SBF)) and cytotoxicity (MG63 osteoblast cell line). The samples were tested for antibacterial property against Staphylococcus aureus. The formation of apatite layer on 1.5 SBF-immersed samples confirms the better bioactivity of the nanocomposite than that of pure TiO_2. High surface area, appropriate hydroxyapatite formation, specific antibacterial action, increased cell viability in osteoblast cell line, controlled swelling and degradation rates were favorably achieved in this nanocomposite. This study shows in situ synthesised TiO_2–CH nanocomposites as the promising biomaterial for orthopedic and tissue engineering applications.

V. Rajendran, P. Prabu and K.E. Geckeler (eds.)
Nano Biomaterials, pp. 309–314 (2012).

INTRODUCTION

Biomaterials have wide applications in bone tissue engineering and orthopedic applications, even though search toward the biomimic, bioactive and biocompatible potential material is a thrust area. Nanostructured amorphous titania (TiO_2) is a well-known biomaterial due to its unique physiochemical nature.[1] However, pure nano-TiO_2 shows toxic nature during biomedical applications. Generally, particle size, specific surface area (SSA), dosage, and transport of nano-TiO_2 are important in determining the toxicity of nanoparticles.[2,3] Therefore, attempts have been made to overcome the adverse effects of TiO_2 by preparing a composite with biocompatible polymers.[4] Chitosan (CH) has a considerable resemblance to extracellular matrix elements with less mechanical property. The defects of organic and inorganic materials are overcome by *in situ* sol–gel derived TiO_2–CH nanocomposite. To the best of our knowledge, no report has been published about the combined properties, i.e. bioactivity, antibacterial activity and biocompatibility, of such biodegradable materials, particularly *in situ*–synthesised TiO_2–CH nanocomposite.

The objective of this work is to synthesise *in situ* sol–gel TiO_2–CH nanocomposites and to characterise the same using conventional techniques. Phosphate buffered saline (PBS), 1.5 mM simulated body fluid (1.5 SBF), bone-infecting microorganisms, and osteoblast cell line (MG63) were used to analyse the swelling capacity, bioactivity, antibacterial activity, and cytotoxicity of the TiO_2–CH nanocomposite, respectively, with a view to explore the necessity of the composites for biomedical application.

EXPERIMENTAL DETAILS

Nano-TiO_2 particles were synthesised by the precipitation method.[5] The precipitates were washed and dried at 393 K for 1 h (hereafter termed as TA). Titanium isopropoxide was diluted with isopropyl alcohol followed by the addition of acetyl acetone (1:4:0.7) to prepare *in situ* TiO_2–CH nanocomposite.[6] After 1 h stirring, 2% acetic acid-dissolved chitosan was added dropwise to become semisolid and then dried at 433 K for 2–3 h (hereafter termed as TAC). The prepared powders were characterised by X-ray diffractometer (XRD), Fourier transform infrared spectrometer (FTIR), Scanning electron microscopy with Energy-dispersive X-ray spectroscopy (SEM–EDX), Transmission electron microscopy (TEM) and X-ray fluorescence spectrometry (XRF). The SSA was measured using the Brunauer–Emmett–Teller surface area analyser.

Swelling Study and *In vitro* Bioactivity

The samples were incubated in PBS for 7 days.[7] At defined days (1, 4, and 7), the wet weight of the samples was measured and compared with the initial weight. Moreover, the swelling ratio was calculated as described by Rani *et al.* (2011). To carry out the *in vitro* bioactivity study, we immersed 1.5 g samples in 1 L of 1.5 SBF solutions [8] and incubated them for 21 days at 310±1 K. The dry weight of the pellets was measured and formation of hydroxyapatite (HAp) was confirmed by XRD, FTIR, and XRF studies.

Antibacterial Activityand Cytotoxicity Study

To explore the antibacterial activity, we added 100 mg of the prepared sample in disk on *S. aureus* swabbed plates by disc diffusion method. The MG63 cell lines were used to analyse the toxicity of the prepared samples. Filter-sterilised nanosamples at three concentrations, namely 50, 20 and 10 μg ml^{-1}, were used for MTT (3-[4, 5-dimethylthiazol-2-yl]-2,5 diphenyltetrazolium bromide) assay. Statistical analysis was carried out to explore the significance of toxicity.

RESULTS AND DISCUSSION

The XRD pattern and FTIR spectra of the prepared nano-TiO_2 and TiO_2–CH nanocomposite before (A) and after *in vitro* incubation (B) are shown, respectively, in Figure 1 and Figure 2. All the XRD peaks are well indexed with anatase phase of TiO_2 (JCPDS file no. 21–1276; Figure 1A). The average crystallite size obtained using Scherrer equation [9] is shown in Table 1. The addition of chitosan in TiO_2 decreases the crystallite size and crystalline nature. The presence of functional group in prepared TiO_2 and composites is observed at 400–800, 1106, 1233, and 1392 cm^{-1}, respectively, for Ti–O–Ti, Ti–OH, Ti–O–C, and C–O (Figure 2A).[10,11] SEM-EDX and TEM images (Figure 3) show the spherical and irregular morphology of the TiO_2 and TiO_2–CH nanocomposite with high purity. The obtained particle size is in the range of 20 nm.

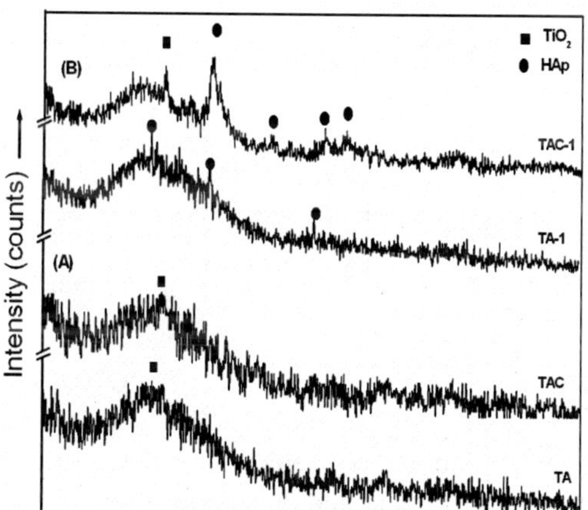

Fig. 1: XRD Pattern of Nanosamples

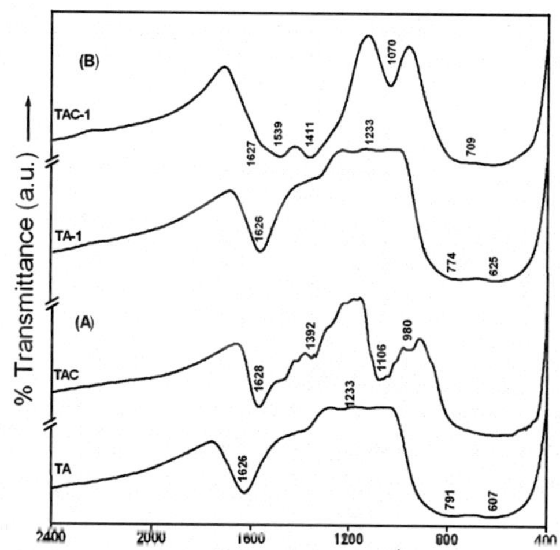

Fig. 2: FTIR Spectra of Nanosamples

The SSA of the prepared nanoparticles is 208.4 and 265.33 $m^2 g^{-1}$, respectively, for TA and TAC. In addition, the total pore volume and average pore diameter are calculated according to Barrett *et al.* (1951) and are shown in Table 1. The higher SSA and pore size of TAC helps in cell growth during the implantation and tissue engineering.

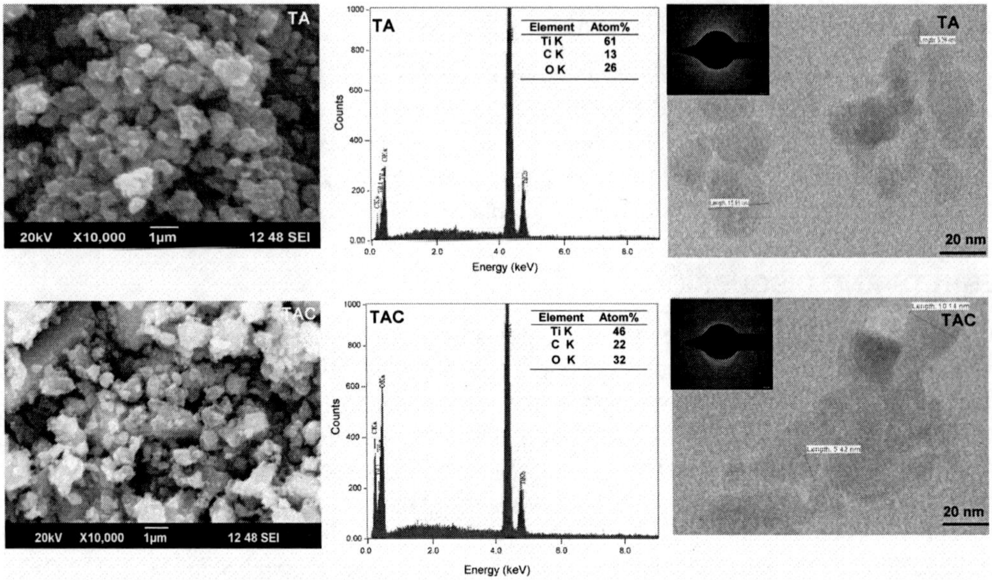

Fig. 3: SEM-EDX and TEM Images of the of TiO_2–CH Nanocomposite

Table 1: Physicochemical Property of Prepared TiO_2–CH Nanocomposites

Sample Name	*TiO_2/Chitosan Concentration ($g\ L^{-1}$)*	*Crystallite Size (nm)*	*Particle Size (nm)*	*Surface Area ($m^2\ gm^{-1}$)*	*Total Pore Volume ($cc\ gm^{-1}$)*	*Average Pore Diameter (nm)*
TA	1:0	3.60	10.58	208.41	0.0945	1.262
TAC	1:0.5	2.13	5.23	265.33	0.130	1.966

Figure 1(B) shows a distinct peak for the HAp layer formation (JCPDS file no. 090432) on the surface of samples after incubation in 1.5 SBF. An irregular increase in crystallite size is observed (Table 2). This is due to the swelling behaviour and agglomeration of the sample. Figure 2(B) shows the presence of CO_3^{2-} characteristic peak and the formation of HAp layer.[12] The XRD and FTIR results confirm the well-formed HAp layer on TiO_2–CH nanocomposites, which indicates good bioactivity.

The chitosan content in the TiO_2 increases the swelling capacity (Table 2; TAC - 42.89% on day 7). The swelling avails to increases the pore size and surface area of the particles, facilitating cell attachment and more nutrient absorption.[8] The weight loss percentage after *in vitro* study (Table 2) indicates that the chitosan containing TiO_2 (TAC) gain the weight instead of loss due to the swelling behavior. Ca/P ratio of TA and TAC shows carbonate-substituted HAp and oxy-HAp, respectively, as 1.73 and 1.5 (Table 2).[9] It is well known that oxy-HAp is an appropriate ratio of accepted stoichiometry for bone formation.[9]

Table 2: Biological Analysis of the Prepared TiO_2–CH Nanocomposite

Sample Name	*Swelling Ratio in Weight*			*After In vitro Bioactivity*			*Zone of Inhibition (mm)*
	Day 1 (%)	*Day 4 (%)*	*Day 7 (%)*	*Crystallite Size (nm)*	*Weight Loss (%)*	*Ca/P Ratio (%)*	
TA	19.02	10.32	12.90	17.53	–2.51	1.73	0
TAC	36.14	42.82	42.89	61.77	1.41	1.58	10

Composites show bigger zone of inhibition against *S. aureus* than the pure TiO_2 due to the presence of the chitosan (Table 2). This facilitates the prevention of infection during the implantation and tissue engineering.

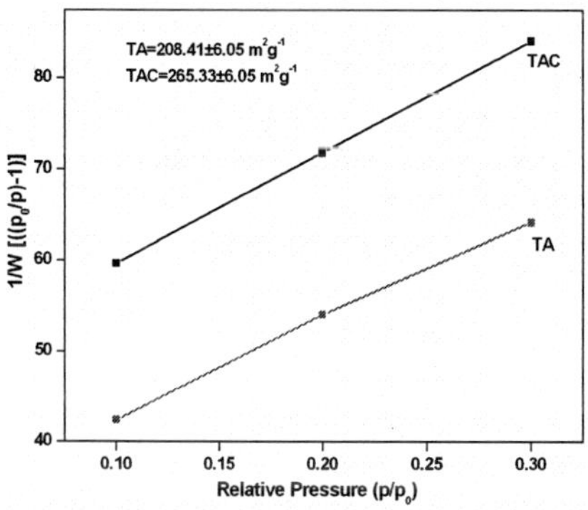

Fig. 4: BET Surface Area of the Nanosamples

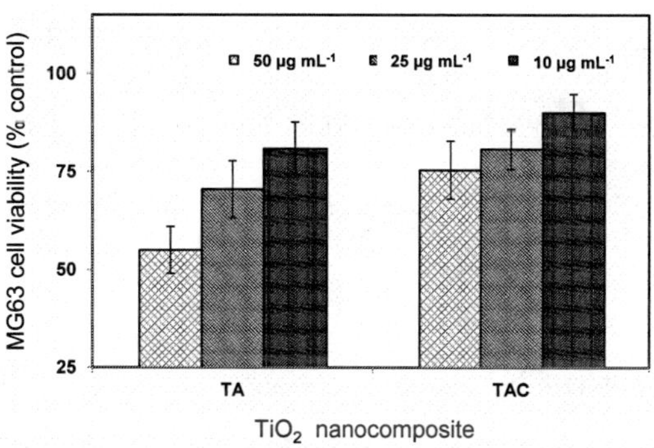

Fig. 5: MTT Assay of TiO_2 Nanocomposite

Figure 5 shows the MTT assay of MG63 treated with nanocomposite. TCA shows higher cell viability than the TA in all the concentrations. Concentrations of samples and cell viability are directly proportional to and in turn indicate that there is no significant toxicity. The statistical analysis indicates the nonsignificant toxicity of nanocomposite. It also implies that the high surface area to volume ratio and swelling property of TAC facilitate ideal reactivity on bone forming osteoblast cell surface [13] and promote cell growth.

CONCLUSIONS

Biocompatible TiO_2–CH nanocomposites are successfully synthesised using *in situ* method. The obtained nanocomposite are anatase in nature with high surface area of 265.33 $m^2\ g^{-1}$ and good swelling property, which improve cell attachment and avoid transport and agglomeration of the particles in the body. The high SSA, specific antibacterial activity, favorable swelling property, stoichiometric ratio of HAp formation, weight loss, and biocompatibility of the prepared sample collectively reveal that *in situ*–synthesised TCA composite material (1:0.5) is a good candidate for regeneration and reconstruction of bone.

REFERENCES

[1] Hua, D., Cheuk, K., Wei-ning, Z. and Chen, W., Chang–fa X. Trans. *Naonferrous Met. Soc. China* 2007, 17, 700–703.

[2] Singh, S., Shi, T., Duffin, R., Albrecht, C., Berlo, D.V., Hohr, D., Fubini, B., Martra, G., Fenoglio, I., Borm, P.J.A. and Schins, R.P.F., *Toxicol. Appl. Pharmacol.* 2007, 222, 141–151.

[3] Kroll, A., Dierker, C., Rommel, C., Hahn, D., Wohlleben, W., Schulze-Isfort, C., Gobbert, C., Voetz, M., Hardinghaus, F., Schnekenburger, J., *Part. Fibre Toxicol.* 2011, 8, 9.

[4] Luo, Y.B., Wang, X.L., Xu, D.Y. and Wang, Y.Z., *Appl. Surf. Sci.* 2009, 255, 6795–6801.

[5] Choi, H.C., Ahn, H.J., Jung, Y.M., Lee, M.K., Shin, H.J., Kim, S.B. and Sung, Y.E. *Appl. Spectrosc,* 2004, 58, 598–602.

[6] Yang, D., Li, J., Jiang, Z., Lu, L. and Chen, X., *Chem. Eng. Sci.*, 2009, 64, 3130–3137.

[7] Rani, V.V.D., Ramachandran, R., Chennazhi, K.P., Tamura, H., Nair, S.V. and Jayakumar, R., *Carbohydr. Polym.,* 2011, 83, 858–864.

[8] Gerhardt, L.C., Jell, G.M.R. and Boccaccini, A.R., *J. Mater. Sci.*: *Mater. Med.*, 2007, 18, 1287–1298.

[9] Rajkumar, M., Meenakshi, Sundaram N. and Rajendran, V., *Digest J. Nanomater. Biostruct.*, 6, 169–179.

[10] Al-Sagheer, F.A. and Merchant, S., *Carbohydr. Polym.*, 2011, 85, 356–362.

[11] Nakayama, N. and Hayashi, T., *Polym. Degrad. Stab.,* 2007, 92, 1255–1264.

[12] Barrett, E.P., Joyner, L.G. and Halenda, P.P., *J. Am Ceram. Soc.*, 1951, 73, 373–380.

[13] Saquib, Q., Al-Khedhairy, A.A., Siddiqui, M.A., Tarboush, F.M.A. and Azam, A., Musarrat, J., *Toxicol. In Vitro* 2012, 26, 351–361.

ZIRCONIUM-SUBSTITUTED NANOBIOACTIVE GLASS PARTICLES FOR BONE IMPLANT APPLICATIONS

M. PRABHU, K. KAVITHA, S. SUTHA, P. MANIVASAKAN, V. RAJENDRAN AND P. KULANDAIVELU[1]

Centre for Nanoscience and Technology, K.S. Rangasamy College of Technology, Tiruchengode, Tamil Nadu
[1]Department of Mechanical Engineering, Centre for Nanoscience and Technology, K.S. Rangasamy College of Technology, Tiruchengode, Tamil Nadu
E-mail: veerajendran@gmail.com

ABSTRACT

In this study, zirconium-substituted nanobioactive glass particles with compositions, namely, $58SiO_2$–$33CaO$–$9P_2O_5$ and $58SiO_2$–$28CaO$–$9P_2O_5$–$5ZrO_2$ were synthesised using simple sol–gel method. The prepared samples were characterised before and after in vitro studies using different techniques, such as X-ray diffraction (XRD), Fourier transform infrared spectroscopy, transmission electron microscopy and scanning electron microscopy. The compositions of the prepared samples revealed amorphous nature and spherical shape with an average particle size of less than 100 nm. Specific surface areas were found to be 90.69 and 81.20 m^2g^{-1} for $58SiO_2$–$33CaO$–$9P_2O_5$ and $58SiO_2$–$28CaO$–$9P_2O_5$–$5ZrO_2$, respectively. The formation of hydroxyapatite layer on the surface of the glass during in vitro studies after 21 days of immersion in simulated body fluid (AZT0 sample) is confirmed by the XRD pattern. The initial study reveals the requisite for the optimization of the composition for need-based applications, such as regenerative dentistry and bone implant.

INTRODUCTION

In recent years, biocompatible materials have been of increasing interest in view of their enhanced bioactivity for bone implant applications. Bioactive glasses and glass ceramics play an

V. Rajendran, P. Prabu and K.E. Geckeler (eds.)
Nano Biomaterials, pp. 315–320 (2012).

important role in the replacement of damaged or diseased bone.[1–2] It is mainly used as a filler material in dental prosthetic and orthopaedic implants. Among the biomaterials, bioactive glass nanoparticles have become an essential biomaterial in view of their bone bonding ability and load-bearing applications. Even though different methods are available, the sol–gel method has unique advantages such as low processing temperature, higher purity and better homogeneity, micro-porosity, and higher surface area.[3] The accessibility of a rate of physico-chemical and mechanical properties of bioactive glasses widens the range of options available for clinical applications.[4]

The poor mechanical strength of bioactive glass is a limiting factor for load bearing applications. In order to overcome this problem, metal oxides, such as MgO, TiO_2, ZrO_2, $Al2O_3$, and Li_2O, are added to glasses to improve their physico-chemical properties and inturn enhance their bioactivity. The influences of MgO, TiO_2, ZrO_2 and Al_2O_3 on the bioactive glasses are revealed extensively.[5–8] However, the work on zirconia doped nanobioactive glass is found to be scanty.

The objective of the present investigation is to synthesise and characterise the zirconium substituted bioactive glass nanoparicles ($58SiO_2$-33CaO-$9P_2O_5$) and ($58SiO_2$-33CaO-$9P_2O_5$-$5ZrO_2$). All the prepared glass nanaoparticles were immersed in a simulated body fluid (SBF) for 21 days at 310 K to explore their ability to form a hydroxyapatite (HAp) layer on the glass surface. The structural properties and bioactivity of ZrO_2 substituted nanobioactive glass are revealed through different characterisation techniques in order to understand their biocompatibility for bone implant applications.

MATERIALS AND METHODS

Synthesis of Nanobioactive Glass

Tetraethyl orthosilicate (TEOS; 99% assay; Sigma-Aldrich), triethylphosphate (TEP; 99.5% assay; HiMedia), calcium nitrate (Ca(NO3)2·4H2O; 98% assay; Merck), zirconium dioxide (ZrO2; 99% assay; Merck), 2 N nitric acid (HNO3; 69% assay; Merck), ethanol, 1 M ammonia (25% assay; Merck) and ultrapure water (Arium 611UF; Sartorius AG) were used without any further purification for the preparation of nanobioactive glass powders. The nanobioactive glass particles of compositions 58SiO2–33CaO–9P2O5 and 58SiO2–33CaO–9P2O5–5ZrO2 (hereafter termed as ZT0 and ZT5) were prepared by sol–gel method.[9] Initially, TEOS was dissolved in water/ethanol and stirred at room temperature for 30 min. After complete hydrolysis of TEOS, the chemicals such as TEP, Ca(NO3)2. 4H2O and ZrO2 were continuously added in sequential order at 30 min interval. After complete dissolution, the solution was condensed by adding an ammonia solution until gel. The gel was dried in a hot-air oven at 353 K for 48 h and then calcined at 773 K for 3 h in a muffle furnace to get the nanobioactive glass particles.

Characterisation

After in vitro studies, structural nature, particle size, surface morphology and specific surface area of zirconium-doped nanobioactive glass particles along with the formation of HAp layer on their surface were characterised by X-ray diffraction (XRD) studies using an X-ray diffractometer (X'PertPro; PANalytical, the Netherlands), Fourier transform infrared (FTIR) spectrometer (Spectrum 100; PerkinElmer, USA), transmission electron microscopy (TEM, CM 200; Philips, USA), scanning electron microscopy coupled with an energy-dispersive X-ray analysis (SEM-EDS; JSM 6360; JEOL, Japan), and BET surface area pore size analyzer (Autosorb AS-1MP; Quantachrome, USA).

In Vitro Bioactivity

The bioactivity of the prepared nanobioactive glass particles was measured by in vitro bioactivity test using freshly prepared SBF, revealing the formation of HAp layer on the surface of the prepared glass samples. The SBF was prepared using standard protocol [10] in which the pH value was maintained as 7.4, which is equivalent to that of human blood plasma. The prepared nanobioactive glass particles (250 mg) were transferred separately into a polyurethane bottle containing 50 mL SBF and were then incubated at 37°C for 21 days. The ionic changes in SBF solution were monitored by measuring the changes in pH value every day for 21 days, using a pH electrode (Orion 720A, 3 Star; Thermo, USA). The samples were removed from SBF after 21 days of in vitro studies and were then gently washed with distilled water and dried at 100°C for 1 h. The dried powders were used for further characterisation studies such as XRD, FTIR and SEM to reveal the formation of HAp layer on the glass surface.

RESULTS AND DISCUSSION

The XRD patterns of the prepared nanobioactive glass particles before (ZT0 and ZT5) and after (AZT0 and AZT5) 21 days of immersion in SBF are shown in Figure 1a. These patterns show an amorphous nature of both samples and confirm that the composition of ZrO_2 in the glass system does not influence the crystallisation of bioactive glasses. The FTIR spectra of the prepared nanobioactive glass samples before (ZT0 and ZT5) and after (AZT0 and AZT5) 21 days of immersion in SBF are shown in Figure 1b. Silicate network absorption bands are observed at 1094, 804 and 472 cm^{-1}, which correspond to asymmetric stretching mode, symmetric stretching mode and the Si–O–Si stretching vibration, respectively. The average pore diameters of ZT0 and ZT5 samples are 18.67 and 20.20 nm, respectively. The pore size analysis reveals that all the nanobioactive glass samples are in the mesoporous range. The specific surface areas of ZT0 and ZT5 samples are 90.69 and 81.20 m^2g^{-1}, respectively. It can be concluded that the high surface area facilitates the formation of HAp layer more on the surface of sol–gel derived nanobioactive glass samples than on the surface of melt-derived glasses.[11]

Size and morphology of the prepared nanobioactive glass particles are confirmed through TEM analysis (Figure 2). ZT0 and ZT5 glass samples show a rough spherical shape with an agglomerated particle size of less than 100 nm. However, the EDX pattern of ZT0 and ZT5 samples reveal high purity with negligible carbon content.

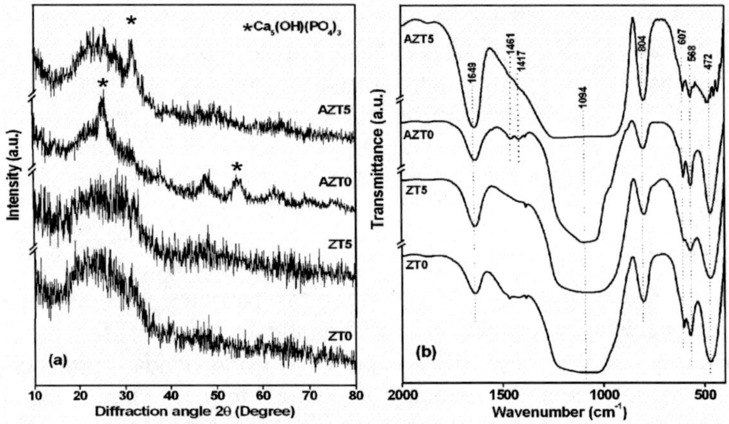

Fig. 1: (a) XRD Pattern, and (b) FTIR Spectrum of Prepared Glass Nanoparticles Before and After *in vitro* Studies

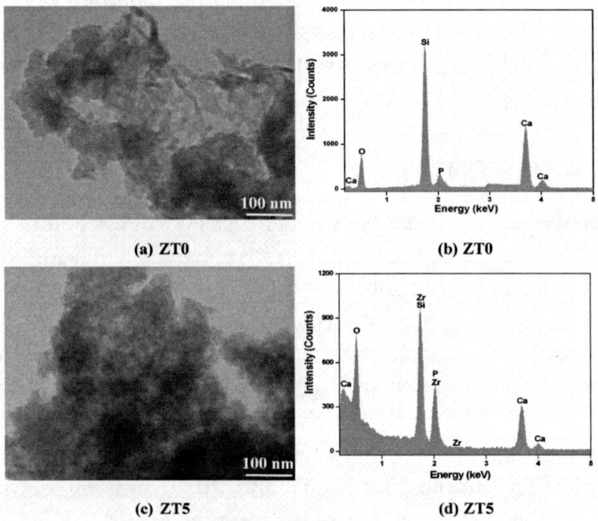

Fig. 2: TEM Images (a and c) and Corresponding EDS Patterns (b and d) of the Prepared Nanobioactive Glass Particles

The XRD pattern of the glass powders (AZT0 and AZT5) after 21 days of immersion in SBF is shown in Figure 1a. The diffraction peaks observed at 25.8°, 31.9° and 53.35° (JCPDS no. 09-0432) correspond to the crystalline peaks of HAp layer formed on the surface [12] of the nanobioactive glass particles. The observed results clearly indicate the formation of thick HAp layer on the AZT0 sample when compared with the AZT5 sample. A small quantity of zirconium is added to base glass; it acts as an intermediate that suppresses dissolution rate and also reduces the rate of formation of HAp layer.

The FTIR spectra of samples (AZT0 and AZT5) after soaking in SBF for 21 days are shown in Figure 1b. The absorption of the phosphate band observed at 607 and 568 cm^{-1} confirms the formation of a calcium phosphate layer in the crystalline phase.[13] In addition, carbonate absorption peaks observed at 1461 and 1417 cm^{-1} are detected in the AZT0 sample. However, it is concluded that the evolution of carbonate band is higher in the AZT0 sample than in the AZT5 samples after 21 days. The well-observed carbonate and phosphate bands indicate the formation of an apatite layer on the surface of the bioactive glass samples.

Fig. 3: SEM Images of the Prepared Glass Nanoparticles Before (a and c) and After (b and d) *in vitro* Studies

The SEM micrographs of all the glass samples before (ZT0 and ZT5) and after (AZT0 and AZT5) 21 days of immersion in SBF are shown in Figure 3. The ZT5 glass samples show an agglomerated and rough spherical surface with an irregular morphology when compared to the ZT0 glass samples. The surface of substituted glass sample (ZT5) became denser with increasing ZrO_2 concentration in silica glasses. An obvious difference in the surface morphology of the glass samples is observed before and after immersion in SBF, indicating the formation of a new apatite layer on the surface. The glass with ZrO_2 content (AZT5) shows a lesser formation of HAp layer on its surface after *in vitro* studies. A thick layer of HAp is noticed on the surface of AZT0 glass samples. These results conclude that the reactivity of AZT0 is high with SBF and hence, the glass surface is fully covered with a crystalline HAp layer without any interface. When a small amount of zirconium is added to the glass system, it acts as an intermediate and hence suppresses the bioactivity. Therefore, a decrease in the formation of HAp layer on the surfaces is more effective due to the lower reactivity of the crystalline state and increased resistance to the ion exchange reaction. Thus, it is concluded that less than 5 mol% of zirconium is required to achieve better bioactivity and biocompatibility.

CONCLUSION

Nanobioactive glass particles of compositions 58SiO2–33CaO–9P2O5 and 58SiO2–33CaO–9P2O5–5ZrO2 are prepared using a sol–gel method. The bioactivity of these samples is evaluated and compared with that of base glass. The presence of 5 mol% of zirconium in the glass composition decreases the formation of apatite layer when compared with base glass. The above studies confirm that optimisation of the zirconium compositions is required to enhance their bioactivity and biocompatibility for specific applications such as regenerative dentistry and bone implant.

ACKNOWLEDGEMENTS

The authors acknowledge the financial assistance provided by Department of Science and Technology (DST), New Delhi, to carry out this research project (SR/S2/CMP-0054/2009, dated 27 September 2010).

REFERENCES

[1] Hench, L.L., *J. Am. Ceram. Soc.,* 74 (1991), 1487.

[2] Maria, A. Lopes and Rui F. Silva, *Fernando J. Monteiro*, José D. Santos, *Biomaterials* 21 (2000) 749.

[3] Vallet-Reg, M., Salinas, A.J., Roman, J. and Gil, M., *J. Mater. Chem.,* 9 (1999), 515.

[4] Chenglin, Chu, Xiaoyan, Xue, Jingchuan, Zhu and Zhongda, Yin, *Mater. Sci. Eng. A* 429 (2006) 18.

[5] Manupriya, K.S., Thind, G., Sharma, K., Singh, V. and Rajendran, S. *et al., J. Am. Ceram. Soc.,* 90 (2007), 467.

[6] Navarro, M., Ginebra, M.P., Clément, J., Martínez, S. and Avila, G. *et al., J. Am. Ceram. Soc.*, 86 (2003), 1345.

[7] Yu., Liping, Hanning, Xiao and Yin, Cheng, *Ceram. Inter*. 34 (2008) 63.

[8] Rajendran, V., Gayathri, Devi A.V., Azooz, M. and El-Batal, F.H., *J. Non Cryst. Solids,* 353 (2007), 77.

[9] Xia, W. and Chang, J., *Mater. Lett.,* 61 (2007), 3251.

[10] Kokubo, T. and Takadama, H., *Biomaterials,* 27 (2006), 2907.

[11] Greenspan, D.C., Zhong, J.P. and Wheeler, D.L., *Bioceramics* 11 (1998), 345.

[12] Lin, K.S.K., Yao-Hung, T., Mou, Y., Yu-Chuan, H. and China-Min, Y. *et al., Chem. Mater.* 17 (2005) 4493.

[13] Costa, H.S., Rocha, M.F., Andrade, G.I., Barbosa-Stancioli, E.F. and Pereira, M.M. *et al., J. Mater. Sci. Mater. Med.* 43 (2008) 494.

ANTIMICROBIAL PROPERTIES OF NANO-SIZED SILVER-DOPED HYDROXYAPATITE COMPOSITE FOR BIOMEDICAL APPLICATIONS

S. Sutha, G. Karunakaran, M. Prabhu and V. Rajendran

Centre for Nano Science and Technology, K.S. Rangasamy College of Technology, Tiruchengode, Tamil Nadu
E-mail: veerajendran@gmail.com

ABSTRACT

In this study, the antimicrobial nano-sized hydroxyapatite (HAp) particles doped with different concentrations of silver (0, 0.2 and 0.5 mol%) with uniform morphology and narrow distribution were synthesised using sonic process. The size of prepared nanoparticles was in the range of 20–42 nm. The results of phase analysis confirmed that the prepared nanoparticles were crystalline in nature with hexagonal phase structure. The morphology of the particles was spherical with reduced particle diameter with an increasing silver concentration in the HAp structure. The specific surface area of the Ag-HAp nanocomposites was analysed by BET surface area analyser, which shows the addition of dopend in HAp improves its specific surface area from 19.1 to 36.7 m^2g^{-1}. The antimicrobial activity of the Ag-HAp nanoparticles for the various bacterial strains such as Bacillus subtilis, Staphylococcus aureus, Klebsiella pneumonia and Escherichia coli was evaluated. The result of antimicrobial analysis indicates that the antimicrobial activity of HAp increases with the increase in silver concentration in HAp structure.

INTRODUCTION

Hydroxyapatite (HAp; ($Ca_{10}(PO_4)_6OH_2$)) is a natural bone mineral composite that shows excellent biocompatibility and bone-bonding ability with the surrounding tissues.[1] The properties of synthetically prepared HAp depend mainly on its morphology, crystallite size,

V. Rajendran, P. Prabu and K.E. Geckeler (eds.)
Nano Biomaterials, pp. pp. 321–326 (2012).

purity and stoichiometry. Owing to its large scope in orthopaedic and dental implant applications, synthetically prepared HAp has been widely studied. The synthesis method highly influences the physico-chemical properties of the HAp, which alters its biological behaviour.[2] The synthetic HAp was prepared using several methods including hydrothermal, solid-state reaction, sol-gel and ultrasonication.[3–5] In all the methods, ultrasonication is the best one for the production of nano-sized HAp particles with controlled morphology.

Recently, the use of inorganic antibacterial agents in biomaterials has attracted interest.[6] The function of these agents is to improve the stability of biomaterials. The most commonly used antibacterial inorganic materials are the ceramics immobilizing antibacterial metals such as silver and copper.[7] The ionic substitution in the synthetic HAp has been widely used because of its excellent ion exchange rate. Silver is an excellent material with antimicrobial activity against most of the gram-positive and -negative bacterial strains and it also shows low toxicity towards mammalian cells.[8] Although Ag in small percentages shows excellent antimicrobial effect, larger amounts can be toxic.[9]

In this study, Ag-doped HAp (Ag-HAp) nanoparticles were prepared using ultrasonication process. Morphological characterisation of the prepared Ag-HAp nanoparticles was carried out. The antimicrobial activity of the prepared Ag-HAp nanoparticles against gram-positive and negative bacterial strains was analysed to ensure optimum antibacterial effect.

MATERIALS AND METHODS

The HAp nanocomposites were prepared by mixing commercial calcium nitrate ($Ca(NO_3)_2$; 99.9%; Merck AR), diammonium hydrogen phosphate ($(NH_4)_2HPO_4$; 99.9%; Merck AR), silver nitrate ($Ag(NO_3)_2$; 99.9%; Aldrich) and ammonium hydroxide (NH_4OH; 25%; Merck GR, USA). The chemicals were used as received. The ultrapure water (Arium 611 Ultrafilter; Sartorius AG, Germany) was used throughout the experiment. In this method, $Ca(NO_3)_2$ solution was prepared by dissolving 0.66 M Ca containing $Ca(NO_3)_2$ in ultrapure water, using a sonochemical reactor (VC 505; Sonics, USA) with a frequency of 20 MHz for 10 min at room temperature. To this solution, we added 0, 0.2, and 0.5 mol% Ag containing $Ag(NO_3)_2$ salt and then added drop-wise the required amount of $(NH_4)_2HPO_4$ dissolved in ultrapure water under sonication. Then, using ammonium hydroxide, the pH of the solution was adjusted to 10.2. The HAp particles prepared with 0, 0.2 and 0.5 mol% Ag concentrations were dried in a hot-air oven at 353 K for 12 h.

The size of the prepared nanoparticles (1–500 nm) was characterised by dynamic light scattering (Nanophox; Sympatec, Germany) with the scattering angle of 90°. The structure of the prepared HAp and Ag-HAp samples was assessed using X-ray diffraction (XRD; X'PertPRO; PANalytical, the Netherlands) with monochromatic CuKα as the radiation source (λ = 1.540 Å). Morphology of the powders was confirmed by transmission electron microscopy (TEM).

The antimicrobial activity of the Ag-HAp nanocomposites was measured using Kirby–Bauer disc diffusion method. Various gram-positive (*Bacillus subtilis* and *Staphylococcus aureus*) and gram-negative (*Klebsiella pneumonia*, *Salmonella typhi* and *Pseudomonas aeruginosa*) microorganisms (National Collection of Industrial Microorganisms, National Chemical Laboratory, Pune, India) were used to analyse the antimicrobial activity. The sub-cultured bacterial strains were incubated at 310 K for 1 h. The antimicrobial tests were performed on the previously prepared Mueller–Hinton agar plates. The discs were loaded with different concentrations of Ag-HAp and incubated at 310 K for 24 h. After the incubation period, the inhibitory zone was measured and a graph was plotted. The microbial resistance of the composites was measured from the inhibitory zone formed around the disc.

RESULTS AND DICUSSION

The variation in particle size distribution for HAp doped with different concentrations of Ag is shown in Figure 1. Incorporation of Ag into HAp structure reduces the particle size from 30–100 to 15–32 nm with a maximum distribution of 24 nm, indicating that the silver ions greatly affect the particle growth of HAp

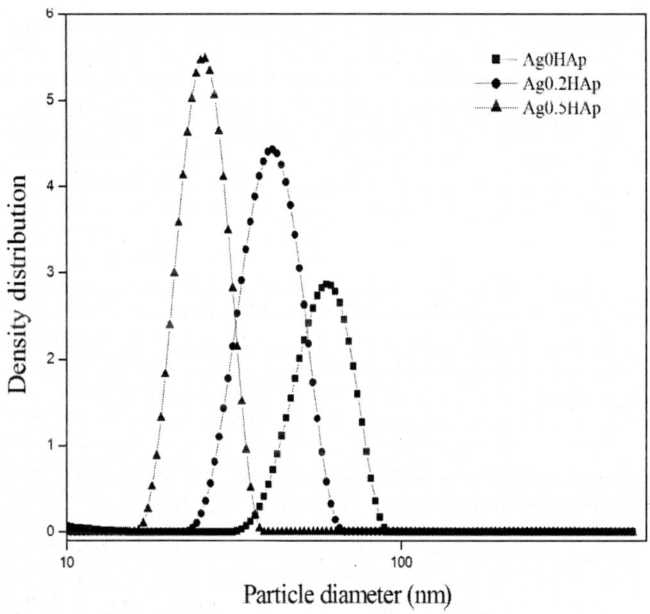

Fig. 1: Particle Size Distribution of Sonochemically Prepared Pure and Silver-Doped Hydroxyapatite

The phase analyses of the pure and Ag-HAp nanocomposites are shown in Figure 2. The obtained XRD patterns are in good agreement with the standard JCPDS file no. 09–0432. These patterns for the undoped and doped HAp show a sharp and well resolved peak with hexagonal phase structure. The observed results indicate that the obtained HAp composites are well crystallised and phase pure. The increasing concentration of Ag into the HAp lattice reduces its degree of crystallinity, which may be due to the substitution of silver for calcium. From this observation, it is assumed that the calcium ions are replaced by the silver ions in the HAp lattice.

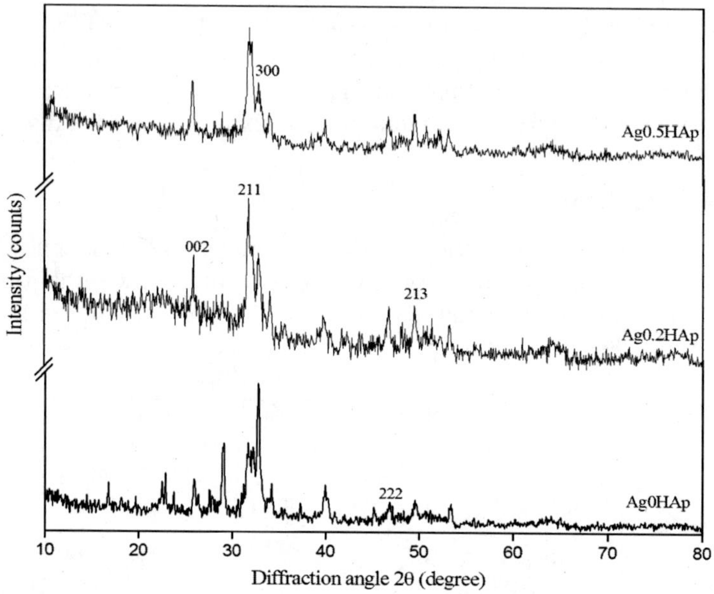

Fig. 2: XRD Pattern of Sonochemically Prepared Pure and Silver-Doped Hydroxyapatite

The change in surface area, pore size and pore volume with the incorporation of Ag ions into the HAp structure was analysed using BET surface area analyser. The obtained results of the Ag-HAp nanocomposites have been summarised in Table 2. The increasing Ag concentration in HAp highly influences the formation of microporous Ag-HAp with increased total pore diameter. Thus, an increase in surface area with increase in Ag concentration in HAp is noticed. From this result, it is clear that 0.5 mol% Ag-substituted HAp shows maximum structural modification and above this concentration the composite reaches its stability.

Table 1: Shows the TEM Micrograph of all the As-synthesised HAp and Ag-HAp Samples

Sample Code	*Particle Size (nm)*	*Surface Area* (m^2g^{-1})	*Pore Volume* $\times 10^{-2}\ ccg^{-1}$	*Pore Diameter (nm)*
Ag0HAp	24	19.1	1.49	1.84
Ag0.2HAp	41	30.94	1.51	1.95
Ag0.5HAp	60	36.83	1.84	2.01

It can be seen that the particles are more uniform in size with the spherical morphology. The average diameter of the particles for all samples lies between the particle size range of 15 and

100 nm. Thus, it is concluded that an increase in silver ion concentration during doping significantly reduces the size of the particles without any morphological change.

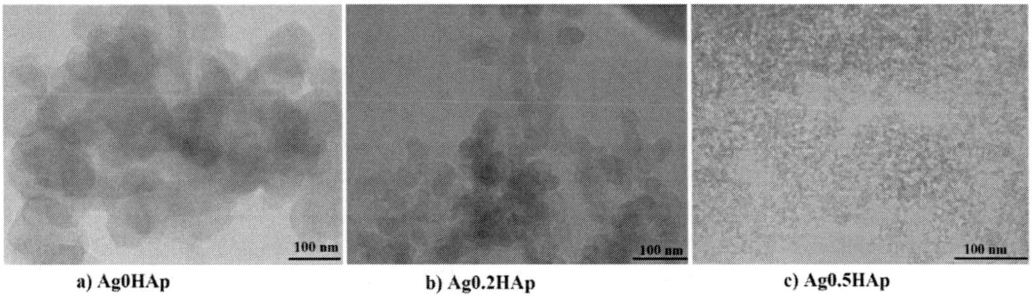

Fig. 3: TEM Micrograph of the Pure and Silver-Doped Hydroxyapatite

The antimicrobial activities of the Ag-HAp samples tested with the most common gram-positive and negative microorganisms, which are responsible for the implant infections, are shown in Figure 4. The results demonstrate that the Ag-HAp samples show a visible microbial inhibition zone around the disc. The formation of this zone increases with an increase in Ag concentration in HAp, suggesting dependence of antimicrobial efficiency on the concentration of the silver ions. The observed higher inhibition zone value of 2–2.8 nm for the gram-positive bacterial stain is reduced for the gram-negative strains to 1.1–1.7 mm. This result suggests that the Ag-HAp samples are having high antimicrobial resistance for the gram-positive bacterial strains, which is reduced for the gram-negative strains.

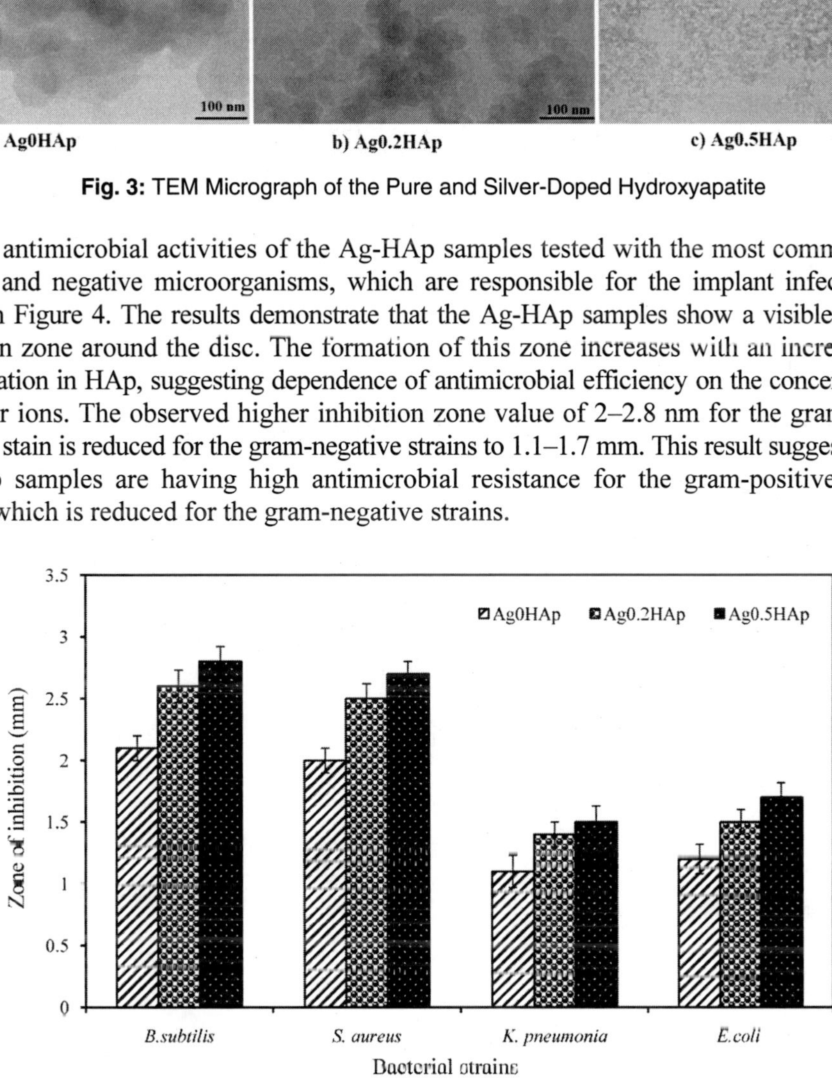

Fig. 4: Antimicrobial Activity of the Ag-HAp Nanoparticles with Various Bacterial Strains

The results of antimicrobial analysis reveal that the undoped HAp samples also show antimicrobial zone formation around the disc. This may be due to the adhesion of microorganisms on the surface of the HAp samples, causing cell damage due to the direct contact of the silver ions with the microorganisms.

CONCLUSION

The nano-sized HAp and Ag-HAp composites are prepared through the sonochemical process. The phase analysis conforms the phase purity with the hexagonal phase structure. The surface area and pore diameter variation with respect to the silver ion concentration is increased. The antimicrobial study reveals that the Ag-HAp nanocomposites show an excellent antimicrobial activity, which is higher for the gram-positive bacterial strains. Thus, the result reveals that the Ag-HAp composites are better choice for the antimicrobial applications in the field of biomaterials.

REFERENCES

[1] Cheng, R.J., Hsieh, M.F., Huang, K.C., Perng, L.H., Chou, F. and Chin, T.S., *J. Sol-gel Sci. Technol.* 33, 229, 2005.

[2] Hench, L.L., *J. Am. Ceram. Soc.* 81, 1705, 1998.

[3] Ramachandra, R.R., Roopa, H.N. and Kannnan, T.S., *J. Mater. Sci. Mater. Med.* 8, 511, 1997.

[4] Liu, H.S., Chin, T.S., Lai, L.S., Chiu, S.Y., Chung, K.H., Chang, C.S. *et al.*, *Ceram. Int.* 23, 19, 1997.

[5] Chai, C.S., Gross, K.A. and Nissan, B., *Biomaterials* 19, 2291, 1998.

[6] Ge, X., Leng, Y., Bao, C., Xu, S.L., Wang, R. and Ren, F., *J. Biomed. Mater. Res.* 95A, 588, 2010.

[7] Percivel, S.L., Bowler, P.G. and Ressell, D., *J. Hosp. Infect.* 60, 1, 1997.

[8] Chen, W., Liu, Y., Courtney, H.S., Bettenga, M., Agrawal, C.M., Bumgardner, J.D., *et al.*, Biomaterials 27, 5512, 2006.

[9] Kim, Y.S., Song, M.Y., Park, J.D., Song, K.S., Ryu, H.R., Chung, Y.H. *et al.*, *Part. Fibre Toxicol.* 7, 20, 2010.

AUTHOR INDEX

Abimanyu, S. 47
Abiraman, T. 29
Akhoon, S.A. 197
Alameh, K. 9
Anitha, R. 53, 59
Anukaliani, A. 251
Aravindan, S. 235
Arun, A.B. 125
Arunkumar, S. 89
Ashokkumar, T. 173, 279, 289

Balakumaran, M.D. 29
Balamurugan, M. 161
Balashanmugam, Pannerselvam 299
Balasubramanian, S. 29, 35
Begum, A. Nishara 235

Chaudhri, Sanjay G. 137
Choi, Soon Mo 191

Das, Nayan Mani 131
Deepa, C. 235
Devi, P. Renuka 13
Dhanalakshmi, C.P. 243
Dharmaraj, Nallasamy 23
Dhodapkar, Rita 107
Dinda, Amit Kumar 83

Emayavaramban, Perumal 23
Eswaramma, S. 143
Ezhilarasu, T. 35

Firdhouse, M. Jannathul 179

Geetha, R. 173, 279, 289
Giribabu, K. 243
Gnanamani, A. 5
Gokhale, Shubha M. 149
Gokilam, R. Vasantha 95, 101
Gomathinayagam, Sivasankari 211
Govindaraju, K. 173, 279, 289
Gupta, P.S. 131

Han, Sung Soo 191
Hemananthan, E. 13
Hussain, M.A. 125

Inbakumar, S. 251

Jagan, K. 227
Jagan, R. 113
Jayakumar, T. 309
Jeevanesan, Vinoth 211
John, Rohith P. 167
Joshi, Mangala 149
Josphine, Sugania 227

Kakade, Amol 203
Kalaichelvan, P.T. 29
Kalaichelvan, Puthupalayam Thangavelu 299
Kanakalakshmi, A. 95, 101
Kandasamy, S. 9
Kansara, Ankit M. 137

Karunakaran, G. 321
Kavitha, K. 309, 315
Khalid, P. 125
Krishnaswami, Venkateshwaran 65
Kulandaivelu, P. 315
Kumar, A. 41
Kumar, Anupama 107
Kumar, Dhiraj 83
Kumar, M. Anis 71, 77
Kumar, Ramadhar 1
Kundu, Aniruddha 219
Kuppusamy, Vanitha 211

Lah, Myoung Soo 167
Lalitha, P. 179
Lapierre, F. 9
Latha, S. 71, 77
Layek, Rama K. 219
Lee, Yong Ill 143
Lokina, S. 119
Luthra, Vandna 149

Madhavi, Lakshmi 53, 59
Maheshvaran, K. 89
Malathi, S. 29, 35
Manchanda, Seema 203
Manivasakan, P. 315
Marimuthu, K. 89
Maurya, D.K. 9
Mitra, Susmita 83
Mitra, Tapas 5
Moon, Dohyun 167
Murugan, E. 259, 269

Nagiah, Naveen 53, 59
Nair, M. Anjali 95, 101
Namrata 149
Nandagopal, M. 47
Nandi, Arun K. 219
Narayanan, R. Lakshmi 155
Narayanan, V. 119, 243
Natarajan, Tirupattur Srinivasan 53, 59
Natesan, Subramanian 65, 211
Nethaji, S. 227

Padmavathy, R. 113
Pardeshi, Sushma 107
Ponnusamy, Chandra Sekar 65
Poornachari 17
Prabhu, M. 309, 315, 321
Prabu, C. 71
Prabu, P. 309
Pradhan, Subashis 167
Prasad, S. Siva 143
Prashant, C.K. 83
Rajendran, V. 309, 315, 321
Rajeshwari 149
Rajiv, Sheeja 1
Ramanathan, Giriprasath 59
Rani, D. Vijula 47
Rani, D.P. Geetha 259, 269
Rani, N. Jamuna 47
Rao, K. Venkateswara 155
Rao, K.S.V. Krishna 143
Reddy, N. Sivagangi 143
Reddy, P. Ramasubba 143
Rekha, P.D. 125
Remy 149
Roy, Dhrubojyoti 131
Rubab, S. 197
Rubia, R. 149

Sadiq, A. Mohamed 173, 279, 289
Sailakshmi, G. 5
Saravanan, S. 161
Sathya, D. 113
Sekar, P. Chandra 47
Selvakumar, P. Mosae 185

Selvamani, P. 71, 77
Shah, M.A. 197
Shanthi, K. 95, 101
Sharavanan, S.P. 47
Sharon, Madhuri 203
Sharon, Maheshwar 203
Sidhwani, Indu Tucker 149
Silambarasi, T. 77
Singaravelu, G. 173, 279, 289
Singaravelu, Sivakumar 53
Singh, Deepti 191
Singh, Dolly 191
Singh, Puyam S. 137
Sivakumar, K. 113
Sivakumar, M. 155
Sivasamy, A. 227
Srinivasan, Natarajan Thirupathur 1
Subramanian, N. 47
Subramanian, P.S. 185
Sugumaran, Abimanyu 65
Sukumar, C. 95, 101
Suman, V.B. 125
Suresh, E. 185
Suresh, R. 243
Suriyaprabha, R. 309
Sutha, S. 315, 321

Tamilselvan, S. 173, 279, 289
Thangaraju, Elakkiya 1
Thangaraju, Vaishnavi 1

Uma, Tiruchirapalli Sivagnanam 53, 59

Vasdev, Kavita 149
Verma, Anita K. 41
Vijayalakshmi, L. 243

Yashika 149
Yogaraj, V. 269

Zhu, Y. 9